Biological and Bioenvironmental Heat and Mass Transfer

Biological and Bioenvironmental Heat and Mass Transfer

Ashim K. Datta
Cornell University
Ithaca, New York

MARCEL DEKKER, INC. NEW YORK · BASEL

Typesetting in LaTeX by Ashim K. Datta. Illustrations by Kevin C. Hodgson and Ashim K. Datta.

ISBN: 0-8247-0775-3

This book is printed on acid-free paper.

Headquarters
Marcel Dekker, Inc.
270 Madison Avenue, New York, NY 10016
tel: 212-696-9000; fax: 212-685-4540

Eastern Hemisphere Distribution
Marcel Dekker AG
Hutgasse 4, Postfach 812, CH-4001 Basel, Switzerland
tel: 41-61-261-8482; fax: 41-61-261-8896

World Wide Web
http://www.dekker.com

The publisher offers discounts on this book when ordered in bulk quantities. For more information, write to Special Sales/Professional Marketing at the headquarters address above.

Copyright © 2002 by Marcel Dekker, Inc. All Rights Reserved.

Neither this book nor any part may be reproduced or transmitted in any form or by any means, electronic or mechanical, including photocopying, microfilming, and recording, or by any information storage and retrieval system, without permission in writing from the publisher.

Current printing (last digit):
10 9 8 7 6 5 4 3 2 1

PRINTED IN THE UNITED STATES OF AMERICA

To my parents

Atindra Nath Dutta
and
Bela Rani Dutta

ACKNOWLEDGMENTS

The author sincerely acknowledges the assistance of many individuals in preparing this manuscript; He hopes that the names included here cover most of them. Jean Hunter, Douglas Haith, Louis Albright, Larry Geohring, Ronald Pitt, Roger Spanswick, Gerald Rehkugler, John Cundiff, Kifle Gebremedhin, Thomas Goldstick, and Arthur Johnson generously brought to bear their particular expertise in discussing transport issues with me. Comments from the instructors who used the prepublication version of this text— Dean Steele, Josse De Baerdemaeker, Guido Wyseure, Sean Kohles, Sudhir Sastry, Bradley Marks, John Nieber, Fu-Hung Hsieh, and Christopher Choi, have been invaluable in the development process. Students of ABEN 350, in the Department of Biological and Environmental Engineering (formerly Agricultural and Biological Engineering) at Cornell University, during the period 1990-1998, as well as students from Katholieke Universiteit Leuven, Belgium, provided feedback that made the text more relevant, manageable, and clear. Development of example problems was facilitated by the Teaching Assistants Kelley Bastian, Haitao Ni, Annie Chi, Steven Lobo, Katherine DeBruin, Jordan Dolande, Indranil Mukerji, and Wenjie Hu. No less important has been the help given to me over the years in typing, drawing, LaTeX, proofreading and related activities from Kevin Hodgson, Sharon Hobbie, Cindy Robinson, Ivan Dobrianov, Andy Ruina, Mark Schroeder, Krishanu Saha, Sue Fredenburg, Jifeng Zhang, Donna Burns, Sandy Bates, and Richard Krizek. The long development process needed encouragement, and the author deeply appreciates everyone who provided it— just a few of whom are Andy Rao, Robert Cooke, Ronald Furry, and Norman Scott. Most important, I want to give my profound thanks to my wife, Anasua, and daughters, Ankurita and Amita, for their patience with my preoccupation for such a long time.

The author also acknowledges financial support from the U.S. Department of Agriculture, subcontracted through Roger Garrett of the University of California, Davis. Also appreciated is various support from the Department of Biological and Environmental Engineering at Cornell University.

PREFACE

It is very important to give the undergraduate engineer a fundamental education in the context of his/her likely application areas. Transport of energy and mass is fundamental to many biological and environmental processes (see pages xi to xx). Areas, from food processing to thermal design of buildings to biomedical devices to pollution control and global warming, require knowledge of how energy and mass can be transported through materials. These wide-ranging applications have become part of emerging curricula in biological engineering, and societies such as the Institute of Biological Engineering and the American Society of Agricultural Engineers have recognized the need for a course (and a text) that presents fundamentals while integrating the diverse subject matter.

The basic transport mechanisms of many of these processes are diffusion (or diffusion-like, such as capillary and dispersion) and bulk flow. Additionally, there is radiative heat transfer. It is crucial for the student to see these concepts as comprehensive and unified subject matter (much like fluid mechanics); they are the building blocks for lifelong learning in many of their interest areas. Such fundamentals-based approach will replace the more empirical and ad-hoc teaching that sometimes exists.

Although the concept of teaching transport processes as a unified subject has existed for over forty years in some engineering disciplines, only in recent years have we seen adequate quantitative studies to make such teaching possible in biological and bioenvironmental processes. This book attempts to bring together under one umbrella the unique content, contexts and parameter regimes of biologically related processes and to emphasize principles and not just mathematical analysis. *Content*, such as bioheat transfer, thermoregulation, freezing, global warming, capillary flow, and dispersion, are some of the topics not typically included in the undergraduate-level teaching of transport phenomena. *Context*, such as plants, animals, water, soil, and air is important at this level, because without this information students have an unnecessarily hard time relating to real physical processes. Context also helps students learn about the physical processes themselves in a quantitative way. For example, studying convective transfer of water vapor over a leaf includes a quantitative introduction to transpiration. (The present text was created by distilling the content of hundreds of research papers and textbooks on similar biological and environmental applications.) The *parameter regimes* of biological processes are also different from those of typical mechanical and chemical processes. For example, biological processes often involve a source term of heat generation or oxygen consumption. Presence of the source or sink term changes the nature of the solution and is emphasized in this text.

How This Book Fits in a Biological Engineering Curriculum

This text is intended for a junior-level engineering science course in curricula that emphasize biology and the environment. The course would build on the prerequisites of partial differential equations and fluid mechanics. Prior knowledge of biological and environmental science, although not required, would be useful. For example, this course can readily build on a course such as Thermodynamics of Biological Systems that has been discussed in the context of a biological engineering curriculum. Mass and heat transfer, much like fluid flow, are just as much building blocks for many of the upper-level courses. Thus, specialized design courses and advanced courses such as bioprocessing, biomedical engineering, food process engineering, environmental processes and their control, and waste management can build on a course that uses this text, greatly reducing the need to teach basic engineering science of mass and heat transfer in these upper-level courses. This text was developed at Cornell University for a junior-level engineering science course.

Approach and Organization of the Book

The overall organization of the book follows the well-tested transport phenomena approach. The chapters and their content on heat and mass transfer are made to follow an almost exact parallel, as shown in the table below. The first two chapters in each part (Part I, "Energy Transfer" and Part II, "Mass Transfer") develop the two build-

	Chapter numbers	
	Energy	Mass
Conservation	1	9
Rate laws	2	10
Governing equation	3	11
Diffusion, steady-state	4	12
Diffusion, transient	5	13
Diffusion (and dispersion) with bulk flow	6	14

ing blocks of conservation laws and rate laws, and the next chapter (Chapters 3 and 11) combines them to build the general governing equations and boundary conditions. The next two chapters in each part (as shown in the table above) cover steady-state or transient diffusion, without any flow, while the last one adds the effect of flow. Chapter 7 covers heat transfer with change of phase, and Chapter 8 covers radiative energy transfer. Porous media flow and simple kinetics of zero and first order are included for completeness as they relate to transport. Effort has been made to clarify important processes such as dispersion. Different application areas in biology and environment are included within this framework of chapters, when they are relevant.

PREFACE

How to use This Book

Students frequently follow individual topics well but have difficulty seeing their relatedness, i.e., the big picture. Thus, major effort has been made to distill the concepts presented here and to show the connections among chapters. Each chapter begins with a small list of major concepts to be covered, together with important terminologies introduced in that chapter, and each chapter has a map showing how all chapters are interconnected in terms of the topic under consideration. Each chapter has a summary at the end that puts every major concept and equation at the reader's fingertips, providing page numbers for easy access, and a set of descriptive questions checks the reader's understanding of concepts and facts. Summary maps in Appendix A (pages 316-318) show the integration of all the scenarios covered in the text. The first-time reader of the subject is strongly encouraged to use these features.

As curricula in biological and related engineering programs evolve, the core of such curricula will include mass and heat transfer as essential building blocks in students' instruction in a natural and obvious way. The author sincerely hopes that this text will serve the needs of these curricula. He also believes that the text must evolve with the curricula. Additional materials helpful for teaching this subject matter can be seen on the Internet at www.ashimdatta.net. Please do not hesitate to contact the author if you have comments on any aspect of the book.

Ashim K. Datta
akd1@cornell.edu

INTRODUCTION

Problem Formulation in the Transport Processes

This book centers around problem formulation, shown schematically in Figure 1—taking a biological problem and, after making sufficient assumptions, formulating it as a mathematical problem that one can solve using standard undergraduate calculus. The mathematical model thus built provides a more quantitative insight into the biological process by seeing how the various parameters of the process influence it. Having started from a general and fundamental description (e.g., mass and energy balances) of the process, it will be easier for students to formulate more complex processes.

The various application areas in biological engineering can be grouped as follows: plant systems, mammalian systems, the bioenvironment, and the industrial processing

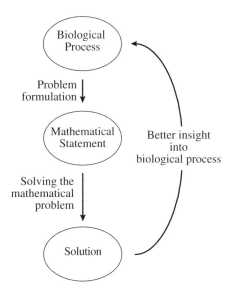

Figure 1: Schematic showing the steps of problem formulation and solution.

of biomaterials and foods. Schematics of examples and problems in each of these four areas are shown on the following pages, together with a brief description of transport processes in each application area and a list of reading materials covering specialized transport in that area. These pages will provide the reader with a glimpse of the biological processes that are discussed in this book.

In addition to the specialized texts covering transport processes in various application areas mentioned in the following pages, several well-known books on general transport phenomena can provide the reader with information on governing equations and solutions for various situations. Some of these general texts are:

Bird, R. B., W. E. Stewart, and E. N. Lightfoot. 1960. *Transport Phenomena*. John Wiley & Sons, New York.

Cussler, E. L. 1997. *Diffusion Mass Transfer in Fluid Systems*. Cambridge University Press, Cambridge, UK.

Deen, W. M. 1998. *Analysis of Transport Phenomena*. Oxford University Press, New York.

Eckert, E. R. G. and R. M. Drake. 1987. *Analysis of Heat and Mass Transfer*. Hemisphere Publishing Corporation, New York.

Fahien, R. W. 1983. *Fundamentals of Transport Phenomena*. McGraw-Hill, New York.

Geankoplis, C. J. 1993. *Transport Processes and Unit Operations*. P T R Prentice Hall, Inc., Englewood Cliffs, New Jersey.

Geankoplis, C. J. 1972. *Mass Transport Phenomena*. C. J. Geankoplis, Columbus, OH.

Incropera, F. P. and D. P. Dewitt. 1996. *Fundamentals of Heat and Mass Transfer*. John Wiley & Sons, New York.

Middleman, S. 1998. *An Introduction to Mass and Heat Transfer*. John Wiley & Sons, New York.

Rosner, D. E. 1986. *Transport Processes in Chemically Reacting Flow Systems*. Butterworths, Boston.

Slattery, J. C. 1999. *Advanced Transport Phenomena*. Cambridge University Press, Cambridge, UK.

Welty, J. R., C. E. Wicks, R. E. Wilson, and G. Rorrer. 2001. *Fundamentals of Momentum, Heat, and Mass Transfer*. John Wiley & Sons, New York.

Transport in the Mammalian System

In mammalian systems, transport processes occur at the cellular, tissue, organ, and whole-body levels (Figure 2). At the cellular level, transport across the cell membrane is driven by passive diffusion of solutes and water, hydraulic and osmotic transport of water, carrier-mediated transport, passive ion transport and active transport. This text includes diffusion, as well as hydraulic and osmotic transport, but not the other important but complex transport mechanisms just mentioned. An extensive treatise of cellular transport is provided in Weiss (1996). At the tissue or organ level, diffusion of oxygen is an important transport process. One of many examples of such transport being the oxygen diffusion from air to the blood stream in alveoli of the lungs. At the whole body level, there is thermoregulation, whereby heat production in the body and heat dissipation through behavioral changes (movements of the whole body) or autonomic, reflex-like changes (such as sweating or shivering) are modified. Thermal therapy, either the use of heat (hyperthermia) or the use of freezing temperatures (cryosurgery) to destroy tissue, demonstrates heat transfer in clinical applications. Transport in artificial organs such as the dialyzer is another important group of problems. The reader can consult several specialized books on transport in biomedical systems for further

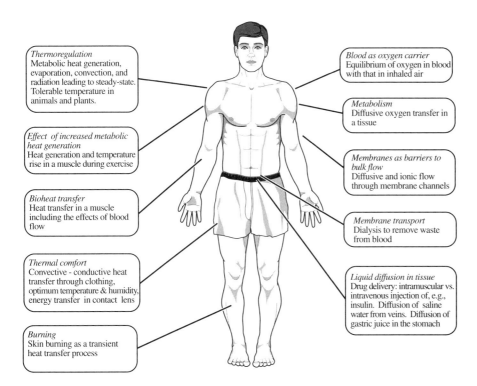

Figure 2: Schematic showing examples of transport in mammalian systems covered in this text.

details. Some titles follow:

Berger, S. A., W. Goldsmith, and E. R. Lewis. 1996. *Introduction to Bioengineering.* Oxford University Press, Oxford, UK.

Charny, C. K. 1992. Mathematical models of bioheat transfer. Advances in Heat Transfer 22:19-153.

Cooney, D. O. 1976. *Biomedical Engineering Principles. An Introduction to Fluid, Heat and Mass Transport Processes.* Marcel Dekker, Inc., New York.

Evans, D. H. 1998. *The Physiology of Fishes.* CRC Press, Boca Raton, Florida.

Fanger, P. O. 1972. *Thermal Comfort: Analysis and Applications in Environmental Engineering.* McGraw-Hill, New York, NY.

Fournier, R. L. 1998. *Basic Transport Phenomena in Biomedical Engineering.* Taylor & Francis, Philadelphia, Pennsylvania.

Lightfoot, E. N. 1974. *Transport Phenomena in Living Systems.* John Wiley & Sons, New York.

Lih, M. M. 1975. *Transport Phenomena in Medicine and Biology.* John Wiley & Sons, New York.

Middleman, S. 1972. *Transport Phenomena in the Cardiovascular System.* Wiley-Interscience.

Shitzer, A. and R. C. Eberhart. 1985. *Heat Transfer in Medicine and Biology: Analysis and Applications.* Plenum Press, New York.

Weiss, T. F. 1996. *Cellular Biophysics. Volume 1: Transport.* The MIT Press, Cambridge, Massachusetts.

Yang, W. 1989. *Biothermal-Fluid Sciences.* Hemisphere Publishing Corporation, New York.

INTRODUCTION

Transport in Plant Systems

From an engineering standpoint,[1] an annual crop plant may be regarded as a self-replicating structure. The first structures produced, leaves and roots, function primarily in transport and energy acquisition. The leaves intercept solar energy for the fixation of carbon dioxide in photosynthesis, acquiring the carbon dioxide by diffusion through the boundary layer over the leaf and through the stomatal pores, while inevitably losing water by evaporation from the cell surfaces of the wet leaf. The roots extract water from the soil, replacing that lost by the leaves and providing the water that constitutes the bulk of growing plant tissue. For many mineral nutrients essential for plant growth, roots are also the initial site of transport into the plant. The properties of soil-root interface also determine whether pollutants in the soil enter the food chain. Within the plant, transport across the cell membranes involves both diffusion, for water and solutes, and carrier-mediated processes, for ions and organic molecules but not water. Bulk flow of water, ions, and some nitrogen compounds from the roots to the shoot occurs in the specialized vascular tissue called the xylem. Some of the examples of transport in plant systems covered in this text are shown in Figure 3. (The topic of carrier-mediated transport is outside the scope of this text, and the reader is referred to specialized texts such as Marschner, 1995). Some texts and reference books covering transport in plant systems are noted below:

Baker, D. A. 1978. *Transport Phenomena in Plants*. Chapman and Hall, London.

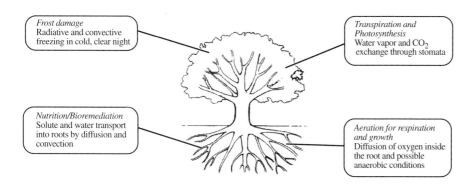

Figure 3: Schematic showing examples of transport plant systems covered in this text.

---

[1] Paragraph based on contribution from Prof. Roger Spanswick, Dept. of Biological and Environmental Engineering, Cornell University.

Buchanan, B., W. Gruissem, and R. L. Jones. 2000. *Biochemistry and Molecular Biology of Plants* (Chapters 3, 15 and 23). American Society of Plant Biologists, Rockville, MD.

Cundiff, J. S. 1999. Simulation of biological systems. Coursenotes for BSE 4144, Virginia Polytechnic and State University, Blacksburg, Virginia.

Flowers, T. J. and A. R. Yeo. 1992. *Solute Transport in Plants.* Blackie Academic & Professional, London.

Marschner, H. 1995. *Mineral Nutrition of Higher Plants.* Academic Press, San Diego.

Merva, G. E. 1995. *Physical Principles of the Plant Biosystem.* American Society of Agricultural Engineers, St. Joseph, MI.

Nobel, P. S. 1999. *Physiochemical and Environmental Plant Physiology.* Academic Press, San Diego.

Siau, J. F. 1984. *Transport Processes in Wood.* Springer-Verlag, New York.

Transport in Industrial Food and Biological Processing

Transport is at the core of industrial (and domestic) processing of food and biomaterials (see examples in Figure 4). In most food processes heat and moisture transport influ-

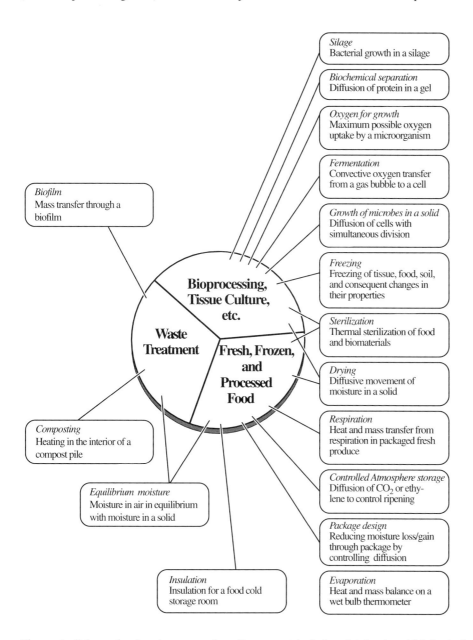

Figure 4: Schematic showing examples of transport in industrial food and biological processing that are covered in this text.

ences chemical and microbiological changes. Sterilization of food to extend its storage life is done primarily using heat, making heat transfer an extremely important transport process. Freezing of food involves heat transfer with a change of phase. Drying, and related processes such as baking and frying, are also important and involve the phase change of water to vapor and the transport of water and vapor through the food matrix. The diffusion of gases through a packaging material is important in packaging food so that the optimum gas composition (controlled atmosphere) is maintained around the food. Fresh produce respires, and here again heat transfer is important, as heat generation from respiration is important in designing the right storage temperature.

Bioprocessing, such as a fermentation process, involves oxygen transport, for example, from a liquid fermentation medium to the cells. The design of waste treatment facilities also requires knowledge of heat and mass transfer, particularly mass transport. For example, a composting pile must be designed so that the heat generated within the pile is released appropriately and so that temperatures do not become so high as to be lethal to the microbes that are responsible for biological activity. Some specialized texts on engineering aspects of food and bioprocessing are noted here for further reading:

Bailey, O. E. and D. F. Ollis. 1986. *Biochemical Engineering Fundamentals*. McGraw-Hill, New York.

Gekas, V. 1992. *Transport Phenomena of Foods and Biological Materials*. CRC Press, Boca Raton, Florida.

Hallström, B., C. Skjöldebrand, and C. Trägårdh. 1987. *Heat Transfer and Food Products*. Elsevier Applied Science Publishers Ltd., Barking, Essex, England.

Heldman, D. R. and R. P. Singh. 1981. *Food Process Engineering*. AVI Publishing, Westport, Conn.

Heldman, D. R. and D. B. Lund. 1992. *Handbook of Food Engineering*. Marcel Dekker, New York.

Johnson, A. 1999. *Biological Process Engineering*. John Wiley & Sons, New York.

Mujumdar, A. S. 1987. *Handbook of Industrial Drying*. Marcel Dekker, New York.

Transport in the Bioenvironmental System

Transport in the bioenvironmental system (Figure 5) can be seen in three major environmental media—water, air, and soil. Transport in soil can involve bulk flow and dispersion of water and chemicals present in it, eventually reaching groundwater. Such

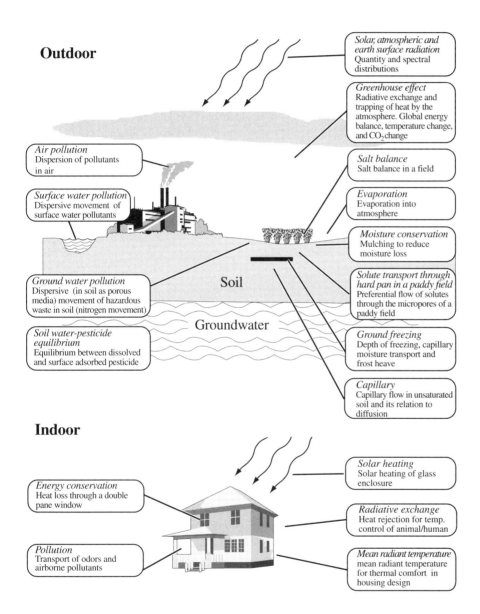

Figure 5: Schematic showing examples of transport in the bioenvironment covered in this text.

chemicals could be fertilizers or pesticides that are routinely applied. Transport in soil is traditionally covered in textbooks on soil physics, although some specialized transport texts covering transport in soil have appeared in the recent years (e.g., Schnoor, 1996). In the air, transport can be of airborne particles such as smoke from a chimney, or of evaporated water, such as from wet soil or a water surface. Specialized texts on air pollution and evaporation also exist, as noted below. The movement of pollutants in surface water occurs in bodies such as streams and lakes and is important in the study of environmental transport. Heat transport in the outdoor environment can involve solar energy— its incidence, as well as its reflection and absorption in the atmosphere, vegetation, and soil. In the indoor environment, heat transport can occur in heat loss through building walls. Heat exchange between humans (or animals) and room air or the interior walls of a building is also important for indoor thermal comfort. Mass transfer indoors can involve transport of odors and indoor pollutants such as smoke, and indoor air quality is a subject of significant interest. Some specialized texts covering transport in the environment are included here for follow-up reading:

Brutsaert, W. 1982. *Evaporation into the Atmosphere: Theory, History and Applications.* Kluwer Academic Publishers, Boston.

Campbell, G. S. 1985. *Soil Physics with Basic Transport Models for Soil-Plant Systems. Developments in Soil Science.* Elsevier Science Publishers, Amsterdam.

Clark, M. M. 1996. *Transport Modeling for Environmental Engineers and Scientists.* John Wiley & Sons, New York, NY.

Ghildyal, B. P. and R. P. Tripathi. 1987. *Soil Physics.* Wiley Eastern Limited, New Delhi, India.

Hillel, D. 1980. *Movement of Solutes and Soil Salinity. Fundamentals of Soil Physics.* Academic Press, New York.

Jumikis, A. R. 1966. *Thermal Soil Mechanics.* Rutgers University Press, New Brunswick, N.J.

Loucks, D. P., J. R. Stedinger, and D. A Haith. 1981. Water quality prediction and simulation. In: *Water Resource Systems Planning and Analysis.* Prentice-Hall, Inc., Englewood Cliffs, N.J.

Monteith, J. L. and M. H. Unsworth. 1990. *Principles of Environmental Physics.* Edward Arnold, New York.

Neprin, S. V. and A. F. Chudnovskii. 1984. *Heat and Mass Transfer in the Plant-Soil-Air System.* Oxonian Press Pvt. Ltd., New Delhi, India.

Schnoor, J. L. 1996. *Environmental Modeling: Fate and Transport of Pollutants in Water, Air and Soil.* John Wiley & Sons, New York.

Stern, A. C. 1976. *Air Pollution*, 3rd Ed., Vol. I. Academic Press, New York.

Thibodeaux, L. J. 1979. *Chemodynamics: Environmental Movement of Chemicals in Air, Water, and Soil.* John Wiley & Sons, New York.

CONTENTS

ACKNOWLEDGMENTS	v
PREFACE	vii
INTRODUCTION	xi
LIST OF SYMBOLS	xxix

I Energy Transfer — 1

1 EQUILIBRIUM, ENERGY CONSERVATION, AND TEMPERATURE — 3
    1.1 Thermal Equilibrium and the Laws of Thermodynamics 5
        1.1.1 Laws of Thermodynamics 5
        1.1.2 Thermal Equilibrium 5
        1.1.3 Energy Conservation 5
    1.2 Non-Equilibrium Thermodynamics and the Transport of Energy ... 6
    1.3 Temperature in Living Systems 6
        1.3.1 Temperature Response to Human Body 7
        1.3.2 Temperature Sensation in Humans 7
        1.3.3 Thermal Comfort of Human and Animals 9
    1.4 Temperature in the Environment 10
        1.4.1 The Greenhouse Effect 10
    1.5 Temperature Scales 11
    1.6 Chapter Summary—Energy Conservation and Temperature 12
    1.7 Concept and Review Questions 13
    1.8 Further Reading 13

2 MODES OF HEAT TRANSFER — 15
    2.1 Conductive Heat Transfer 15
        2.1.1 Thermal Conductivity of Biological and Other Materials ... 20
        2.1.2 Thermal Diffusivity 20
        2.1.3 Density and Specific Heat 21
    2.2 Convective Heat Transfer 22
    2.3 Radiative Heat Transfer 23

2.4	Chapter Summary—Modes of Heat Transfer	23
2.5	Concept and Review Questions	24
2.6	Further Reading	24
2.7	Problems	25

3 GOVERNING EQUATION AND BOUNDARY CONDITIONS OF HEAT TRANSFER   27

3.1	Governing Equation for Heat Transfer Derived	29
	3.1.1 Meaning of Each Term in the Governing Equation	31
	3.1.2 Examples of Thermal Source (Generation) Term in Biological Systems	31
	3.1.3 Utility of the Energy Equation	32
3.2	General Boundary Conditions	34
3.3	The Bioheat Transfer Equation for Mammalian Tissue	36
3.4	Governing Equation Derived in Cylindrical Coordinates	38
3.5	Governing Equations in Various Coordinate Systems	40
3.6	An Algorithm to Solve Transport Problems	42
3.7	Chapter Summary—Governing Equations and Boundary Conditions	43
3.8	Concept and Review Questions	43
3.9	Further Reading	44
3.10	Problems	44

4 CONDUCTION HEAT TRANSFER: STEADY-STATE   45

4.1	Steady-State Heat Conduction in a Slab	47
	4.1.1 Thermal Resistance and Overall Heat Transfer Coefficient	48
4.2	Steady-State Heat Conduction in a Cylinder	50
	4.2.1 Thermal Resistance Term for the Hollow Cylinder	52
	4.2.2 Comparison with the Solution for the Slab	52
	4.2.3 Thickness of Fur in Small and Large Animals	53
4.3	Steady-State Heat Conduction in a Slab with Internal Heat Generation	55
	4.3.1 Thermoregulation–Maintaining the Core Body Temperature of Human and Animals	56
4.4	Steady-State Heat Transfer from Extended Surfaces: Fins	58
	4.4.1 Fins and Bio-heat Transfer	60
4.5	Chapter Summary—Steady-State Heat Conduction	61
4.6	Concept and Review Questions	62
4.7	Further Reading	63
4.8	Problems	64

5 CONDUCTION HEAT TRANSFER: UNSTEADY STATE   69

5.1	Transient Heat Transfer with No Internal Resistance	71
5.2	Biot Number: Deciding When to Ignore the Internal Resistance	72
5.3	Transient Heat Transfer with Internal Resistance	75
	5.3.1 How Temperature Changes with Time	77
	5.3.2 Temperature Change with Position and Spatial Average	77

	5.3.3	Temperature Change with Size	78
	5.3.4	Charts Developed from the Solutions: Their Uses and Limitations	78
	5.3.5	Internal and External Resistance	80
	5.3.6	Numerical Methods as Alternatives to the Charts	81
5.4	Transient Heat Transfer in a Finite Geometry—Multi-Dimensional Problems		82
5.5	Semi-infinite Region		82
5.6	Chapter Summary—Transient Heat Conduction		86
5.7	Concept and Review Questions		86
5.8	Further Reading		86
5.9	Problems		87

6 CONVECTION HEAT TRANSFER — 95

- 6.1 Governing Equation for Convection 95
- 6.2 Temperature Profiles and Boundary Layers Over a Surface 97
- 6.3 Laminar and Turbulent Flows 99
- 6.4 Convective Heat Transfer Coefficient Defined 100
- 6.5 Significant Parameters in Convective Heat Transfer 101
- 6.6 Convective Heat Transfer Coefficient Calculations 104
  - 6.6.1 Flat plate, Forced Convection 106
  - 6.6.2 Flat plate, Natural Convection 108
  - 6.6.3 Flow Over Cylinder, Natural Convection 110
  - 6.6.4 Flow Over Cylinder, Forced Convection 110
  - 6.6.5 Flow Through Cylinder, Forced Convection 111
  - 6.6.6 Flow Over Sphere, Natural Convection 111
  - 6.6.7 Flow Over Sphere, Forced Convection 111
  - 6.6.8 Laminar vs. Turbulent Flow 111
  - 6.6.9 Orders of Magnitude for Heat Transfer Coefficient Values . . 112
  - 6.6.10 Coefficients for Air Flow Over Human Subjects 112
  - 6.6.11 Wind Chill Factor and Boundary Layer Thickness 112
- 6.7 Chapter Summary—Convective Heat Transfer 115
- 6.8 Concept and Review Questions 115
- 6.9 Further Reading 116
- 6.10 Problems 117

7 HEAT TRANSFER WITH CHANGE OF PHASE — 123

- 7.1 Freezing and Thawing 125
- 7.2 Freezing of Pure Water 126
  - 7.2.1 Freezing Process 126
  - 7.2.2 Property Changes During Freezing 127
- 7.3 Freezing of Solutions and Biomaterials 128
  - 7.3.1 Solutions 128
  - 7.3.2 Cellular Tissues 129
  - 7.3.3 Cooling Rates and Success of Freezing 130
- 7.4 Temperature Profiles and Freezing Time 131

		7.4.1	Freezing Time for an Infinite Slab of Pure Liquid	131
		7.4.2	Freezing Time for Biomaterials	133
	7.5	Evaporation		135
		7.5.1	Evaporation from Wet Surfaces	136
		7.5.2	Evaporation Inside a Solid Matrix	138
		7.5.3	Evaporation of Solutions	138
	7.6	Chapter Summary—Heat Transfer with Change of Phase		139
	7.7	Concept and Review Questions		140
	7.8	Further Reading		140
	7.9	Problems		142
8	**RADIATIVE ENERGY TRANSFER**			**145**
	8.1	The Electromagnetic Spectrum		147
	8.2	Reflection, Absorption and Transmission of Waves at a Surface		147
		8.2.1	Transmissivity of a Leaf and Photosynthesis	148
		8.2.2	Transmissivity of the Atmosphere: Greenhouse Effect	149
		8.2.3	Albedo–Reflection from Soil	149
		8.2.4	Absorption and Transmission in Biomaterials	150
	8.3	Thermal Radiation from an Ideal (Black) Body		150
	8.4	Fraction of Energy Emitted		153
	8.5	Thermal Radiation from a Real Body: Emissivity		154
	8.6	Emission from Human Bodies		156
	8.7	Solar, Atmospheric, and Earth Surface Radiation		158
		8.7.1	Solar Radiation–Magnitude and Spectral Distribution	158
		8.7.2	Emission from Earth's Surface	159
		8.7.3	Atmospheric Emissions	159
		8.7.4	Global Energy Balance	160
	8.8	Radiative Exchange Between Bodies		160
		8.8.1	Radiative Heat Transfer Coefficient	163
		8.8.2	Radiative Exchange Between a Leaf and Surroundings	163
		8.8.3	Radiative Exchange Between Human (or Animal) and Its Surroundings	165
	8.9	Chapter Summary—Radiative Energy Transfer		165
	8.10	Concept and Review Questions		167
	8.11	Further Reading		168
	8.12	Problems		169

II Mass Transfer 173

9	**EQUILIBRIUM, MASS CONSERVATION, AND KINETICS**		**175**
	9.1	Concentration	177
		9.1.1 Concentrations in a Gaseous Mixture	177
	9.2	Species Mass Balance (Mass Conservation)	179
	9.3	Equilibrium	179
		9.3.1 Equilibrium Between a Gas and a Liquid	180

CONTENTS   XXV

      9.3.2   Equilibrium Between a Gas and a Solid (with Adsorbed Liquid) 182
      9.3.3   Equilibrium Between Solid and Liquid in Adsorption 185
  9.4 Chemical Kinetics: Generation or Depletion of a Mass Species 185
      9.4.1   Rate Laws of Homogeneous Reactions 186
      9.4.2   Zeroth Order Reaction 186
      9.4.3   First Order Reaction 187
      9.4.4   *nth* Order Reaction 189
      9.4.5   Effects of Temperature 189
  9.5 Chapter Summary—Equilibrium, Mass Conservation, and Kinetics .. 189
  9.6 Concept and Review Questions 190
  9.7 Further Reading 190
  9.8 Problems 191

10 MODES OF MASS TRANSFER   195
  10.1 A Primer on Porous Media Flow 197
      10.1.1 Physical Interpretation of Hydraulic Conductivity K and Permeability k 198
      10.1.2 Capillarity and Unsaturated Flow in a Porous Media 201
      10.1.3 Osmotic Flow in a Porous Media 202
  10.2 Diffusive Mass Transfer 203
      10.2.1 Molecular Diffusion 203
      10.2.2 Capillary Diffusion 211
  10.3 Dispersive Mass Transfer 213
  10.4 Convective Mass Transfer 214
      10.4.1 Convection-Diffusion Mass Transfer Over a Surface 214
      10.4.2 Convection-Dispersion Mass Transfer 216
  10.5 Comparison of the Modes of Mass Transfer 216
  10.6 Chapter Summary—Modes of Mass Transfer 218
  10.7 Concept and Review Questions 219
  10.8 Further Reading 219
  10.9 Problems 220

11 GOVERNING EQUATIONS AND BOUNDARY CONDITIONS OF MASS TRANSFER   223
  11.1 Modified Fick's Law for Bulk Flow or Convection 224
      11.1.1 Velocity and Mass Average Velocity 225
      11.1.2 Flux Equation for a Convective Situation 225
  11.2 Governing Equation for Mass Transfer 227
      11.2.1 Meanings of Each Term in the Governing Equation 229
  11.3 General Boundary Conditions 230
  11.4 Governing Equations for Mass Diffusion in Various Coordinate Systems 233
  11.5 Chapter Summary—G.E. & B.C. of Mass Transfer 234
  11.6 Concept and Review Questions 234
  11.7 Further Reading 234
  11.8 Problems 235

12 DIFFUSION MASS TRANSFER: STEADY STATE   237

12.1 Steady-State Mass Diffusion in a Slab 238

    12.1.1 One-Dimensional Mass Diffusion Through a Composite Slab—Overall Mass Transfer Coefficient 240

12.2 Steady-State Diffusion in a Slab with Chemical Reaction 244

    12.2.1 Photosynthesis and the Transport of Water Vapor and CO_2 in a Leaf 247

12.3 Chapter Summary—Steady-State Mass Diffusion with and Without Chemical Reaction 248

12.4 Concept and Review Questions 249

12.5 Further Reading 249

12.6 Problems 250

13 DIFFUSION MASS TRANSFER: UNSTEADY-STATE   253

13.1 Transient Mass Transfer with No Internal Diffusive Resistance 254

13.2 Transient Mass Transfer with Internal Diffusive Resistance 257

    13.2.1 How Concentration Changes with Time 259

    13.2.2 Concentration Change with Position and Spatial Average . . . 259

    13.2.3 Concentration Change with Size 260

    13.2.4 Charts Developed from the Solutions: Their Uses and Limitations 260

    13.2.5 When Both Internal and External Resistances are Present: Convective Boundary Condition 261

13.3 Transient Diffusion in a Finite Geometry—Multi-Dimensional Problems 262

13.4 Transient Diffusion in a Semi-Infinite Region 266

13.5 Chapter Summary—Transient Diffusive Mass Transfer 269

13.6 Concept and Review Questions 269

13.7 Further Reading 270

13.8 Problems 270

14 CONVECTION MASS TRANSFER   277

14.1 Convection-Dispersion in an Infinite Fluid 279

14.2 Convection-Dispersion in a Semi-Infinite Porous Solid 284

14.3 Convection-Dispersion in a Semi-Infinite Porous Solid: Inclusion of Sorption 289

14.4 Convection-Diffusion in a Stagnant Gas 291

14.5 Convection-Diffusion Over a Surface 295

    14.5.1 Concentration Profiles and Boundary Layers Over a Surface . 295

    14.5.2 Convective Mass Transfer Coefficient Defined 298

    14.5.3 Significant Parameters in Convective Mass Transfer 299

    14.5.4 Calculation and Physical Implications of Convective Mass Transfer Coefficient Values 300

    14.5.5 Convection-Diffusion of Heat and Mass (Simultaneous) Over a Surface: Example of a Wet Bulb Thermometer 304

14.6 Chapter Summary—Convective Mass Transfer 305

14.7 Concept and Review Questions 306

14.8 Further Reading 307
14.9 Problems 308

III APPENDIX 313

A SUMMARY 315
- A.1 Generic Transport Equation and the Physical Meanings of the Constituent Terms 316
- A.2 Generic Boundary Conditions and Initial Condition Needed to Solve the Transport Equation in A.1 316
- A.3 Solution Map for Heat Conduction and Mass Diffusion Problems (Without Bulk Flow) 317
- A.4 Solution Map for Mass Diffusion or Dispersion with Bulk Flow (Convection-Diffusion/Dispersion) 318
- A.5 Basic Heat and Mass Transfer Parameters and Fluxes 319
- A.6 Heat and Mass Transfer Governing Equations and Boundary Conditions 320

B PHYSICAL CONSTANTS, UNIT CONVERSIONS, AND MATHEMATICAL FUNCTIONS 321
- B.1 Physical Constants 321
- B.2 Some Useful Conversion Factors 322
- B.3 Error Function Tabulated 325
- B.4 Charts for Unsteady Diffusion 326

C HEAT TRANSFER AND RELATED PROPERTIES 331
- C.1 Basal Metabolic Rate for a Few Animals 332
- C.2 Typical Metabolic Rate for Various Human Activities 333
- C.3 Blood Flow, Oxygen Consumption, and Metabolic Rate for Various Human Organs 334
- C.4 Thermal Properties of Animal Materials 335
- C.5 Thermal Conductivities of Some Animal Hair Coats 337
- C.6 Thermal Properties of Some Agricultural Materials 338
- C.7 Thermal Properties of Food Materials (Representative Values) . . . 339
- C.8 Thermal Properties of Air at Atmospheric Pressure 340
- C.9 Psychrometric Chart 341
- C.10 Vapor Pressure of Liquid Water from 0 to 100°C 342
- C.11 Steam Properties at Saturation Temperatures 343
- C.12 Thermophysical Properties of Saturated Water 344

D MASS TRANSFER PROPERTIES 345
- D.1 Apparent Diffusivities in Solids 346
- D.2 Apparent Diffusivities in Liquids 347
- D.3 Diffusivities in Air at 1 atm 348
- D.4 Representative Values of Resistances to Water Vapor Transport Out of Leaves 349

	D.5	Approximate Ranges of Dispersion Coefficients in Surface Water . . .	350
	D.6	Approximate Ranges of Dispersion Coefficients in Porous Media . . .	351
	D.7	Kinetic Parameters for Various First Order Rate Processes in Biological and Environmental Systems	352
	D.8	Surface Tension of Water	353
	D.9	References to Data on Heat and Mass Transfer and Related Properties	354
E	**MISCELLANEOUS ENVIRONMENTAL DATA**		**357**
	E.1	Atmospheric Temperature, Pressure, and Other Parameters as Function of Altitude	358
	E.2	National (U.S.) Primary Ambient Air Quality Standards	359
F	**EQUATIONS OF MOTION IN VARIOUS COORDINATE SYSTEMS**		**361**
	F.1	The Equation of Motion in Rectangular Coordinates $(\mathbf{x}, \mathbf{y}, \mathbf{z})$	361
	F.2	The Equation of Motion in Cylindrical Coordinates $(\mathbf{r}, \theta, \mathbf{z})$	362
	F.3	The Equation of Motion in Spherical Coordinates $(\mathbf{r}, \theta, \phi)$	362
G	**SOME USEFUL MATHEMATICAL BACKGROUND**		**365**
	G.1	Series Solution to the One-Dimensional Heat Equation	365
	G.2	Similarity Transformation of the Heat Equation	369
	G.3	Lumped Parameter Analysis as Related to the Energy Equation	371
	G.4	Transformation of the Convection-Diffusion Equation to the Diffusion Equation	372
	G.5	Solution to Steady One-Dimensional Diffusion in a Slab with Chemical Reaction	373
INDEX			**375**

LIST OF SYMBOLS

A	area, m^2
Bi	$= hL/k$, Biot number, dimensionless
Bi_m	$= h_m K^* L / D_{AB}$, mass transfer Biot number, dimensionless
c	total concentration (sometimes abbreviated c_A), kg/m^3
c_A	concentration of component A, kg of A/m^3
$c_{A,s}$	concentration of component A at a surface, kg of A/m^3
$c_{A,\infty}$	concentration of component A in the bulk fluid, kg of A/m^3
$c_{A,i}$	initial concentration, kg/m^3
c_{av}	average concentration, kg/m^3
c_p	specific heat at constant pressure, kJ/kg·°C
d	diameter, m
D_{AB}	diffusivity of species A in species B, m^2/s
erf	error function
E	total emissive power, W/m^2
$\dot{E}$	energy of flowing fluid per unit time, kJ/s
E_z	dispersion coefficient in the z direction, m^2/s
E_a	activation energy, J/mole
f	frictional coefficient in Stokes equation for diffusivity
F	energy flux, W/m^2
Fo	$= \alpha t / L^2$, Fourier number, dimensionless
g	gravity, m/s^2
Gr	$= g\beta \Delta T \rho^2 L^3 / \mu^2$, Grashof number, dimensionless
Gr_{AB}	$= g\rho \Delta \rho_A L^3 / \mu^2$, mass transfer Grashof number, dimensionless
h	convective heat transfer coefficient, W/m^2·°C
h	Planck's constant, 6.625×10^{-34} J·s
h	sum or pressure and matric potential, m
h_m	convective mass transfer coefficient, m/s
H	enthalpy per unit mass, kJ/kg
H	Henry's law constant, atm/mole fraction
$\bar{H}$	humidity of air, kg of vapor/ kg of dry air
$\mathcal{H}$	hydraulic head, m
$j_{A,z}$	diffusive or dispersive mass flux of A in the z direction, kg/m^2·s
j^v	volumetric flux, m^3/m^2·s or m/s

k	thermal conductivity, W/m·°C
k	permeability, m²
k''	reaction rate constant for first order reaction, 1/s
k_f	thermal conductivity of fluid, W/m·°C
K	hydraulic conductivity, m/s
K^*	distribution coefficient or partition coefficient, units vary
L	half thickness of a slab or characteristic length, m
L_p	membrane permeability, m/Pa·s
m	non-dimensional parameter in Chapters 4 and 12
m	mass, kg
m_s	mass of solids, kg
m	non-dimensional (inverse Biot) number k/hL
M	molecular weight, kg
$\mathcal{M}$	molality or moles of solute per unit mass solvent mol/kg
$\dot{m}$	mass flow rate, kg/s
n	non-dimensional distance
n	order of reaction, dimensionless
$n_{A,z}$	mass flux of species A in the z direction, kg/m²·s
n^v	volumetric flux, m³/m²·s
$N_{A,z}$	mass flow rate of species A in the z direction, kg/s
Nu	Nusselt number, dimensionless
Nu_x	Nusselt number at a location x, dimensionless
Nu_L	Average Nusselt number based on characteristic length L, dimensionless
P	total pressure, N/m² or Pa
p_A	partial pressure of component A, N/m² or Pa
p_m	permeability, m/s
P	perimeter, m
Pr	$=\mu c_p/k$, Prandtl number, dimensionless
q_x	heat flow in the x direction, W
q_x''	heat flux in the x direction, W/m²
Q	volumetric heat generation, W/m³
r	radial direction
r_A	rate of generation of A per unit volume, kg of A/m³·s
RH	Relative humidity, fraction
R_g	universal gas constant = 8.315 kJ/kmol·K
Ra	$= Gr \times Pr$, Rayleigh number, dimensionless
Ra_m	$= Gr_{AB} \times Sc$, mass transfer Rayleigh number, dimensionless
Re	$= \rho u_\infty L/\mu$, Reynolds number, dimensionless
Sc	$= \mu/\rho D_{AB}$, Schmidt number, dimensionless
Sh	$= h_m L/D_{AB}$, Sherwood number, dimensionless
t	time, s
$t_{1/2}$	half life, s
T	temperature
T_s	surface temperature
T_i	initial or inlet temperature
T_f	freezing point, K

LIST OF SYMBOLS

T_∞	fluid or ambient temperature
u	velocity in the x direction, m/s
u_∞	free stream velocity in the x direction, m/s
U	thermal energy per unit volume, J/m^3
U	overall heat transfer coefficient, W/m$^2\cdot{}^\circ$C
U_m	overall mass transfer coefficient, m/s
v	velocity in y direction, m/s
V	volume, m^3
w	moisture content, kg of water/kg of dry solids
x	x coordinate, m
x_A	mole fraction of species A in liquid phase
y	y coordinate, m
y_A	mole fraction of species A in vapor phase
z	z coordinate, m

Greek Letters

α	thermal diffusivity, m^2/s; also absorptivity, dimensionless
β	coefficient of thermal expansion
δ	penetration depth of microwaves
δ_{conc}	concentration boundary layer thickness, m
δ_{vel}	velocity boundary layer thickness, m
$\delta_{thermal}$	thermal boundary layer thickness, m
Δ	finite change
ΔH_f	latent heat of fusion, kJ/kmol or kJ/kg
ΔH_{vap}	latent heat of vaporization, kJ/kmol or kJ/kg
$\Delta \beta_i$	volume fraction of pores with radius r_i, dimensionless
ϵ	emissivity, dimensionless
η	non-dimensional quantity in Chapter 5
γ	surface tension, N/m
κ	Boltzmann's constant, 1.380×10^{-23} J/K
λ	wavelength, m
μ	viscosity, kg/m$\cdot$s
ν	kinematic viscosity, m^2/s
Π	osmotic pressure, Pa
ρ	density or mass concentration, kg/m^3; also reflectivity, dimensionless
σ	Stefan-Boltzmann constant, 5.676×10^{-8} W/m$^2\cdot$K^4
σ_{AB}	collision diameter, $^\circ$A
$\Omega_{D,AB}$	dimensionless function defined in Chapter 10
τ	transmissivity, dimensionless
θ	non-dimensional temperature

Biological and Bioenvironmental Heat and Mass Transfer

Part I

Energy Transfer

Chapter 1

EQUILIBRIUM, ENERGY CONSERVATION, AND TEMPERATURE

CHAPTER OBJECTIVES

After you have studied this short introductory chapter, you should be able to:

1. Explain thermal equilibrium and how it relates to energy transport.

2. Understand temperature ranges important for biological systems and temperature sensing in mammals.

3. Understand temperature ranges important for the environment.

KEY TERMS

- first and second law of thermodynamics
- thermal equilibrium
- energy transfer
- tolerable temperatures
- deep body temperature
- cold receptor
- heat receptor
- greenhouse effect

This is the first chapter in the part on energy transfer. It discusses energy transfer as the logical next step after what you have studied in thermodynamics. Relationship of this chapter to other chapters is shown in Figure 1.1.

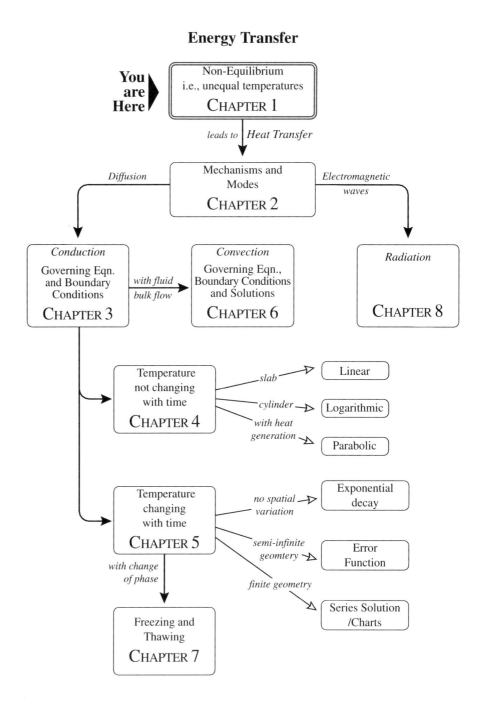

Figure 1.1: Concept map of energy transfer showing how the contents of this chapter relates to other chapters on energy transfer.

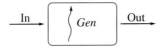

Figure 1.2: A control volume for energy conservation showing different components.

1.1 Thermal Equilibrium and the Laws of Thermodynamics

1.1.1 Laws of Thermodynamics

Thermodynamics deals with interchanges between various forms of energy. The first law of thermodynamics states that energy is conserved. Total energy of the system plus surroundings remain constant. Different forms of energy can inter-convert but their sum remains constant. The second law of thermodynamics states that the total entropy of a system plus surroundings never decreases. This is equivalent to the statement that heat spontaneously flows from a body of higher temperature to one at lower temperature.

1.1.2 Thermal Equilibrium

Two systems are said to be in thermal equilibrium when their temperatures are equal. This is different from *steady state*. In steady state, temperatures do not change with time. In equilibrium, there are no heat flows. This will be explained further in Chapter 4.

1.1.3 Energy Conservation

Conservation of energy is the first law of thermodynamics, as was just mentioned. It is one of the two pillars on which the subject of energy transfer stands. The other pillar being the laws describing the rate of energy transfer, as will be discussed later. To apply the conservation of *thermal* energy to an arbitrary system as shown in Figure 1.2, we can write a word equation as

$$\begin{matrix} \text{Rate of} \\ \text{Energy In} \end{matrix} - \begin{matrix} \text{Rate of} \\ \text{Energy Out} \end{matrix} + \begin{matrix} \text{Rate of} \\ \text{Energy Generation} \end{matrix} = \begin{matrix} \text{Rate of} \\ \text{Energy Storage} \end{matrix} \quad (1.1)$$

Note that other forms of energy can convert into thermal energy. When this happens, we think of this source of thermal energy as part of "Energy Generation." Increase in energy storage is manifested as increase in the temperature. Equation 1.1 will be used in Chapter 3 over a control volume to develop general equations of energy transfer.

For an illustration of energy conservation using the Eqn. 1.1, consider a solid and a liquid at two different temperatures as shown in Figure 1.3. At some point in time,

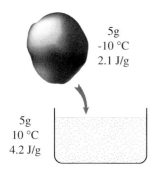

Figure 1.3: Illustration of energy conservation.

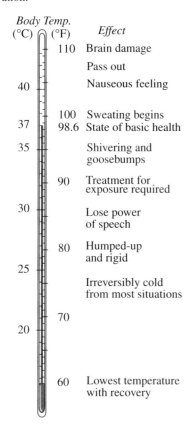

Figure 1.4: Effect of various deep body temperatures on humans. Data from Egan (1975).

the solid is put in the liquid. We are interested in knowing the final temperature of the system. Consider no heat loss to the surroundings at any time.

Let m_1, c_{p_1}, T_1 and m_2, c_{p_2}, T_2 be the mass, specific heat and temperature, respectively, of bodies 1 and 2. Recall that the energy content of a mass m and specific heat c_p at temperature T is $mc_p(T - T_{ref})$ where T_{ref} is a reference temperature. In using Eqn. 1.1 with the liquid as the system, we can write the individual terms as

$$\begin{aligned}
\text{Energy In} &= m_1 c_{p_1}(T_1 - 0) \\
\text{Energy Out} &= 0 \\
\text{Energy Generation} &= 0 \\
\text{Change in Energy Storage} &= \underbrace{(m_1 c_{p_1} + m_2 c_{p_2})(T - 0)}_{\text{final}} - \underbrace{m_2 c_{p_2}(T_2 - 0)}_{\text{initial}}
\end{aligned}$$

Plugging in Eqn. 1.1, we get

$$m_1 c_{p_1}(T_1 - 0) = (m_1 c_{p_1} + m_2 c_{p_2})(T - 0) - m_2 c_{p_2}(T_2 - 0) \tag{1.2}$$

Using the numerical values as shown in Figure 1.3, we get

$$5(2.1)(-10 - 0) = (5(2.1) + 5(4.2))(T - 0) - 5(4.2)(10 - 0) \tag{1.3}$$

from which we can calculate the final temperature $T = 3.33°C$. Thus, using energy conservation, we can find the final or equilibrium temperature or the final state of the system. Note that energy conservation alone is unable to predict how long it will take to reach the final or equilibrium temperature.

1.2 Non-Equilibrium Thermodynamics and the Transport of Energy

As was just illustrated, the laws of thermodynamics deal with the final or equilibrium state of a process. They provide no information on the nature of the interaction (what type of energy transfer?) or the rate of the process (how long does it take to reach the equilibrium states?). Since energy transfer takes place only when two bodies are not in equilibrium, the subject of energy transfer is sometimes described as non-equilibrium thermodynamics. As thermodynamics does not provide rate information, additional rate laws are defined in non-equilibrium thermodynamics to study the rates of energy transfer. The two rate laws that will be defined in this text are the Fourier's law of energy diffusion (Chapter 2) and Fick's law of mass diffusion (Chapter 10).

1.3 Temperature in Living Systems

Most organisms live within a narrow temperature range, with a maximum temperature becoming more disastrous than a minimum temperature. As shown in Figure 1.5, most biological activity is confined to a rather narrow temperature range of 0-60°C.

1.3. TEMPERATURE IN LIVING SYSTEMS

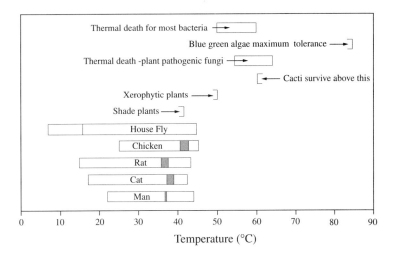

Figure 1.5: Tolerable temperature regimes for many plants and animals. Shaded areas represent the normal temperature for the organism and the unshaded areas represent the extreme tolerances. Adapted from The energy environment in which we live, by D. M. Gates, 1963. Printed by permission of *American Scientist*, Journal of Sigma Xi, The Scientific Research Society.

Although many simple organisms and a few higher forms can remain viable after exposure to absolute zero, most plants and animals do not carry on biological activity below 0°C. Effects of freezing will be described in detail in Chapter 7. Some of the higher animals maintain a very narrow range in body temperature, as shown in Figure 1.5, through complex physiological controls. Hence, for both plants and animals, the temperature of an organism resulting from environmental influence is of critical importance. The biological systems studied in this text often refer to higher animals and therefore the temperature change in this context is within a narrow range.

1.3.1 Temperature Response to Human Body

As an example of how temperature affects the state of a living system, consider the effect of deep body temperature on humans as shown in Figure 1.4 on the previous page. Man is a constant temperature animal with deep body temperature of about 98.6°F (37°C). It is obvious that temperature needs to be controlled within a very narrow range for effective biological functions.

1.3.2 Temperature Sensation in Humans

The human being can perceive different gradations of cold and heat, as shown in the top portion of Figure 1.6. Thermal gradations are discriminated by two types of temperature receptors—heat receptor and cold receptor. In addition, there are pain receptors

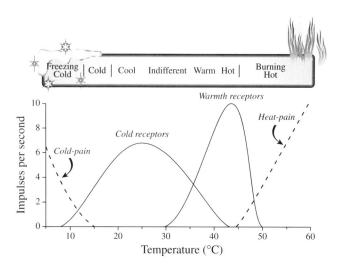

Figure 1.6: Strength of signal from cold, heat, and pain receptors at different temperatures.

that are stimulated only by extreme degrees of heat or cold, as in "burning hot" or "freezing cold" sensations. The brain interprets the input from the different combinations of these receptors as a particular temperature sensation. The heat receptor and cold receptors are free nerve endings located immediately under the skin at discrete but separated points. It is believed that the cold and heat receptors are stimulated by changes in their metabolic rates, these changes resulting from the fact that temperature alters the rates of intracellular chemical reactions more than twofold for each $10°C$ change.

As shown in Figure 1.6, the heat receptors are most sensitive to a range of temperatures about $30 - 45°C$. The cold receptors, on the other hand, are sensitive to the range of temperatures between $10 - 40°C$. Note that the strength of the signal, measured in terms of nerve impulses per second, is different at different levels of temperature. In the extreme hot or cold region, only pain fibers are stimulated. Extreme degrees of cold or heat can both be painful and both these sensations, when intense enough, may give almost the same quality of sensation— that is, freezing cold and burning hot sensations are almost alike; they are both painful.

Both hot and cold receptors rapidly adapt, so that within about a minute of continuous stimulation, the sensation of heat or cold begins to fade. Thus thermal senses respond markedly to *changes in temperature* in addition to being able to respond to steady states of temperature. Thus when the temperature of the skin is actively falling, a person feels much colder than when the temperature remains cold at that same level. This explains why, on a cold day, it feels much colder when first coming out of a warm house into the outdoors than staying in the same outdoors for longer time. Conversely, when temperature is rising, the person feels much warmer than if that same temperature remained constant. A tub of warm water feels much warmer when first entering in it than if staying in the tub for a longer time.

1.3.3 Thermal Comfort of Human and Animals

Human skin surface should be about 92°F for comfort. The human body maintains a balance with its environment through minor physiological changes (i.e., by increasing or decreasing the flow of blood to the skin). Thermoregulation is the physiological mechanism by which mammals and birds attempt to balance heat gain and loss in order to maintain a constant body temperature. This is described in detail in Chapter 4. When surrounded by air, body heat losses are primarily by convection, evaporation, and radiation as shown in Figure 1.7.

Body heat losses are affected by air temperature, humidity, and velocity, and other factors as shown in Figure 1.8. These are the thermal comfort factors. Since evaporation (sweating) has a cooling effect, higher evaporation is required at higher temperature. Lower humidity of air facilitates larger evaporation, which leads to temperature-humidity combination for human comfort as shown in Figure 1.8.

Information such as in Figure 1.8 is valuable in the design of buildings. Building surface temperatures are important factors for achieving thermal comfort since the body exchanges radiative heat with the walls around it. This is is discussed in more detail in

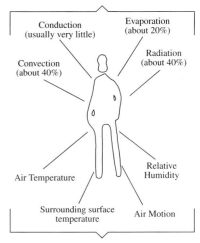

Figure 1.7: Heat loss and thermal comfort factors for human and animals. Adapted from Concepts in thermal comfort by M. D. Egan, ©1975 by Prentice Hall. Reprinted by permission.

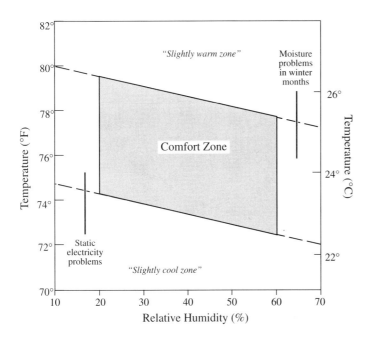

Figure 1.8: Temperature and humidity combinations for human comfort. Adapted from Concepts in thermal comfort by M. D. Egan, ©1975 by Prentice Hall. Reprinted by permission.

Chapter 8. We can also see from Figure 1.8 that in the comfort zone, human tolerance to humidity is much greater than human tolerance to temperature. Consequently, air temperatures need to be more carefully controlled.

1.4 Temperature in the Environment

1.4.1 The Greenhouse Effect

Gases such as CO_2, methane, and Chloroflurocarbons (CFCs) in our atmosphere have the property to let energy from the sun pass through to earth but stops the energy reflected from the earth surface from escaping into space. This trapping of energy by the atmosphere, much like the heating up of the insides of cars or greenhouses on a sunny day, is the well known greenhouse effect. The more these gases are present in our atmosphere, the higher is the earth's temperature. For example, the carbon dioxide concentration and surface temperature have been shown to be closely related, as shown in Figure 1.9. As can be seen from Figure 1.9, during the past 100 years there has been about 0.5°C of **real warming**. The question is how much the temperature

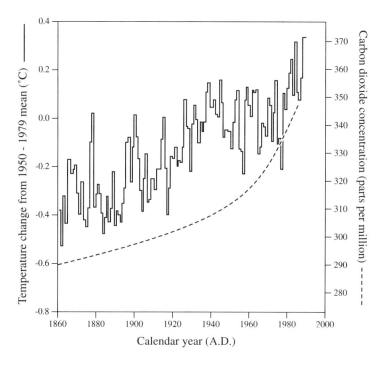

Figure 1.9: History of carbon dioxide concentration and global temperature change for recent times. From *The Changing Climate* by S. H. Schneider. Copyright ©1989 by Scientific American, Inc. All rights reserved.

will be raised, not whether the temperature is raised by the greenhouse gases. Several research centers throughout the world are working on an accurate prediction of the rise in temperature as a function of human activity. **A few degrees of warming** can raise the sea level between 0.2 and 1.5 meters because of wide-spread melting at the polar icecaps, leading to disastrous consequences.

1.5 Temperature Scales

To refresh our knowledge, some common temperature scales are provided in Table 1.1.

Table 1.1: Various temperature scales and the values of some reference temperatures

	Celsius (°C)	Kelvin (K)	Fahrenheit (°F)	Rankine (°R)
Steam point	100	373.15	212	671.67
Ice point	0	273.15	32	491.67
Absolute zero	-273.15	0	-459.67	0

The relationships between the scales are given as

$$T(°F) = 1.8 T(°C) + 32 \quad (1.4)$$
$$T(K) = T(°C) + 273.15 \quad (1.5)$$
$$T(°R) = T(°F) + 459.67 \quad (1.6)$$

Whenever temperature differences are involved, one may use either the absolute scale or the relative scale. For example, a temperature difference of 15°C is equal to 15 K. Thus

$$\Delta T(K) = \Delta T(°C) \quad (1.7)$$
$$\Delta T(°R) = \Delta T(°F) \quad (1.8)$$
$$\Delta T(°F) = 1.8 \Delta T(°C) \quad (1.9)$$

Note that wherever a quantity is to be multiplied or divided by temperatures (not temperature differences), absolute scales for temperature must be used except in empirical equations. A feel for various temperature magnitudes is provided in Table 1.2.

Table 1.2: Some temperature values

Lowest achieved temperature	10^{-12} K
Liquid Helium	4.2 K
Liquid Nitrogen	$-196.15°$C
Liquid CO_2	$-73°$C
Temperate rise in atmosphere due to the greenhouse effect	$0.5°$C
Home refrigerator	
(freezer section)	$-24°$C
(refrigerator section)	$0°$C
Weather	
New York (winter)	$-18°$C($\approx 0°$F)
Miami (winter)	$2°$C($\approx 35°$F)
New York (summer)	$35°$C($\approx 95°$F)
Miami (summer)	$33°$C($\approx 91°$F)
Normal human body	$37°$C
Surface gases in the sun	$6000°$C

1.6 Chapter Summary— Energy Conservation and Temperature

- **Laws of Thermodynamics (page 5)**

  1. First law of thermodynamics states that energy is conserved.
  2. Second law of thermodynamics is equivalent to the condition that heat spontaneously flows from a body of higher temperature to a body of lower temperature.
  3. Laws of thermodynamics provide no information on the rate of the energy transfer process.

- **Temperature and the Living Systems (page 6)**

  1. Most living organisms can tolerate only narrow ranges of temperatures.
  2. Temperature sensation in humans is provided by heat and cold receptors located immediately under the skin.

- **Temperatures in the Environment (page 10)**

  1. The greenhouse effect has lead to about $0.5°$C of real warming. A few degrees of warming can be disastrous.

2. The temperature rise due to the greenhouse effect has been related to increase in gases such as CO_2, methane and Chlorofluorocarbons in our atmosphere.

1.7 Concept and Review Questions

1. What additional information is needed beyond the knowledge of thermodynamic equilibrium to decide *how long* it takes for a heating or cooling process?

2. What is the typical temperature range below or above which cell death occurs in human tissue?

3. What has been the approximate temperature rise of the earth's surface in the last 100 years?

4. Is the temperature sensation for a given person always over a fixed temperature range?

5. From a comfort standpoint, which one is a more critical variable—temperature or humidity?

1.8 Further Reading

Egan, M. D. 1975. *Concepts in Thermal Comfort*. Prentice Hall, Englewood Cliffs, N.J.

Gates, D. M. 1963. The energy environment in which we live. American Scientist. 51:327-348.

Guyton, A. C. and J. E. Hall. 1996. *Textbook of Medical Physiology*. W. B. Saunders Company, Philadelphia.

Nobel, P. S. 1991. *Physiochemical and Environmental Plant Physiology*. Academic Press, San Diego.

Rich, L. G. 1973. *Environmental Systems Engineering*. McGraw-Hill Book Company, New York.

Schneider, S. H. 1989. The Changing Climate. Scientific American, September, 1989.

Shitzer, A. and R.C. Eberhart. 1985. *Heat Transfer in Medicine and Biology*, Volumes I and II. Plenum Press, New York.

Chapter 2

MODES OF HEAT TRANSFER

CHAPTER OBJECTIVES

After you have studied this chapter, you should be able to:

1. Understand the physical processes and the rate laws describing the three modes of heat transfer, conduction, convection, and radiation.
2. Understand the material properties that affect heat conduction in a material.

KEY TERMS

- **diffusion**
- **conduction**
- **Fourier's law**
- **thermal conductivity**
- **thermal diffusivity**
- **heat flux**
- **heat flow rate**
- **convection**
- **bulk flow**
- **convective heat transfer coefficient**
- **convection**
- **radiation**

In this chapter we will study the fundamental ways energy can be transported. Figure 2.1 shows how the contents of this chapter relates to the overall subject of energy transfer.

2.1 Conductive Heat Transfer

Conductive heat transfer is the movement of thermal energy through a medium from its more energetic particles to the less energetic. The temperature at any location in a

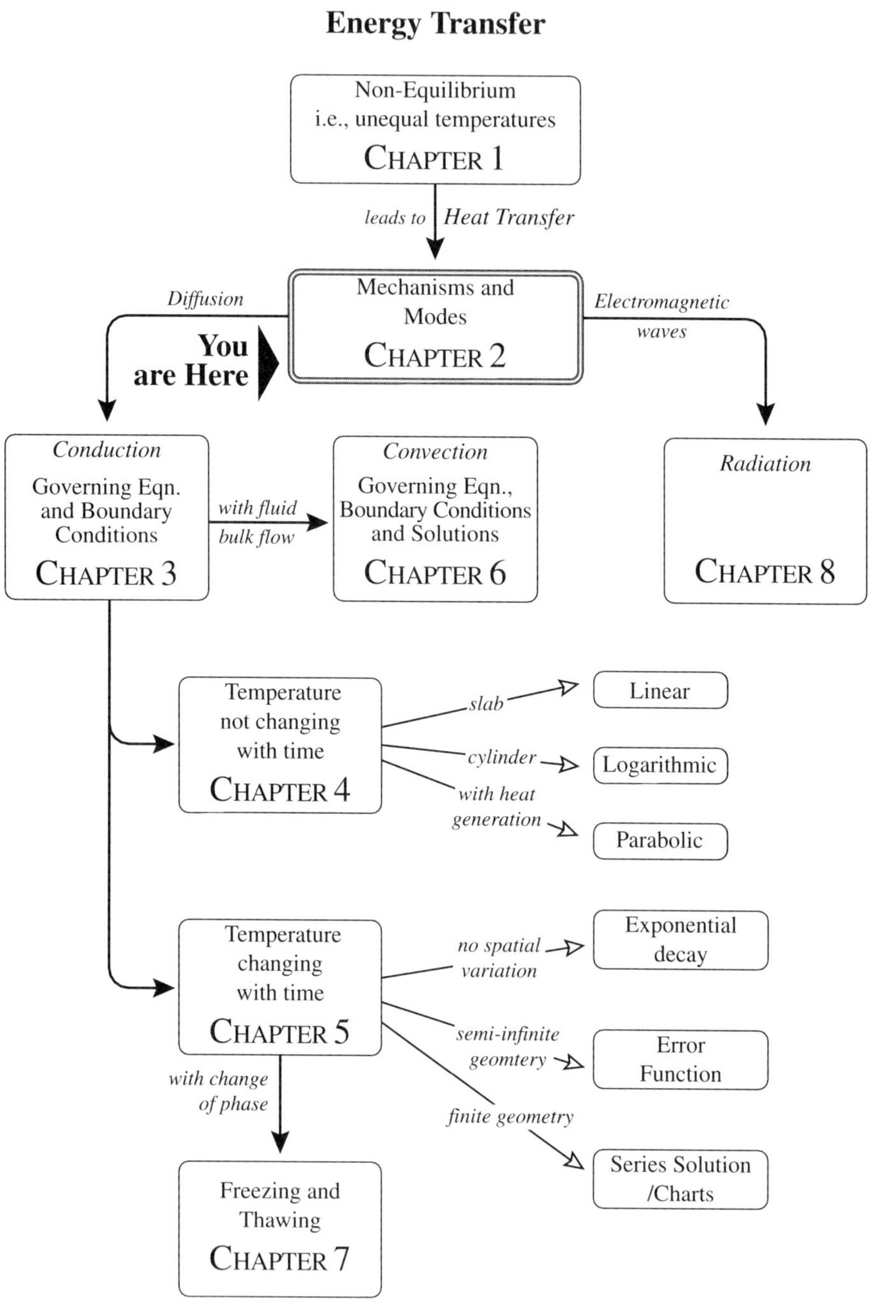

Energy Transfer

Figure 2.1: Concept map of energy transfer showing how the contents of this chapter relates to other chapters on energy transfer.

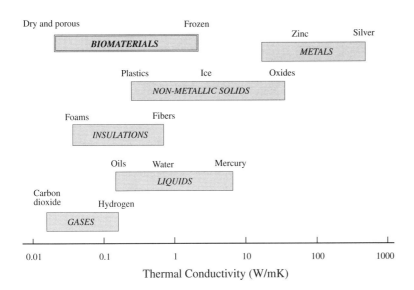

Figure 2.2: Range of thermal conductivity values (at normal temperature and pressure) for various materials compared with those for biomaterials. Adapted from Fundamentals of Heat and Mass Transfer by F. P. Incropera and D. P. DeWitt, ©1990 John Wiley & Sons, Inc. Reprinted by permission of John Wiley & Sons, Inc.

material is associated with the energy of the molecules around that location. Higher temperature corresponds to higher molecular energy. In a gas, this energy can be translational, rotational or vibrational. As the molecules are constantly colliding, they are transferring energy from the more energetic molecules to the less energetic. Thus, energy is transferred from higher to lower temperature. This net transfer of energy due to random molecular motion is termed as the *diffusion* of energy. In a solid, translational and rotational motions are restricted. As the temperature of one area of a material increases, the molecules in that area vibrate more and bump into neighboring molecules. This contact between molecules imparts some of the vibrational motion of the first molecule to the second molecule, which then begins to vibrate to a greater extent. This trend continues throughout the material, spreading heat energy by the introduction of increased vibrational motion. In metals there are unbound electrons that translate freely in the material. These electrons allow for a faster and more efficient means of heat conduction in metallic materials than in non-metallic materials. Since thermal conductivity (explained later) is a measure of the efficiency of heat conduction, it is higher for metallic solids than for non-metallic solids, as shown in Figure 2.2.

In a liquid, the intermolecular spacings are larger than in solids. Molecules move more randomly. This leads to less effective energy transport and lower values of conductivity, as shown in Figure 2.2. In a gas, the molecules are farther apart. Molecular movements are most random and interactions are less frequent. Thus the lowest thermal conductivity values are found in gases (Figure 2.2).

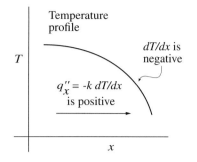

Figure 2.3: Direction of heat flux as related to the temperature gradient.

Conduction heat transfer can be described by the Fourier's equation (rate law):

$$\frac{q_x}{A} = -k\frac{dT}{dx} \quad (2.1)$$

where q_x is the rate of heat flow in the x direction, A is area perpendicular to the x direction through which the heat flows, k is the thermal conductivity of the medium (solid, liquid, or gas), and T is the temperature at a location x. The significance of the negative sign is that heat flows in the direction of decreasing temperature, as shown in Figure 2.3. The quantity q_x/A is called heat flux, which is heat flow per unit time *per unit area*, and will be denoted by the symbol q_x''. The two primes on the symbol q_x'' are simply for convenience and do not carry any special meaning. Thus, in terms of the notation for heat flux, Fourier's law is

$$q_x'' = -k\frac{dT}{dx} \quad (2.2)$$

Note that the heat flux q_x'' or heat flow rate q_x always has direction associated with it. Thus, if the subscript x is omitted for simplicity, the appropriate direction is given by the direction of the gradient in temperature. The units of thermal conductivity k are given by:

$$[k] = \frac{[q_x]}{[A]\left[\frac{dT}{dx}\right]} = \frac{[W]}{[m^2]\left[\frac{°C}{m}\right]} = \frac{W}{m\,°C} \text{ or } \frac{W}{mK}$$

A note on units— throughout this book, units with a denominator will be written using a slash. For example, unit for thermal conductivity just described will be W/m·K. When multiple units are present in the denominator, as m and K in case of thermal conductivity, the units in the denominator are separated by a dot (·) and the reader needs to be careful in their use. An alternative is to use a bracket to group the terms in the denominator, such as W/(mK) in case of thermal conductivity. For a different system of units, thermal conductivity is expressed in

$$[k] = \frac{[Btu/hr]}{[ft^2]\left[\frac{°F}{ft}\right]} = \frac{Btu}{hr\,ft\,°F}$$

Only for very simple materials such as simple gases, can thermal conductivities be predicted from molecular knowledge (e.g., kinetic theory of gases). For almost all real materials, available thermal conductivity data is from measurements.

Consider a simple heat transfer situation of one-dimensional flow of energy at steady state through a flat surface such as through a wall, as shown in Figure 2.4. To maintain steady state, the same amount of energy has to flow at any location x, making q_x'' a constant. For such a situation, Equation 2.1 can be integrated over finite distances (between locations 1 and 2) to obtain

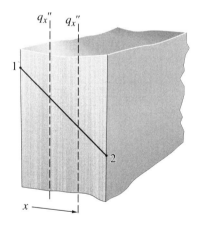

Figure 2.4: Schematic showing same value of heat flux at two locations in a slab at steady state.

$$\int_1^2 dT = -\frac{q_x''}{k}\int_1^2 dx \quad (2.3)$$

integrating

2.1. CONDUCTIVE HEAT TRANSFER

$$\Delta T = -\frac{q_x''}{k}\Delta x$$

$$q_x'' = -k\frac{\Delta T}{\Delta x} \quad (2.4)$$

where $\Delta T = T_1 - T_2$ and $\Delta x = x_1 - x_2$. Equation 2.4 can also be viewed as an approximation of Eqn. 2.2 for small distances or when temperature changes linearly with distance as in this steady-state situation.

Example 2.1.1 Heat Flux Through a Wall

Consider a 50 mm thick wall whose inner and outer surface temperatures are 20°C and 5°C, respectively. Compare the heat fluxes for two different materials of the wall, brick and wood. The thermal conductivity of the two materials are 0.69 W/m·K and 0.208 W/m·K, respectively.

Known: Thickness of a plane wall and its surface temperatures

Find: Compare heat fluxes through two wall materials, brick and wood

Schematic and Given Data: Schematic of the problem is shown in Figure 2.5. The given data are:

1. surface temperatures of the wall are 5°C and 20°C, respectively
2. thickness of the wall is 50 mm
3. thermal conductivity of brick is 0.69 W/m·K and that of wood is 0.208 W/m·K

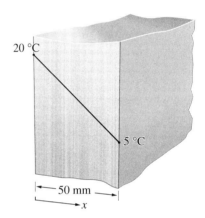

Figure 2.5: Schematic for Example 2.1.1.

Assumptions:

1. Variation of thermal conductivity with temperature is negligble
2. Heat transfer is one-dimensional, along the thickness

Analysis:
For a brick wall,

$$\begin{aligned} q_x'' &= -k\frac{\Delta T}{\Delta x} \\ &= -0.69\left[\frac{W}{m\cdot K}\right]\frac{(5-20)\,[K]}{50\times 10^{-3}\,[m]} \\ &= 207\left[\frac{W}{m^2}\right] \end{aligned}$$

For a wood wall,

$$q''_x = -k\frac{\Delta T}{\Delta x}$$
$$= -0.208 \left[\frac{W}{m \cdot K}\right] \frac{(5-20)\,[K]}{50 \times 10^{-3}\,[m]}$$
$$= 62 \left[\frac{W}{m^2}\right]$$

Comments: Because of brick's higher thermal conductivity, the brick wall loses heat faster than the wood wall.

2.1.1 Thermal Conductivity of Biological and Other Materials

Thermal conductivity of biomaterials as compared to other materials is shown in Figure 2.2. Biomaterials are often mostly water. Therefore, thermal conductivities of unfrozen wet biomaterials are often close to the values for water. It is useful to think of a biomaterial as a composite of its primary ingredients water, ice (for frozen materials), and air (for dry materials), which have large variations in conductivity values as shown in Figure 2.6. Dry biomaterials have a thermal conductivity much less than water (somewhat closer to air), whereas frozen biomaterials have thermal conductivity values closer to that of ice. See, for example, data for food materials in Appendix C.7 on page 339. Thermal properties for various materials are provided in Appendices C.4-C.8 on pages 335-340.

2.1.2 Thermal Diffusivity

Fourier's law (Eqn. 2.2) can be rewritten as:

$$q''_x = -k\frac{dT}{dx} = -\frac{k}{\rho c_p}\frac{d(\rho c_p T)}{dx} = -\alpha\frac{dU}{dx} \qquad (2.5)$$

Here ρ is the density, c_p is the specific heat of the material and

$$U = \rho c_p T \qquad (2.6)$$

is the thermal energy per unit volume. The proportionality constant α is called the thermal diffusivity. Thus, the thermal diffusivity is defined as

$$\alpha = \frac{k}{\rho c_p} \qquad (2.7)$$

The units of thermal diffusivity α can be calculated as

$$[\alpha] = \frac{[k]}{[\rho c_p]} = \frac{W/m \cdot K}{kg/m^3 \; J/kg \cdot K} = \frac{m^2}{s} \qquad (2.8)$$

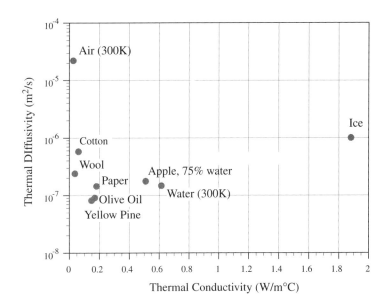

Figure 2.6: Thermal conductivity and thermal diffusivity of some biomaterials.

Since dU/dx is the gradient in energy, α is the proportionality constant between energy flux and energy gradient, i.e.,

$$\begin{matrix} \text{Flux} \\ \text{of Energy} \end{matrix} = \alpha \times \begin{matrix} \text{Gradient} \\ \text{in Energy} \end{matrix} \qquad (2.9)$$

To further interpret thermal diffusivity, note that thermal conductivity provides the rate of heat flow. This alone does not decide temperature change. The amount of energy needed for every degree of temperature change is also a factor in deciding temperature rise. Whereas thermal conductivity gives an indication of the ease of heat flow, thermal diffusivity gives an indication of the ease of temperature change in a transient process.

2.1.3 Density and Specific Heat

Density ρ and specific heat c_p are parts of "thermal mass" of the system. They affect temperature change in the system as was just discussed under thermal diffusivity. The higher the density and specific heat, the larger is the energy it takes to change the temperature by one degree. The quantity ρc_p has the units $J/m^3 \cdot K$ and is called the volumetric heat capacity, as can be seen from its units. Examples of data on density and specific heat can be seen in Appendix C.8.

Two types of densities are used—solid density and bulk density. Distinction between these two densities is important when the material is porous. Solid density (also called mean particle density) is mass per unit volume of just the solid portion in a porous media. Dry bulk density is the mass of the dried solid to its total volume (solids and pores together). The porosity of a material is defined as the volume of pores occupied by air and water (if present) divided by the total volume of the solid.

2.2 Convective Heat Transfer

Convective heat transfer is the movement of heat through a medium as a result of the net motion of a material in the medium (e.g. fluid flow over a surface). This is shown in Figure 2.7. Forced convection is due to an external force such as a fan, while free or natural convection is driven by a density difference in the material. Convection over a surface is described by:

$$q_{1-2} = hA(T_1 - T_2) \tag{2.10}$$

where q_{1-2} is the heat flow rate from 1-2 (in W or Btu/hr), A is the area normal to the direction of heat flow (m^2 or ft^2), $T_1 - T_2$ is the temperature difference between surface and fluid, and h is the convective heat transfer coefficient, also called the film coefficient. The units of h can be calculated as:

$$[h] = \frac{[q]}{[A][\Delta T]} = \frac{W}{m^2 \, °C} = \frac{Btu}{hr \, ft^2 \, °F}$$

Equation 2.10 is not a law but a defining equation for h. It is important to note that the convective heat transfer coefficient h always includes the effect of conduction in the fluid, in addition to its bulk flow. The effect of conduction, which is a random molecular effect, is always there, in the presence or absence of bulk flow and it cannot be stopped. The convective heat transfer coefficient h is a function of system geometry, fluid and flow properties, and magnitude of ΔT. Convection will be studied in more detail in Chapter 6.

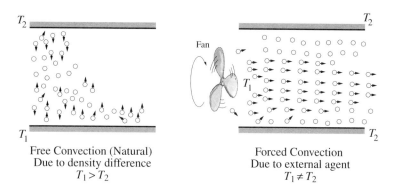

Figure 2.7: A schematic comparing free or natural (on the left) and forced (on the right) convection.

2.3 Radiative Heat Transfer

All matter, when at temperatures above absolute zero, emits radiative energy. This radiation is attributed to changes in the electron configuration of the atoms within the matter and is emitted as electromagnetic waves. Unlike conduction and convection, radiation does not require a medium. The maximum flux at which radiation may be emitted by a body at absolute temperature T is given by the Stefan-Boltzmann law:

$$\frac{q}{A} = \sigma T^4 \tag{2.11}$$

where:

$$\sigma = 5.676 \times 10^{-8} \frac{\text{W}}{\text{m}^2\text{K}^4} = 0.1714 \times 10^{-8} \frac{\text{Btu}}{\text{hr ft}^2 {}^\circ\text{R}^4}$$

Net energy transfer depends on surface and geometric factors. Radiative heat transfer will be studied in detail in Chapter 8.

2.4 Chapter Summary—Modes of Heat Transfer

- **Conductive Heat Transfer (page 15)**

    1. Translational, rotational, and vibrational transfer of energy from one molecule to another through physical contact is termed as conduction or diffusion.
    2. Fourier's law, given by $q''_x = -k\frac{\partial T}{\partial x}$ describes the conduction mode of heat transfer.
    3. Thermal conductivity k is a material property that determines the ease of heat conduction. Higher value of k means higher rate of heat conduction.
    4. Thermal diffusivity α is related to thermal conductivity and determines the ease temperature change at a location

- **Convective Heat Transfer (page 22)**

    1. It is the transfer of energy when there is net motion, i.e., bulk flow in the medium. The effect of bulk flow is *in addition to* the conductive mode of energy transfer. This mode of energy transport can occur due to the presence of a liquid or a gas.
    2. Convective energy transport is described by $q''_{1-2} = h(T_1 - T_2)$

- **Radiative Heat Transfer (page 23)**

    1. Radiative energy transport is due to the spontaneous emission of electromagnetic waves by all matter. Such energy transport does not require a medium.
    2. It is described by the equation $q'' = \sigma T^4$

2.5 Concept and Review Questions

1. Newton's law of viscosity is given by

$$\tau = -\mu \frac{\partial u}{\partial x} \qquad (2.12)$$

   where τ is the shear stress, μ is the viscosity, and u is the velocity. Compare the variables in Fourier's law (Equation 2.2) with this equation and discuss the analogies in the physical quantities and their units. Hint: Multiply and divide u in the above equation by by a constant ρ (density) and compare the resulting equation with Equation 2.5.

2. What is the order of magnitude of the thermal conductivity for gases, liquids, and solids at room temperature and 1 atm pressure?

3. Would you expect wood to have the same thermal conductivity in all three directions? (Hint: see Figure 10.9.)

4. How is the convective heat transfer related to the conductive heat transfer?

5. What is the most fundamental difference between the conductive and the radiative modes of heat transfer?

6. Describe a heat transfer situation in everyday life and discuss what aspects of it you want to know in more detail.

2.6 Further Reading

More references on thermal properties can be found on page 354 in Appendix D.9.

Adiutori, E. F. 1989. *The New Heat Transfer.* Ventuno Press, West Chester, Ohio.

Cengel, Y. A. 1998. *Heat Transfer: A Practical Approach.* McGraw-Hill, New York, NY.

Chato, J. C. 1985. Selected thermophysical properties of biological materials. In: *Heat Transfer in Medicine and Biology: Analysis and Applications, Volume 2.* Plenum Press, New York.

Incropera, F. P. and D.P. Dewitt. 1996. *Fundamentals of Heat and Mass Transfer.* Wiley, New York.

Kittel, C. and H. Kroemer. 1980. *Thermal Physics.* W. H. Freeman and Company, San Francisco.

Rao, M. A. and S.S.H. Rizvi. 1986. *Engineering Properties of Foods.* Marcel Dekker, Inc., New York.

2.7 Problems

2.7.1 Graphical implementation of Fourier's law

Figure 2.8 shows the temperature profile in a cooking pot at some point in time during cooking. Consider the food as solid (no bulk movement or convection). (a) For a thermal conductivity of 0.45 W/m·K, calculate the heat flux graphically as a function of height and plot on the same figure. (b) Show the direction of positive heat flux in the plot. (c) Since the heat flux is lower toward the top, less energy is flowing at an upper location than below. Explain whether the thermal energy is conserved in this example.

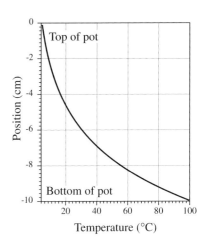

Figure 2.8: Temperature profile in a cooking pot.

2.7.2 Conductive heat flux from human body

Suppose you are sitting on a chair and you are able to maintain the temperature of its wooden surface at the average skin temperature of 33°C. The other side of the wooden chair (2.54 cm thick) remains at the surrounding air temperature of 20°C. What is the conductive heat flux at steady state? The thermal conductivity of wood is 0.208 W/m·K.

2.7.3 Convective heat flux from human body

Calculate the average convective heat flux from a bare body when the average skin surface temperature is 33°C, the surrounding air temperature is 20°C and a gentle breeze is leading to a heat transfer coefficient on the skin surface of 50W/m^2·K. Compare this with the conductive heat flux calculated in problem 2.7.2.

2.7.4 Energy flux at the sun's surface

What is the energy flux at the surface of sun? Assume an average surface temperature of sun to be 5000K. Compare this to average solar flux just outside the earth's atmosphere as 1353 W/m^2 and explain the difference.

2.7.5 Radiative energy flux from human body

Find the average energy flux from the human body due to radiation assuming the average skin surface temperature to be 33°C. Is this the *net* radiative heat loss when the person is inside a room? Explain. Compare this with the convective heat loss calculated in problem 2.7.3.

Chapter 3

GOVERNING EQUATION AND BOUNDARY CONDITIONS OF HEAT TRANSFER

CHAPTER OBJECTIVES

After you have studied this chapter, you should be able to:

1. Identify the terms describing storage, convection, diffusion and generation of energy in the general governing equation for heat transfer.
2. Specify the three common types of heat transfer boundary conditions.
3. Describe heat transfer in a mammalian tissue with blood vessels using the bioheat transfer equation.

KEY TERMS

- **governing equation**
- **control volume**
- **storage term**
- **convection term**
- **diffusion term**
- **generation term**
- **boundary temperature specified**
- **boundary heat flux specified**
- **boundary convective heat transfer**
- **bioheat transfer**

In this chapter we will develop a general equation that describes temperature in a material in any kind of heating or cooling situation. Later, in Chapters 4 through 8, we

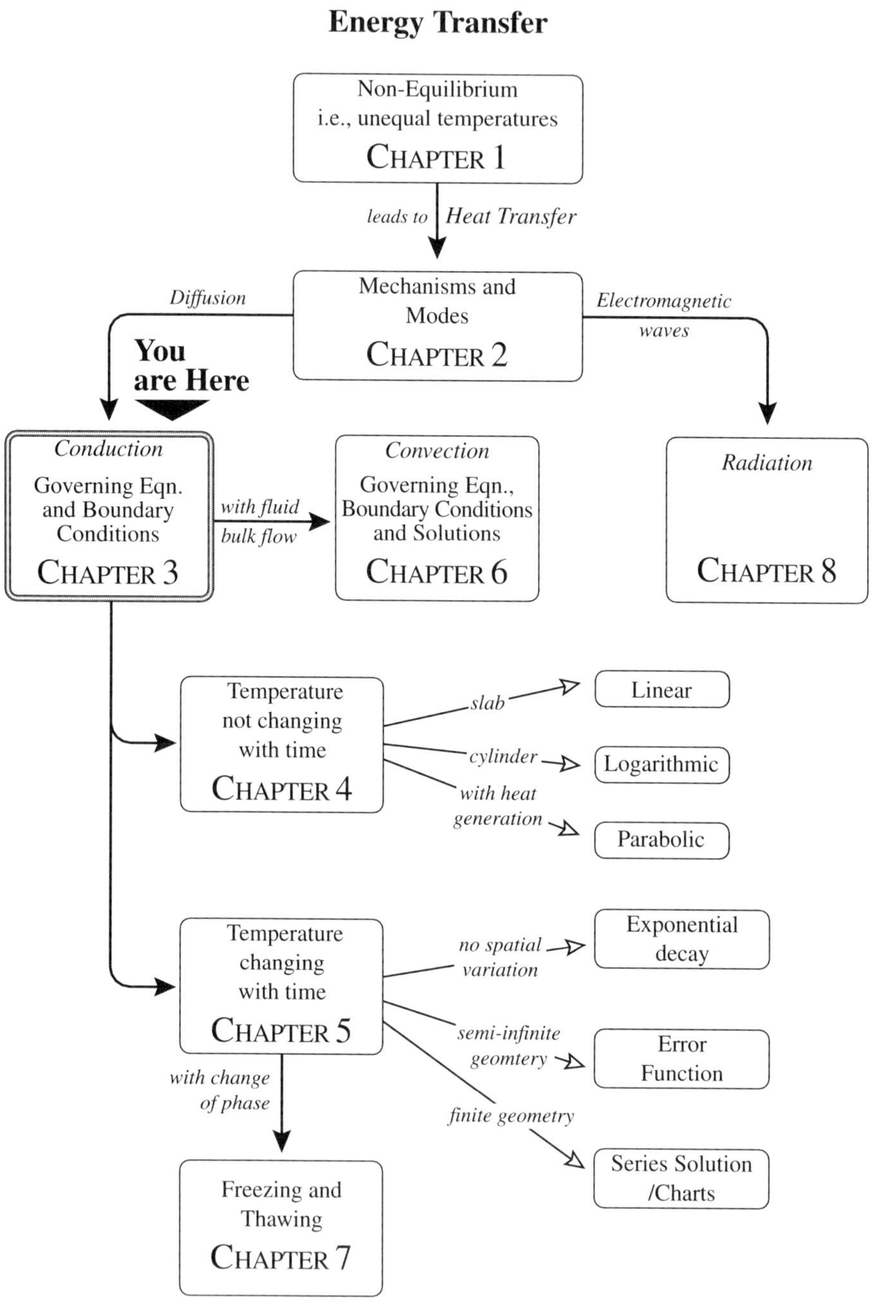

Figure 3.1: Concept map of energy transfer showing how the contents of this chapter relates to other chapters on energy transfer.

3.1 Governing Equation for Heat Transfer Derived from Energy Conservation and Fourier's Law

We want a general equation that describes temperature distribution in all kinds of situations, not just the simple steady state heat transfer (Eqn. 2.4 in Chapter 2). Consider an elemental control volume as shown in Figure 3.2 where q'' is heat flux (rate of heat flow per unit area) in a particular direction. The heat flux is comprised of conductive and convective heat transport into and out of the control volume. Convective or bulk heat transport term arises from the bulk movement of a fluid, if present. As a fluid mass flows into a system, it carries heat (thermal energy) with it at the rate:

$$\begin{aligned} E &= \dot{m} c_p (T - T_R) \\ &= u A \rho c_p (T - T_R) \end{aligned} \qquad (3.1)$$

where u is the fluid velocity and T_R is some reference temperature. Thus, the heat flux due to convection is

$$\frac{E}{A} = u \rho c_p (T - T_R) \qquad (3.2)$$

Within the control volume, heat is generated at the rate of Q per unit volume. Increase in the stored heat would manifest in the increase in temperature of the control volume. Conversely, if the stored heat in the control volume decreases, its temperature drops. Generation of heat is not to be confused with storage of heat. Generation is the transformation of energy from one form into heat. For example, during exercise, mechanical energy is transformed into heat in the muscle. When electrical current is passed through a material, electrical energy is converted into heat. Other examples of generation are described later in this section. Storage of energy (and rise in temperature) is the effect whereas generation and conduction of energy are the causes.

Using the first law of thermodynamics for the control volume in Figure 3.2, which states that the energy in the control volume is conserved (Eqn. 1.1), we can write:

$$\begin{array}{c} \text{Energy} \\ \text{In} \end{array} - \begin{array}{c} \text{Energy} \\ \text{Out} \end{array} + \begin{array}{c} \text{Energy} \\ \text{Generated} \end{array} = \begin{array}{c} \text{Energy} \\ \text{Stored} \end{array} \qquad (3.3)$$

Note that in using the term "energy," we are implicitly assuming only the thermal form of energy or heat. For simplicity in deriving the heat transfer equation, let us consider heat flow only in the x direction. The various quantities in Eqn. 3.3 can be written as:

$$\begin{array}{c} \text{Energy in} \\ \text{during time } \Delta t \end{array} = \left(q''_x \Delta y \Delta z + [u \Delta y \Delta z \rho c_p (T - T_R)]_x \right) \Delta t$$

$$\begin{array}{c} \text{Energy out} \\ \text{during time } \Delta t \end{array} = \left(q''_{x+\Delta x} \Delta y \Delta z + [u \Delta y \Delta z \rho c_p (T - T_R)]_{x+\Delta x} \right) \Delta t$$

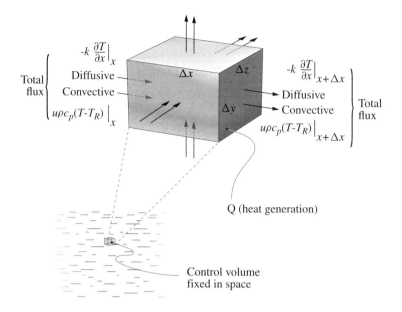

Figure 3.2: Control volume showing energy inflow and outflow by conduction (diffusion) and convection.

$$\text{Energy generated during time } \Delta t = Q \Delta x \Delta y \Delta z \Delta t$$

$$\text{Energy stored during time } \Delta t = \Delta x \Delta y \Delta z \rho c_p \Delta T$$

Substituting in Equation 3.3, we get:

$$\Delta t \left(q''_x \Delta y \Delta z - q''_{x+\Delta x} \Delta y \Delta z \right.$$
$$+ \rho c_p \Delta y \Delta z \left[u(T - T_R)_x - u(T - T_R)_{x+\Delta x} \right]$$
$$\left. + Q \Delta x \Delta y \Delta z \right) = \rho c_p \Delta x \Delta y \Delta z \Delta T$$

Dividing throughout by $\Delta x \Delta y \Delta z \Delta t$ and rearranging:

$$-\frac{q''_{x+\Delta x} - q''_x}{\Delta x} - \rho c_p \frac{(u(T - T_R)_{x+\Delta x} - u(T - T_R)_x)}{\Delta x} + Q = \rho c_p \frac{\Delta T}{\Delta t}$$

Making Δx and Δt go to zero and using the definition of a derivative:

$$-\frac{\partial q''_x}{\partial x} - \rho c_p \frac{\partial}{\partial x}(uT) + Q = \rho c_p \frac{\partial T}{\partial t}$$

Note that T_R has been dropped since it is a constant. Using Fourier's law for heat

3.1. GOVERNING EQUATION FOR HEAT TRANSFER DERIVED

conduction (Eqn. 2.1) to substitute for the heat flux, q_x'':

$$-\frac{\partial}{\partial x}\left(-k\frac{\partial T}{\partial x}\right) - \rho c_p \frac{\partial}{\partial x}(uT) + Q = \rho c_p \frac{\partial T}{\partial t}$$

If k can be assumed constant, this is simplified to:

$$\underbrace{\frac{\partial T}{\partial t}}_{\text{storage}} + \underbrace{\frac{\partial(uT)}{\partial x}}_{\substack{\text{flow or} \\ \text{convection}}} = \underbrace{\frac{k}{\rho c_p}\frac{\partial^2 T}{\partial x^2}}_{\text{conduction}} + \underbrace{\frac{Q}{\rho c_p}}_{\text{generation}} \tag{3.4}$$

Equation 3.4 is the general governing equation for energy transfer in one dimension cartesian coordinate system with constant thermal properties. This equation is also known as the energy equation or the heat equation. Often, the equation of continuity or mass conservation (see Eqn. 11.16 on page 229)

$$\frac{\partial u}{\partial x} = 0$$

is used to write an alternate form of Eqn. 3.4 as

$$\underbrace{\frac{\partial T}{\partial t}}_{\text{storage}} + \underbrace{u\frac{\partial T}{\partial x}}_{\substack{\text{flow or} \\ \text{convection}}} = \underbrace{\frac{k}{\rho c_p}\frac{\partial^2 T}{\partial x^2}}_{\text{conduction}} + \underbrace{\frac{Q}{\rho c_p}}_{\text{generation}} \tag{3.5}$$

Note that the convective term has been simplified.

3.1.1 Meaning of Each Term in the Governing Equation

Although Eqn. 3.4 looks complex with all the different terms, the good news is that we will never try (in this text) to solve it keeping all the terms. Depending on the particular situation, we only keep a few of the terms. However, this means we need to be fully aware of what the terms represent, so that we can ignore the ones that are not relevant for a particular situation. For this purpose, we rearrange Eqn. 3.4 as:

$$\underbrace{\rho c_p \frac{\partial T}{\partial t}}_{\text{storage}} + \underbrace{\rho c_p \frac{\partial}{\partial x}(uT)}_{\text{convection}} = \underbrace{k\left(\frac{\partial^2 T}{\partial x^2}\right)}_{\text{conduction}} + \underbrace{Q}_{\text{generation}} \tag{3.6}$$

Note that in this rearranged format, each term has an unit of W/m^3 and thus represent energy per unit volume per unit time. The meaning of each term is now summarized in Table 3.1.

3.1.2 Examples of Thermal Source (Generation) Term in Biological Systems

Any living thing is a heat producer. A working muscle such as in the heart or limbs produces heat. The metabolic process itself results in a release of energy in the form

Table 3.1: Various terms in the governing equation and their interpretations

Term	What does it represent	When can you ignore it
Storage	Rate of change of stored energy	Steady state (no variation of temperature with time)
Convection	Rate of net energy transport due to bulk flow	Typically in a solid, with no bulk flow through it
Conduction	Rate of net energy transport due to conduction	Slow thermal conduction in relation to generation or convection. For example, in short periods of microwave heating
Generation	Rate of generation of energy	No internal heat generation due to biochemical reactions, etc.

of heat (see Table in Appendix C.1 on page 332). Different activity levels generate body heat to different extents as shown in Table C.2 on page 333. Fermentation, composting, and other biochemical reactions generate heat. Microwaves are absorbed by a water-containing material to produce heat. An example of this is heating of food in a microwave oven.

3.1.3 Utility of the Energy Equation

It is very general.

1. It is useful for any material.

2. It is useful for any size or shape. Similar equations can be derived for other coordinate systems.

3. It is easier to derive the more general equation and simplify

4. It is safer - as you drop terms you are aware of the reasons.

Can we make it more general?

1. To use with compressible fluids.

2. To use when all properties vary with temperature. We need numerical solutions (explained later in the book) to solve such problems.

3. To include mass transfer. For example, the equation cannot predict the temperature inside a steak during cooking in an oven since the equation does not include water loss from the steak.

3.1. GOVERNING EQUATION FOR HEAT TRANSFER DERIVED

Example 3.1.1 Solution to Specific Situations: Need for Boundary Conditions

Consider a system such as a slab with temperature variation only in one dimension. Also, consider the system to be at steady-state and having no heat generation. The following terms drop out from the general governing equation

$$\underbrace{\cancel{\frac{\partial T}{\partial t}}^{0}}_{\text{steady state}} = \frac{k}{\rho c_p} \frac{\partial^2 T}{\partial x^2} + \underbrace{\cancel{\frac{Q}{\rho c_p}}^{0}}_{\text{no generation}} \quad (3.7)$$

leading to the specific governing equation for this problem as:

$$\frac{d^2 T}{dx^2} = 0 \quad (3.8)$$

A general solution to this second order ordinary differential equation can be written as:

$$T = C_1 x + C_2 \quad (3.9)$$

As expected, such general solutions have constants (C_1 and C_2) yet to be determined. More information is needed – need for boundary conditions to evaluate these constants.

Consider the physical situation where the two surfaces (boundaries) of the slab are at two temperatures T_1 and T_2, so that these *boundary conditions* can be written mathematically as

$$T = T_1 \text{ at } x = 0 \quad (3.10)$$
$$T = T_2 \text{ at } x = L \quad (3.11)$$

The conditions 3.10 and 3.11 are substituted in Eqn. 3.9 to obtain

$$T_1 = C_2$$
$$T_2 = C_1 L + C_2$$

and solve for C_1 and C_2. These constants are then replaced in Eqn. 3.9 to obtain the particular solution

$$T = \frac{T_2 - T_1}{L} x + T_1 \quad (3.12)$$

From the mathematical reasoning provided in this section, it follows that the number of boundary conditions needed equal the highest order derivative in the governing ordinary differential equation. Thus, Eqn. 3.8 being a second order ordinary differential equation, we needed two boundary conditions. Instead of the steady-state heat transfer discussed here, if heat transfer is transient, the governing equation would be a partial differential equation, involving also the time derivative. Thus, we would also need one condition for time (the time derivative is first order). This necessary condition for time is generally provided at the initial time of $t = 0$, hence it is referred to as the initial condition. From a physical standpoint, the boundary conditions are needed since they influence the interior solution. Likewise, initial condition is needed for a transient problem since the temperatures at a later time are influenced by the initial temperature. Note that initial conditions are not needed for a steady-state problem.

3.2 General Boundary Conditions

As mentioned in the previous section, conditions at the boundary are necessary for obtaining the full details of the solution. The description of a heat transfer problem in a system is not complete without the information on the thermal conditions on the bounding surfaces of the system. The following are the three most common types of thermal conditions that can occur on the boundary of a system in a heat transfer situation:

1. Surface temperature is specified: One of the simple thermal conditions that can occur on a surface is a specified temperature. For a one-dimensional heat transfer, as shown in Figure 3.3, this boundary condition is expressed as

$$T\bigg|_{x=o} = T_s \tag{3.13}$$

The temperature at the surface, T_s, can be specified as a constant or a function of time. As an example, consider steam condensation on a surface, keeping it at 100°C. This is expressed as

$$T\bigg|_{x=o} = 100$$

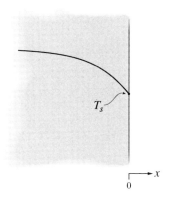

Figure 3.3: A surface temperature specified boundary condition.

2. Surface heat flux is specified Sometimes it is possible to know and specify the rate of heat transfer or *heat flux* on a surface. For a one-dimensional heat transfer, as shown in Figure 3.4, this boundary condition is expressed as

$$-k\frac{dT}{dx}\bigg|_{x=0} = q_s'' \tag{3.14}$$

Here the surface heat flux, q_s'', can be specified as a constant or a function of time. For example, consider shining a handheld infrared heat lamp on your back where the skin is receiving energy at the rate of 4000 W/m². This is expressed as

$$-k\frac{dT}{dx}\bigg|_{x=0} = 4000$$

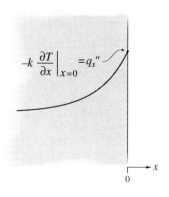

Figure 3.4: A heat flux specified boundary condition.

Note that surface heat flux, q_s'', although a source of energy, is different from internal heat generation that is included in the governing equation. There are two important special cases of specified surface heat flux. These are now discussed.

2a) Special case: Insulated condition

Surfaces are insulated to reduce heat flux. When a surface is highly insulated, the heat flux through the surface is very small and can be approximated as zero, as illustrated in Figure 3.5. This boundary condition is expressed as

$$-k \left.\frac{dT}{dx}\right|_{x=0} = 0 \qquad (3.15)$$

2b) Special case: Symmetry condition

Another common situation arises in a heating or cooling process when the geometry and the boundary conditions are symmetric, as shown in Figure 3.6 for a slab of uniform thickness (therefore symmetric about the centerline) as well as it is cooled symmetrically (same boundary condition on both faces). The resulting temperature profile in the slab will also be symmetric about the centerline, having a zero slope at the centerline. This is expressed as

$$-k \left.\frac{dT}{dx}\right|_{x=L} = 0 \qquad (3.16)$$

Note that this symmetry condition resembles the insulated condition mentioned above. To maintain symmetry, heat flux has to be zero at the line of symmetry.

3. Convection at the surface

Perhaps the most common type of thermal condition that can occur on a boundary is convection of a fluid over it. Thus heat conducted out of the boundary is convected away by the fluid. This condition is written as simply a heat balance at the boundary. Thus, referring to Figure 3.7, the heat balance at the boundary is

$$\underbrace{-k \left.\frac{dT}{dx}\right|_{x=0}}_{\text{heat conduction}} = \underbrace{h(T|_{x=0} - T_\infty)}_{\text{heat convection}} \qquad (3.17)$$

Here h is the heat transfer coefficient mentioned on page 22 and further described in Chapter 6. For example, suppose air at $10°C$ is blowing over the surface in Figure 3.7, leading to a convective heat transfer coefficient, h, of 50 W/m^2. The convective boundary condition for this surface can now be written as

$$-k \left.\frac{dT}{dx}\right|_{x=0} = 50(T|_{x=0} - 10)$$

Note that the surface temperature $T(x = 0)$ is not known and will come out of the solution. This is in contrast with Eqn. 3.13, where the surface temperature was known. As a special case of Eqn. 3.17, when $h \to \infty$, it is approximated to

$$T|_{x=0} = T_\infty$$

which is the temperature specified (first type of) boundary condition mentioned above.

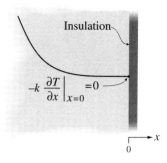

Figure 3.5: An insulated (zero heat flux specified) boundary condition.

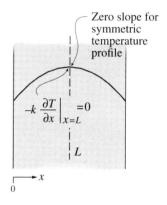

Figure 3.6: A symmetry (zero heat flux specified) boundary condition at the centerline.

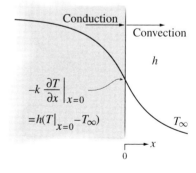

Figure 3.7: A convection boundary condition.

3.3 The Bioheat Transfer Equation for Mammalian Tissue

Although the general governing equation (Eqn. 3.4) and its more general versions are valid for most physical systems, it would be hard to apply the equations for a mammalian tissue that has blood vessels of varying sizes and varying amounts of blood flows. In such a system, thermal conductivity, density and specific heat (the thermal properties) would vary significantly over small distances, among other complications. Researchers have attempted to develop a bioheat transfer equation that is simpler than applying the equations in section 3.5 directly but still captures the essence of the heat transfer process in a mammalian tissue with blood vessels.

Before we discuss the bioheat transfer equation, let us have a simple look at the mammalian tissue system. The circulatory system in our body comprises of two sets of blood vessels— arteries and veins (Figure 3.8) which carry blood from the heart and back. By the pumping action of the heart, blood flows through larger arteries to progressively smaller arteries to arterioles and capillaries to small vein to larger veins and eventually back to heart. Figure 3.9 shows a schematic of the arterial system with different size vessels. Figure 3.10 shows a schematic of temperature equilibration between the blood and the solid tissue, as the blood traverses different size vessels in the systemic circulation. As blood leaves the heart and travels in the large arteries, its temperature remains essentially constant. This is the arterial blood temperature, T_a. Most of the temperature equilibration occurs as the blood passes through vessels

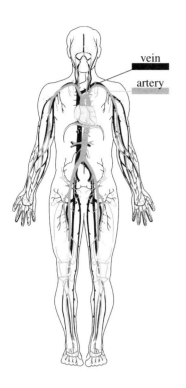

Figure 3.8: Arteries and veins of the circulatory system.

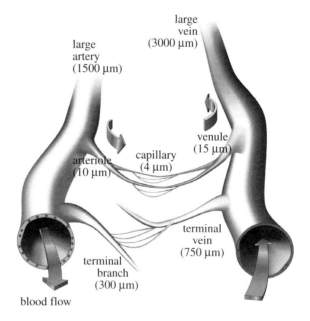

Figure 3.9: Variation of blood vessel sizes.

3.3. THE BIOHEAT TRANSFER EQUATION FOR MAMMALIAN TISSUE

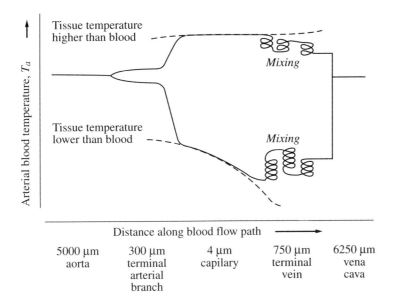

Figure 3.10: Variation of blood temperature in the blood vessels.

whose diameter is between that of the arterial branch and that of the arteriole. As the blood reaches the latter, blood temperature becomes essentially that of the solid tissue (warmer or colder, as shown). Beyond this point, blood temperature follows the solid tissue temperature through its spatial and time variations until blood reaches the terminal veins. At this point the blood temperature ceases to equilibrate with the tissue, and remains virtually constant, except as it mixes with other blood of different temperatures at venous confluences. Finally, the cooler blood from peripheral regions and warmer blood from internal organs mix within the vena cavae and the right atrium and ventricle. Following thermal exchange in the pulmonary circulation and remixing in the left heart, the blood attains the same temperature it had at the start of the circuit.

The bioheat transfer equation can be derived for an idealized tissue system with blood vessels through it, as shown in Figure 3.11 in the same way as Equation 3.4 was derived. Many assumptions that have to be made include 1) homogeneous material with isotropic (same in all directions) thermal properties; 2) large blood vessels are ignored; 3) blood capillaries are isotropic; 4) blood is at arterial temperature but quickly reaches the tissue temperature by the time it reaches the end of the artery system. Using these assumptions, the governing *bioheat equation* for a mammalian tissue that includes the heat carried by the blood vessels is

$$\underbrace{\rho c \frac{\partial T}{\partial t}}_{\substack{\text{change} \\ \text{in storage}}} = \underbrace{k \nabla^2 T}_{\text{conduction}} + \underbrace{\rho_b c_b \dot{V}_b^v (T_a - T)}_{\substack{\text{convection} \\ \text{due to blood flow}}} + \underbrace{Q}_{\substack{\text{metabolic} \\ \text{heat generation}}} \qquad (3.18)$$

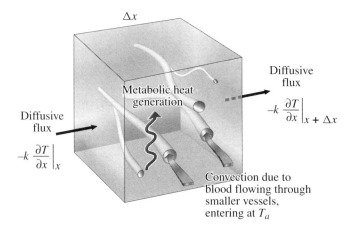

Figure 3.11: Idealized heat transfer in a tissue showing metabolic heat generation Q and convective heat transfer due to the passage of blood.

where $\dot{V}_b^v$ is the flow rate of blood in m^3 of blood / m^3 of tissue per second, k, ρ, c are thermal properties of the tissue, ρ_b and c_b are thermal properties of blood, T_a is arterial blood temperature. Note that the energy carried in the blood is like an additional heat source term of magnitude $\rho_b c_b \dot{V}_b^v (T_a - T)$. This way, Equation 3.18 can be derived from Equation 3.4 by rewriting the source term with the energy carried by the blood fluid. Bioheat transfer is an active research area and alternative equations for bioheat transfer are being formulated. Equation 3.18 is the most widely used bioheat transfer equation today.

The bioheat equation is used generally to solve for tissue temperature distributions. While such considerations as tissue geometry, thermal property values, inhomogenities, and boundary conditions are very important, this equation can be used to predict the effects of frost bite, determine the depth of damage in a burn victim, and find the amount of heat lost through various parts of the body.

3.4 Governing Equation Derived in Cylindrical Coordinates

Depending on the geometry of the body whose energy transfer we are interested in, a different coordinate system may provide a more convenient way to obtain a solution. A geometrical configuration that is of great importance in engineering is the long cylinder (solid or hollow). The cylindrical coordinate is the natural one to use in this case. We will derive the governing equation in cylindrical coordinate system by performing energy balance on a control volume, in a procedure completely analogous to that used to derive Eqn. 3.4 for the cartesian coordinate system. For convenience, we will drop the convection term.

3.4. GOVERNING EQUATION DERIVED IN CYLINDRICAL COORDINATES

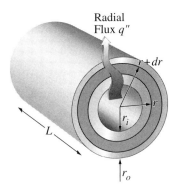

Figure 3.12: Radial heat flow in a long cylinder.

Using energy conservation (first law of thermodynamics)

$$\text{Energy In} - \text{Energy Out} + \text{Energy Generated} = \text{Energy Stored} \tag{3.19}$$

Referring to Figure 3.12, these terms can be substituted as

$$\left[2\pi r L q_r'' - 2\pi(r+\Delta r)L q_{r+\Delta r}'' + 2\pi r \Delta r L Q\right]\Delta t = 2\pi r \Delta r L \rho c_p \Delta T$$

Simplifying and rearranging

$$-\frac{\left((r+\Delta r)q_{r+\Delta r}'' - r q_r''\right)}{r\Delta r} + Q = \rho c_p \frac{\Delta T}{\Delta t}$$

Making Δr and Δt go to zero and using the definition of a derivative:

$$-\frac{1}{r}\frac{\partial}{\partial r}(r q_r'') + Q = \rho c_p \frac{\partial T}{\partial t}$$

Using Fourier's law:

$$k\frac{1}{r}\frac{\partial}{\partial r}\left(r\frac{\partial T}{\partial r}\right) + Q = \rho c_p \frac{\partial T}{\partial t}$$

Rearranging

$$\underbrace{\frac{\partial T}{\partial t}}_{\text{storage}} = \underbrace{\frac{k}{\rho c_p}\frac{1}{r}\frac{\partial}{\partial r}\left(r\frac{\partial T}{\partial r}\right)}_{\text{conduction}} + \underbrace{\frac{Q}{\rho c_p}}_{\text{generation}} \tag{3.20}$$

3.5 Governing Equations for Heat Conduction in Various Coordinate Systems

The governing energy equation is shown here in some familiar coordinate systems. Control volumes for cartesian, cylindrical and spherical coordinate systems are shown in Figures 3.13, 3.14 and 3.15, respectively.

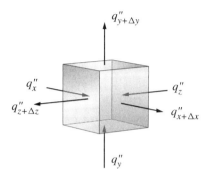

Figure 3.13: Energy balance over a control volume in a cartesian coordinate system.

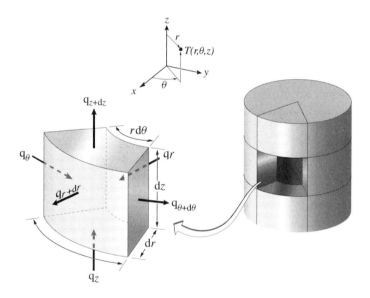

Figure 3.14: Energy balance over a control volume in a cylindrical coordinate system.

3.5. GOVERNING EQUATIONS IN VARIOUS COORDINATE SYSTEMS

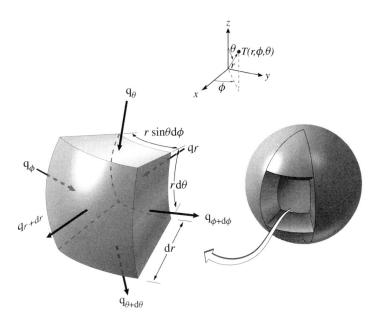

Figure 3.15: Energy balance over a control volume in a spherical coordinate system.

Cartesian

$$\frac{k}{\rho c_p}\left[\frac{\partial^2 T}{\partial x^2} + \frac{\partial^2 T}{\partial y^2} + \frac{\partial^2 T}{\partial z^2}\right] + \frac{Q}{\rho c_p} = \frac{\partial T}{\partial t} \qquad (3.21)$$

Cylindrical

$$\frac{k}{\rho c_p}\left[\frac{1}{r}\frac{\partial}{\partial r}\left(r\frac{\partial T}{\partial r}\right) + \frac{1}{r^2}\left(\frac{\partial^2 T}{\partial \phi^2}\right) + \frac{\partial^2 T}{\partial z^2}\right] + \frac{Q}{\rho c_p} = \frac{\partial T}{\partial t} \qquad (3.22)$$

Spherical

$$\frac{k}{\rho c_p}\left[\frac{1}{r^2}\frac{\partial}{\partial r}\left(r^2\frac{\partial T}{\partial r}\right) + \frac{1}{r^2\sin^2\theta}\frac{\partial^2 T}{\partial \phi^2} + \frac{1}{r^2\sin\theta}\frac{\partial}{\partial \theta}\left(\sin\theta\frac{\partial T}{\partial \theta}\right)\right]$$
$$+ \frac{Q}{\rho c_p} = \frac{\partial T}{\partial t} \qquad (3.23)$$

Symbolically (Any coordinate system)

$$\frac{k}{\rho c_p}\nabla^2 T + \frac{Q}{\rho c_p} = \frac{\partial T}{\partial t} \qquad (3.24)$$

3.6 An Algorithm to Solve Transport Problems

Using the governing equations, a physical problem is translated into a mathematical one. This mathematical problem is subsequently solved. The steps in this solution process can be summarized as shown in Figure 3.16.

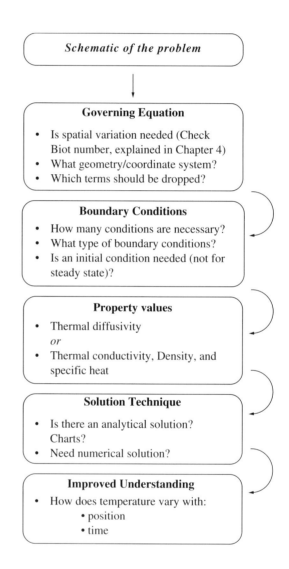

Figure 3.16: A step by step procedure to solve heat and mass transfer problems, showing the steps in case of a heat transfer problem.

3.7 Chapter Summary—Governing Equations and Boundary Conditions

- **Governing Equation (page 29)**

    1. It is a mathematical statement of energy conservation. It is obtained by combining conservation of energy with Fourier's law for heat conduction.
    2. Depending on the appropriate geometry of the physical problem, choose a governing equation in a particular coordinate system from the equations in Section 3.5.
    3. Different terms in the governing equation can be identified with conduction, convection, generation, and storage. Depending on the physical situation, some terms may be dropped.

- **Boundary Conditions (page 34)**

    1. Boundary conditions are the conditions at the surfaces of a body.
    2. Initial conditions are the conditions at time $t = 0$.
    3. Boundary and initial conditions are needed to solve the governing equation for a specific physical situation.
    4. One of the following three types of heat transfer boundary conditions typically exists on a surface:
        (a) Temperature at the surface is specified (Eqn. 3.13)
        (b) Heat flux at the surface is specified (Eqn. 3.14)
        (c) Convective heat transfer condition at the surface (Eqn. 3.17)

- **Problem Formulation (page 42)**

    1. It is the development of mathematical formulation of a physical problem, written in terms of governing equation and boundary conditions.
    2. Follow the steps as shown in Figure 3.16.

3.8 Concept and Review Questions

1. Write Fourier's law of heat conduction using spherical coordinates.
2. What are the assumptions in the bio-heat transfer equation?
3. What is meant by the term *one-dimensional* when applied to conduction problems?
4. How is the generation of energy different from storage?
5. Provide physical reasoning for why the convective boundary condition approaches temperature-specified boundary condition for a high heat transfer coefficient.
6. What geometry (spherical, rectangular, etc.) would you choose to analyze heat transfer in the soil near the earth's surface? Why?

3.9 Further Reading

Ben-Hassan, R. M., A. E. Ghaley, and N. Ben-Abdallah. 1993. Heat generation during batch and continuous production of single cell protein from cheese whey. Biomass and Bioenergy 4(3):213-225.

Bird, R. B., W. E. Stewart, and E. N. Lightfoot. 1960. *Transport Phenomena*. John Wiley & Sons, New York. (Comprehensive source for governing equations).

Carslaw, H. S. and J. C. Jaeger. 1959. *Conduction of Heat in Solids*. Oxford University Press, Oxford, UK. (Comprehensive source for analytical solution of the energy equation).

Charny, C. K. 1992. Mathematical models of bioheat transfer. Advances in Heat Transfer 22:19-153.

Chen, M. 1985. The tissue energy balance equation. In: *Heat Transfer in Medicine and Biology*, edited by A. Shitzer and R. C. Eberhart. Plenum Press, New York.

Diller, K. R. 1992. Modeling of bioheat transfer processes at high and low temperatures. Advances in Heat Transfer, 22:157-357. (Comprehensive source for bioheat transfer processes related to biomedical and related applications).

Eberhart, R. C. 1985. Thermal models of single organs. In: *Heat Transfer in Medicine and Biology*, edited by A. Shitzer and R. C. Eberhart. Plenum Press, New York.

Ozisik, M. N. 1980. *Heat Conduction*. John Wiley & Sons, New York. (Solutions of the energy equation in rectangular, cylindrical and spherical coordinate system).

3.10 Problems

3.10.1 Radial Heat Transfer in a Sphere

Derive the governing equation for radial heat transfer in a sphere (Eqn. 3.23). Use the control volume shown in Figure 3.15.

3.10.2 Equivalence Between Boundary Conditions

You have been introduced the following two boundary conditions at the surface.

$$T|_{\text{surface}} = T_\infty \tag{3.25}$$

$$-k \frac{\partial T}{\partial x}\bigg|_{\text{surface}} = h(T|_{\text{surface}} - T_\infty) \tag{3.26}$$

1) Show that the two boundary conditions are equivalent for large values of h. 2) Give physical reasoning as to why you can't use Equation 3.25 when h is small.

Chapter 4

CONDUCTION HEAT TRANSFER: STEADY-STATE

CHAPTER OBJECTIVES

After you have studied this chapter, you should be able to:

1. Formulate and solve for a conductive heat transfer process in a slab geometry where temperature does not change with time (steady-state).

2. Extend the steady-state heat transfer concept for a slab to a composite slab using the overall heat transfer coefficient.

3. Extend the steady-state heat transfer in a slab to include internal heat generation.

4. Formulate and solve for a steady-state conductive heat transfer process in a cylindrical geometry.

KEY TERMS

- steady-state
- thermal resistance
- R-value
- thermal resistances in series
- overall heat transfer coefficient
- heat transfer with with energy generation
- thermoregulation—behavioral and autonomic processes
- core body temperature
- extended surfaces—fins

This chapter will deal with steady state heat transfer. Steady state is defined as the situation when variables such as temperature are not changing with time. Note that this is not the same as equilibrium, when the system is at the same temperature everywhere. In steady state, heat can continue to flow if there is a temperature difference. It's just

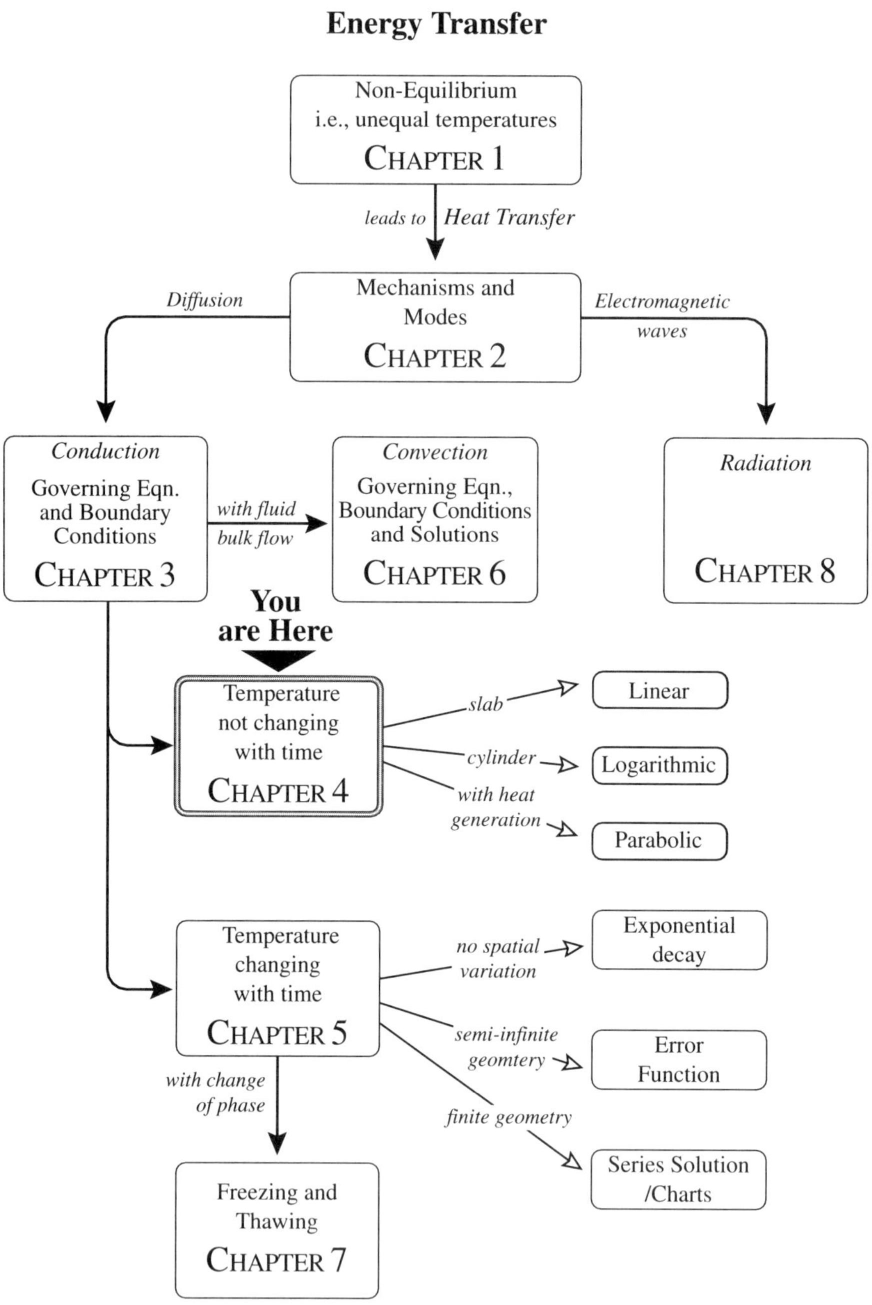

Energy Transfer

Figure 4.1: Concept map of energy transfer showing how the contents of this chapter relates to other chapters on energy transfer.

4.1 Steady-State Heat Conduction in a Slab

Although all geometries are three-dimensional in reality, under certain conditions variations in two of the dimensions can be neglected. In a slab geometry as shown in Fig. 4.2, if the surface temperatures are uniform, at locations not too close to the edges the temperature variation is only along the thickness of the slab. In other words, the heat transfer is one-dimensional. For no heat generation and steady state heat transfer, the general governing equation

$$\underbrace{\cancel{\frac{\partial T}{\partial t}}^{0}}_{\text{steady state}} + \underbrace{\cancel{u\frac{\partial T}{\partial x}}^{0}}_{\text{no bulk flow}} = \underbrace{\frac{k}{\rho c_p}\frac{\partial^2 T}{\partial x^2}}_{\text{conduction}} + \underbrace{\cancel{\frac{Q}{\rho c_p}}^{0}}_{\text{no generation}} \qquad (4.1)$$

becomes

$$\frac{d^2 T}{dx^2} = 0 \qquad (4.2)$$

after eliminating the terms. For the simplest kind of boundary condition where constant temperatures of T_1 and T_2 can be assumed at the two surfaces, we can write the boundary conditions as

$$T(x = 0) = T_1 \qquad (4.3)$$
$$T(x = L) = T_2 \qquad (4.4)$$

The solution to Eqn. 4.2 is given by (see example 3.1.1)

$$T = \frac{T_2 - T_1}{L}x + T_1 \qquad (4.5)$$

which shows a linear change in temperature from T_1 to T_2 at steady state. The heat flow is given by:

$$q_x = -kA\frac{dT}{dx}$$

since the temperature profile is linear (dT/dx is constant),

$$\begin{aligned} q_x &= -kA\frac{T_2 - T_1}{L - 0} \\ &= kA\frac{T_1 - T_2}{L} \end{aligned}$$

which is the same at all locations x along the thickness.

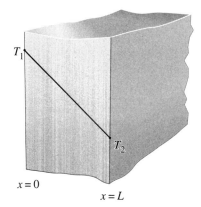

Figure 4.2: A linear temperature profile at steady state in a slab geometry.

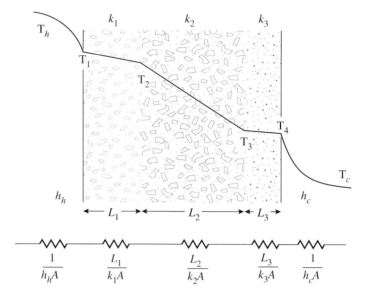

Figure 4.3: Temperature profile and thermal resistances for steady-state conduction through a composite slab with convection at the two surfaces.

4.1.1 One-Dimensional Conduction Through a Composite Slab—Thermal Resistance and Overall Heat Transfer Coefficient

Quite often in practice the slab geometry studied in the previous chapter occurs with two or more materials in series, as shown in Figure 4.3. Examples can be heat transfer through layers of skin, fat, and muscle, all having different properties. Another common example is the wall of a house in a cold climate that can have layers of wood, insulation, sheet rock, etc. Here again if the surface temperatures are uniform at locations not too close to the edges, the temperature variation is one-dimensional. If we ignore heat generation and consider steady state heat transfer, the same constant heat flow q_x appears at any section. For the conduction heat transfer through the solid layers, we can write, using the schematic in Figure 4.3

$$q_x = \frac{T_1 - T_2}{\frac{L_1}{k_1 A}} \qquad (4.6)$$

$$q_x = \frac{T_2 - T_3}{\frac{L_2}{k_2 A}} \qquad (4.7)$$

$$q_x = \frac{T_3 - T_4}{\frac{L_3}{k_3 A}} \qquad (4.8)$$

4.1. STEADY-STATE HEAT CONDUCTION IN A SLAB

For the convective heat transfer at the two surfaces, the same heat flow q_x can be written in terms of the heat transfer coefficients h,

$$q_x = \frac{T_h - T_1}{\frac{1}{h_h A}} \quad (4.9)$$

$$q_x = \frac{T_4 - T_c}{\frac{1}{h_c A}} \quad (4.10)$$

The Eqns. 4.6 through 4.10 can be written as

$$T_1 - T_2 = q_x \frac{L_1}{k_1 A}$$

$$T_2 - T_3 = q_x \frac{L_2}{k_2 A}$$

$$T_3 - T_4 = q_x \frac{L_3}{k_3 A}$$

$$T_h - T_1 = q_x \frac{1}{h_h A}$$

$$T_4 - T_c = q_x \frac{1}{h_c A}$$

Adding the left and the right hand side of the equations, we get:

$$T_h - T_c = q_x \left(\frac{1}{h_h A} + \frac{L_1}{k_1 A} + \frac{L_2}{k_2 A} + \frac{L_3}{k_3 A} + \frac{1}{h_c A} \right)$$

Rewriting the equation in terms of flux, we get

$$q_x = \frac{T_h - T_c}{\underbrace{\frac{1}{h_h A}}_{\text{convective}} + \underbrace{\frac{L_1}{k_1 A} + \frac{L_2}{k_2 A} + \frac{L_3}{k_3 A}}_{\text{conductive}} + \underbrace{\frac{1}{h_c A}}_{\text{convective}}} \quad (4.11)$$

$$= \frac{\text{temperature difference}}{\sum \text{thermal resistance}} \quad (4.12)$$

In order to understand Eqn. 4.12, consider the analogy from electricity where the current flow is given by

$$\text{current flow} = \frac{\text{total potential difference}}{\sum \text{electrical resistance}} \quad (4.13)$$

From the similarity of the two equations, it can be said that if current flow is analogous to heat flow and temperature is analogous to potential, then the terms in denominator in Eqn. 4.12 denote thermal resistances. The terms $1/h_h A$ and $1/h_c A$ represent thermal resistance due to convection in the fluid. Similarly, the term $L_1/k_1 A$ represents thermal resistance due to the first layer of solid. The heat transfer system depicted in Figure 4.3 is analogous to an electrical system with resistances in series. As a further analogy to electricity, we note that the inverse of resistance is conductance. Applying this to heat transfer, we find that the inverse of the resistance $1/hA$, or hA or simply h (the heat transfer coefficient) is like a conductance. Like higher conductance increases the current flow, higher h leads to higher heat flow.

R-value as a common measure of thermal resistance In everyday language, concept of thermal resistance due to a layer of solid is often expressed as *R-value*, defined in terms of the thickness of the layer, L, and its thermal conductivity, k, as

$$R\text{-}value = \frac{L}{k} \quad (4.14)$$

which is the thermal resistance per unit area. Although the *R-value* would be measured in SI units in m²·K/W, in practice (in the U.S.) it is often assumed to have the units of hr·ft²·°F/Btu. For example, a building materials store may tell you that you need a *R-value* of 30 for the ceiling while you need a *R-value* of 11 for the outside walls. It makes sense to talk about the *R-value* or thermal resistance (per unit area) rather than thickness of insulation, because, if you are only told about thickness of insulation, you would also need to know its thermal conductivity to find its resistance. *R-value* is the final parameter, thermal resistance, that includes both thickness and thermal conductivity.

Overall heat transfer coefficient Instead of dealing with the different layers of resistances, as in Eqn. 4.12, it is sometimes useful to talk about a total or overall resistance. If we define an overall heat transfer coefficient U by the equation

$$\begin{aligned} q_x &= UA(T_h - T_c) \\ &= \frac{T_h - T_c}{\frac{1}{UA}} \end{aligned} \quad (4.15)$$

then, in order for Eqn. 4.15 and Eqn. 4.11 to be the same, U has to be given by

$$\frac{1}{UA} = \frac{1}{h_h A} + \frac{L_1}{k_1 A} + \frac{L_2}{k_2 A} + \frac{L_3}{k_3 A} + \frac{1}{h_c A} \quad (4.16)$$

Eqn. 4.16 shows that the overall heat transfer coefficient U is made up of individual conductive and convective resistances. Oftentimes in practice an U value is given for a heat transfer situation, without mentioning the individual convective and conductive components that make up such a value. In such a situation, only Eqn. 4.15 is needed to calculate the total heat transfer q_x.

4.2 Steady-State Heat Conduction in a Cylinder

Like the slab geometry we just studied, another simple geometry that is of considerable importance is a solid or a hollow cylinder, as shown in Figure 4.4. Pipes carrying hot water or steam inside with colder air outside are perhaps the most common example of heat transfer in such a geometry. If the temperatures are uniform on cylinder surface,

4.2. STEADY-STATE HEAT CONDUCTION IN A CYLINDER

temperature gradients are only in the radial direction and therefore heat transfer is one-dimensional. The general governing equation in the cylindrical coordinate (Eqn. 3.22) is simplified for steady state, no bulk flow, no heat generation, and heat transfer only in the radial direction as

$$\underbrace{\rho c_p \cancel{\frac{\partial T}{\partial t}}^0}_{\text{steady state}} = k \left(\frac{1}{r} \frac{\partial}{\partial r} \left(r \frac{\partial T}{\partial r} \right) + \underbrace{\frac{1}{r^2} \cancel{\left(\frac{\partial^2 T}{\partial \phi^2} \right)}^0}_{\text{radial symmetry}} + \underbrace{\cancel{\frac{\partial^2 T}{\partial z^2}}^0}_{\text{no axial variation}} \right) + \underbrace{\cancel{\dot{Q}}^0}_{\text{no heat generation}} \quad (4.17)$$

to obtain the specific governing equation for this problem as

$$\frac{1}{r} \frac{d}{dr} \left(r \frac{dT}{dr} \right) = 0 \quad (4.18)$$

and the boundary conditions when the inner and outer surfaces are at two different temperatures are

$$T(r = r_0) = T_0 \quad (4.19)$$
$$T(r = r_i) = T_i \quad (4.20)$$

To solve Eqn. 4.18, it is rewritten as

$$\frac{d}{dr} \left(r \frac{dT}{dr} \right) = 0$$

and integrated once to obtain

$$r \frac{dT}{dr} = c_1$$

dividing throughout by r, we get

$$\frac{dT}{dr} = \frac{c_1}{r}$$

Integrating for a second time, we get

$$T = c_1 \ln r + c_2 \quad (4.21)$$

where c_1 and c_2 are integration constants. Applying Eqns. 4.19 and 4.20 as the boundary conditions, we get

$$T_0 = c_1 \ln r_0 + c_2$$
$$T_i = c_1 \ln r_i + c_2$$

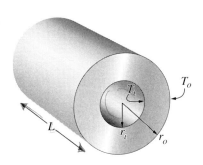

Figure 4.4: Schematic of a hollow cylinder with heat transfer only in the radial direction.

After solving for c_1 and c_2 from these two equations, they are plugged into Eqn. 4.21 to obtain

$$T = T_i - \frac{T_i - T_o}{\ln \frac{r_o}{r_i}} \ln \frac{r}{r_i} \tag{4.22}$$

4.2.1 Thermal Resistance Term for the Hollow Cylinder

Heat flow in Fig. 4.4 is through a cylindrical surface whose surface area increases with radius r. Thus, the effective thermal resistance of the wall is not the same as for a slab of same thickness where the surface area is constant at any position into the slab. To obtain an expression for the thermal resistance of the hollow cylinder, we write the equation for heat flow as

$$\begin{aligned} q_r &= -kA\frac{dT}{dr} \\ &= -2k\pi r L \frac{dT}{dr} \\ &= -2k\pi r L \left(\frac{-(T_i - T_o)}{\ln \frac{r_o}{r_i}} \right) \frac{r_i}{r} \frac{1}{r_i} \\ &= -k\frac{2\pi L (T_o - T_i)}{\ln \frac{r_o}{r_i}} \\ &= \frac{T_i - T_o}{\frac{\ln \frac{r_o}{r_i}}{2\pi k L}} \end{aligned} \tag{4.23}$$

which shows that the thermal resistance term for the hollow cylinder is given by

$$\frac{\ln \frac{r_o}{r_i}}{2\pi k L} \tag{4.24}$$

Note that the thermal resistance term given by expression 4.24 is not the *R-value* for the hollow cylinder. Whereas thermal resistance is defined in terms of heat flow, *R-value* is defined in terms of heat flux or heat flow per unit area (see Eqn. 4.14 for a slab). The *R-value* for the hollow cylinder will be defined in Section 4.2.3 below.

4.2.2 Comparison with the Solution for the Slab

The temperature profile for steady state heat transfer in a hollow cylinder given by Eqn. 4.22 is non-linear, unlike the linear temperature profile in a slab given by Eqn. 4.5. This is illustrated graphically in Figure 4.5. To see why the temperature profiles have to be different, note that the cross-sectional area $2\pi r L$ through which heat flows increases with radius r. Therefore, the temperature gradient dT/dr has to decrease with radius

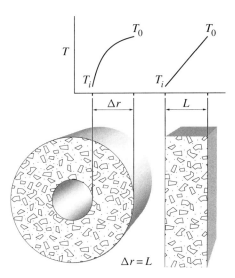

Figure 4.5: Schematic comparing non-linear temperature profile in a hollow cylinder with linear temperature profile in a slab at steady-state.

r to keep the heat flow same at steady state, which is in fact the case since $dT/dr = (T_0 - T_i)/r \ln \frac{r_0}{r_i}$. In other words,

$$\begin{aligned} q_r &= -kA\frac{dT}{dr} \\ &= -k\underbrace{(2\pi r L)}_{\text{area increases}} \underbrace{\frac{T_0 - T_i}{r \ln(r_0/r_i)}}_{\text{gradient decreases}} \\ &= \underbrace{-k\, 2\pi L (T_0 - T_i)/\ln \frac{r_0}{r_i}}_{\text{constant}} \end{aligned}$$

For the slab, the cross sectional area remains constant. This leads to a constant temperature gradient $dT/dx = (T_0 - T_i)/L$ so that the same heat flows at steady state. The temperature profile in the slab is therefore linear.

4.2.3 Thickness of Fur in Small and Large Animals

Fur provides a natural thermal insulation for animals. It is interesting to see how the thermal resistance of fur varies with its thickness. Let us consider different size animals that can be treated as if they were approximately of cylindrical shape. Heat loss per unit surface area of the animal is the most relevant because it closely relates to heat loss per unit volume and can be written using Eqn. 4.23 as

$$q''_{\substack{body \\ surface}} = \frac{T_i - T_o}{(2\pi r_i L)\left(\dfrac{\ln r_o/r_i}{2\pi k L}\right)} \qquad (4.25)$$

where $q''_{body\ surface}$ is the heat flux at the body surface (without the fur), T_i is the temperature at the body surface and T_o is the temperature at the outer surface of the fur. The body surface area of the animal without the fur is $2\pi r_i L$ (r_i is its radius without the fur). The quantity in the denominator of Eqn. 4.25 is the *R-value* for the layer of fur over a cylindrical shaped animal. If the thickness of fur is $\Delta r = r_0 - r_i$, we can write this *R-value* for fur as

$$\begin{aligned} R_{fur} &= 2\pi r_i L \frac{\ln\left(1 + \dfrac{\Delta r}{r_i}\right)}{2\pi k L} \\ &= r_i \frac{\ln\left(1 + \dfrac{\Delta r}{r_i}\right)}{k} \end{aligned} \qquad (4.26)$$

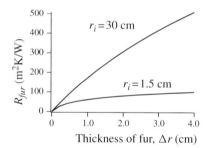

Figure 4.6: Plot of R-value of fur, R_{fur}, as related to thickness of fur, Δr, for an assumed fur thermal conductivity of 0.05 W/m·K.

For small values of $\Delta r/r_i$, the term $\ln(1 + \Delta r/r_i)$ in Eqn. 4.26 can be approximated[1] as $\Delta r/r_i$. Using this approximation,

$$R_{fur} = \frac{\Delta r}{k} \qquad (4.27)$$

Equation 4.27 shows that the R_{fur} increases linearly with the thickness of fur Δr for small values of Δr relative to the radius of the animal, r_i.

Equation 4.26 is plotted in Figure 4.6 for two values of radius of the animal, r_i. It shows that the initial linear increase in R_{fur} with fur thickness, Δr (as predicted by Equation 4.27) continues for a larger value of fur thickness when the animal is bigger (large r_i). While for a small animal, R_{fur} quickly levels off as the fur thickness increases. Thus, increased fur thickness beyond a certain value is not as beneficial for a small animal whereas a large animal would continue to benefit significantly (R_{fur} or thermal resistance per unit surface area increases considerably) over a larger thickness of fur. Looking at it differently, the smaller the animal, the more difficult it is to provide the insulation. Indeed, small animals may only survive in cold climates by behavioral responses that avoid severe stress and by having very high rates of metabolism (Cena and Clark, 1979).

---

[1] For small values of x, $\ln(1 + x)$ can be approximated as

$$\begin{aligned} \ln(1+x) &= x - \frac{1}{2}x^2 + \frac{1}{3}x^3 - \cdots \\ &\approx x \end{aligned}$$

4.3 Steady-State Heat Conduction in a Slab with Internal Heat Generation

All the steady-state heat transfer situations considered in this chapter so far do not have any energy generation. Let us include energy generation in a slab which has temperatures uniform on its two surfaces so that the heat transfer is still one-dimensional, as shown in Fig. 4.7. Examples can be temperature variation in a compost pile where energy is generated due to biochemical processes. The human body with internal metabolic heat generation is another example (see section 4.3.1). We are interested in the steady-state heat transfer in a slab with constant internal heat generation. The general governing equation is simplified for steady-state

$$\underbrace{\frac{\partial T}{\partial t}}_{\text{steady state}}{}^{0} + \underbrace{u\frac{\partial T}{\partial x}}_{\text{no bulk flow}}{}^{0} = \underbrace{\frac{k}{\rho c_p}\frac{\partial^2 T}{\partial x^2}}_{\text{conduction}} + \underbrace{\frac{Q}{\rho c_p}}_{\text{heat generation}} \quad (4.28)$$

to arrive at the specific governing equation

$$\frac{d^2 T}{dx^2} = -\frac{Q}{k} \quad (4.29)$$

The boundary conditions for a constant temperature of T_1 on both surfaces are given by

$$T(x = L) = T_1 \quad (4.30)$$
$$\left.\frac{dT}{dx}\right|_{x=0} = 0 \quad \text{(from symmetry)} \quad (4.31)$$

To solve Eqn. 4.29, it is integrated once to obtain

$$\frac{dT}{dx} = -\frac{Q}{k}x + c_1$$

Using the second boundary condition (Eqn. 4.31),

$$c_1 = 0$$

Integrating again,

$$T = -\frac{Q}{2k}x^2 + c_2 \quad (4.32)$$

Using the first boundary condition,

$$T_1 = -\frac{QL^2}{2k} + c_2 \quad (4.33)$$

Subtracting Eqn. 4.33 from Eqn. 4.32 to eliminate c_2,

$$T - T_1 = -\frac{Q(x^2 - L^2)}{2k}$$
$$= \frac{QL^2}{2k}\left(1 - \frac{x^2}{L^2}\right) + T_1 \quad (4.34)$$

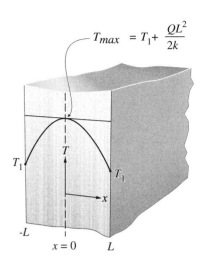

Figure 4.7: Steady state temperature profile in a slab with constant internal heat generation.

This solution is plotted in Figure 4.7. The temperatures inside adjust such that the total energy lost through the surface exactly equals the total energy generated inside, leading to a steady state. The heat lost from the surface at $x = L$ is given by the heat flow at that surface:

$$q = -kA \left.\frac{dT}{dx}\right|_{x=L} \tag{4.35}$$

Substituting for temperature from Eqn. 4.34, we get

$$q = QLA \tag{4.36}$$

Note that QLA is also the heat generated in half of the slab. Steady-state condition is satisfied since the generated heat is equal to the heat lost through the surface.

As an example, Figure 4.8 shows the near steady state temperature profile in a compost pile where biochemical activity produces heat inside the pile and it is cooled from the top and the bottom. Note the similarity between the temperature profile in Figure 4.8 and the temperature profile in Figure 4.7.

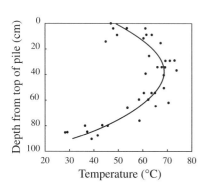

Figure 4.8: Steady state temperature profile in a compost pile (Lynch and Cherry, 1996).

4.3.1 Thermoregulation–Maintaining the Core Body Temperature of Human and Animals

Thermoregulation is maintaining a constant temperature Thermoregulation is the physiological mechanism by which mammals and birds attempt to balance heat gain and loss in order to maintain a constant body temperature when exposed to variations in cooling power of the external medium. Such animals are referred to as warm blooded animals. Some other animals, such as fishes, reptiles, and invertebrates, whose body temperature varies with, and is usually higher than, the temperature of the environment do not have thermoregulation and are referred to as *poikilotherms* or cold blooded animals.

Thermoregulation is an active process Maintenance of a constant body temperature is equivalent to a steady state heat transfer situation achieved through a balance between heat production, gain, and loss. However, it cannot depend totally on diffusion which is a passive process. Rate of metabolic heat generation varies with the animal, as can be seen in Table C.1 on page 332. The environmental temperature can also vary. In a passive system as in cold blooded animals, the steady state temperature would simply be a result of heat balance and would shift to a different value depending on the sustained variation in environmental temperature or metabolic generation. In contrast, in a regulated system as in warm blooded animals, a particular steady state temperature is defended by the activation of appropriate thermoregulatory effectors.

Components of a thermoregulatory system The thermoregulatory processes can be behavioral or autonomic. Behavioral processes of thermoregulation are those which involve the movements of the whole body of the organism relative to the environment such as moving to a preferred temperature or making changes in posture to change heat gain or loss. Autonomic responses are automatic, reflex-like production or loss of heat in response to heat stress or cold stress. Autonomic heat production can come from shivering or supposedly from the metabolic activity of localized deposits of adipose tissue. Autonomic heat loss can be non-evaporative or evaporative. Non-evaporative loss is effected through changing the thermal resistance of the peripheral tissues by changing the extent of constriction of the blood vessels through them. When the environmental temperature is warmer and the heat loss need to be maintained, the pre-capillary arterioles in the layer of fatty tissue beneath the skin are fully dilated and the arterial blood passes through the layer of the fatty tissue unimpeded, keeping the skin temperature quite close to the core and allowing a higher heat loss. When an animal is cold-stressed, the precapillary arterioles are fully constricted and the blood flow to the fatty layer is almost zero. This fatty layer becomes inert with high thermal resistance and the skin temperature is also reduced due to the lack of blood flow, both of which reduce the heat loss. Evaporative heat losses are from panting or sweating. Panting, which is open-mouthed rapid shallow breathing, is evaporation, particularly from the tongue, by the inhaled air. Sweating is an aqueous secretion that spreads over the skin in response to heat stress.

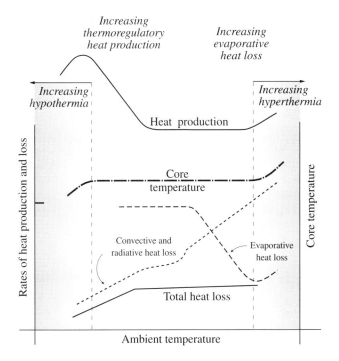

Figure 4.9: Regulation of body core temperature through manipulation of heat production and losses.

Limits of thermoregulation The thermoregulatory functions described above tries to maintain the core body temperature in response to changes in ambient temperature as shown in Figure 4.9. With a fall in ambient temperature, thermoregulatory heat production increases until the animal reaches its maximum ability to generate heat. Any further fall in ambient temperature leads to decline in core temperature, eventually at an increasing rate since the maximum level of heat production may not be sustainable at this lower temperature. This leads to hypothermia. With increase in ambient temperature, the maximum rate of evaporative heat loss is eventually reached. Beyond this point, body temperature can rise and may do so at an increasing rate since the high rate of evaporative loss is not sustainable (heat fatigue) and also the metabolic rate increases significantly at higher temperature. In most species, when temperatures reach $42 - 43°C$, there is damage to the central nervous system, with fatal consequences.

4.4 Steady-State Heat Transfer from Extended Surfaces: Fins

From the definition of convective heat transfer coefficient $q = hA(T - T_\infty)$, we know that to increase heat transfer rate from a surface, one possibility is to increase the convective heat transfer coefficient, h. However, sometimes, situations arise where even the maximum value of h is insufficient or the cost of obtaining pumps, etc., to increase fluid velocity is prohibitive. If we look at this convective heat transfer equation, another choice is to reduce the surrounding temperature, T_∞, which is often impractical. The only other variable that that can possibly be manipulated is the area A. In other words, heat transfer rate can be increased by increasing the surface area over which heat transfer takes place. A *fin* can do precisely this, by extending surfaces from the wall into the surrounding fluid. A schematic of this is shown in Figure 4.10. Everyday examples include radiator fins and fins attached to electronic chips. Examples of fin are also present in the biological world, an example of which is the elephant ear discussed in the next section.

At steady state, the governing equation for a simple fin with constant cross-sectional area (Figure 4.10) is derived as follows:

$$q_x - q_{x+\Delta x} - \Delta q_{conv} = 0$$

where Δq_{conv} is the convective heat loss over the surface area $P\Delta x$ over distance Δx, P being the perimeter. Let A_c be the cross-sectional area and k the thermal conductivity of the fin material. Substituting the heat flux q_x in terms of temperature gradient and the convective heat loss Δq_{conv} in terms of convective heat transfer coefficient, we get

$$-kA_c \left.\frac{\partial T}{\partial x}\right|_x - \left(-kA_c \left.\frac{\partial T}{\partial x}\right|_{x+\Delta x}\right) - hP\Delta x(T - T_\infty) = 0$$

$$kA_c \left(\left.\frac{\partial T}{\partial x}\right|_{x+\Delta x} - \left.\frac{\partial T}{\partial x}\right|_x\right) - hP\Delta x(T - T_\infty) = 0$$

Dividing by $kA_c \Delta x$ and taking limits as $\Delta x \to 0$, the first term becomes the second derivative in temperature and the governing equation for the fin can be written as:

4.4. STEADY-STATE HEAT TRANSFER FROM EXTENDED SURFACES: FINS

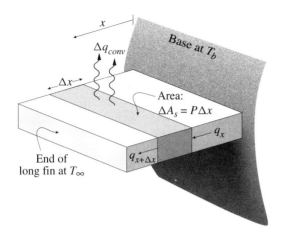

Figure 4.10: Block diagram of a fin showing the base and the extended surface.

$$\frac{\partial^2 T}{\partial x^2} - \frac{hP}{kA_c}(T - T_\infty) = 0 \qquad (4.37)$$

Several boundary conditions are possible. For the special case of a long fin, the end of the fin can be assumed to be at the surrounding temperature. For this situation, the boundary conditions can be written as

$$T(x = 0) = T_b \qquad (4.38)$$
$$T(x \to \infty) = T_\infty \qquad (4.39)$$

To solve the above governing equation, it is easier to transform it using a new temperature variable $\theta = T - T_\infty$ and a transformed base temperature $\theta_b = T_b - T_\infty$. Using these transformations, the new governing equation is

$$\frac{\partial^2 \theta}{\partial x^2} - \frac{hP}{kA_c}\theta = 0 \qquad (4.40)$$

and the new boundary conditions are

$$\theta(x = 0) = \theta_b \qquad (4.41)$$
$$\theta(x \to \infty) = 0 \qquad (4.42)$$

The solution to the above governing equation is given by (see also Section 12.2 on page 244 for another solution):

$$\theta = c_1 e^{mx} + c_2 e^{-mx} \qquad (4.43)$$

where $m^2 = hP/kA_c$ and c_1 and c_2 are constants to be determined using the boundary conditions (Eqns. 4.41 and 4.42). To satisfy the second boundary condition (Eqn. 4.42),

c_1 has to be zero. Using the first boundary condition (Eqn. 4.41), $c_2 = \theta_b$ and the solution is

$$\theta = \theta_b e^{-mx} \qquad (4.44)$$

This shows an exponential drop in temperature.

Heat loss with and without a fin To compare the heat loss in the presence of a fin to the heat loss in its absence, let us first calculate the heat loss with a fin. All of the lost energy has to come through the base of the fin where the rate of heat flow is

$$\begin{aligned} q_{\text{fin}} &= -kA_c \left.\frac{\partial \theta}{\partial x}\right|_{x=0} \\ &= -kA_c\, \theta_b(-m)e^{-mx}\big|_{x=0} \\ &= km A_c \theta_b \\ &= \sqrt{hPkA_c}\,\theta_b \end{aligned} \qquad (4.45)$$

Without a fin, the heat loss is simply what can come out through the surface area equal to the cross-sectional area at the base:

$$q_{\text{no fin}} = hA_c\theta_b \qquad (4.46)$$

Effectiveness of the fin can be defined as

$$\frac{q_{\text{fin}}}{q_{\text{no fin}}} = \sqrt{\frac{kP}{hA_c}} \qquad (4.47)$$

We can choose higher thermal conductivity k of the fin material to increase the fin effectiveness or higher heat loss. We can also choose a higher ratio of perimeter P to cross-sectional area A_c or thin geometry to increase the efficiency or heat loss. Note that under some conditions the ratio on the right hand side of Eq. 4.47 can be less than 1. This means we can actually decrease the heat loss. This makes sense because, for example, if the thermal conductivity of the fin material is very low, it will in fact insulate the surface or reduce heat loss instead of increasing it.

4.4.1 Fins and Bio-heat Transfer

Warm blooded animals appear to have adapted to new or changing environments by varying the size and shape of their bodies and extremities. These physical changes facilitated heat conservation or dissipation as dictated by ambient conditions. Larger animals need to develop means of dealing with great amounts of heat that they produce. The African elephant (Figure 4.11), the largest land mammal, has accordingly developed the largest thermoregulatory organ known in any animal, the pinna or external ear, which it uses as radiator-convector. The pinna is considered to be the main external organ responsible for the temperature regulation of the body.

The combined surface area of both sides of both ears of an African elephant is about 20% of its total surface area. The high surface to volume ratio (see discussion

Figure 4.11: An African elephant with a large external ear or pinna that is important for thermoregulation.

following Eqn. 4.47), large surface area, and extensive vascular network of subcutaneous vessels in the medial side of the ear make it behave like a fin and play a role in temperature regulation. Temperature distribution patterns measured in a pinna are shown in Figure 4.12. Note how the temperature changes from where the ear attaches to the head (base of the fin) to the outer edges, characteristic of a fin. Movement of the pinna (flapping) also increases the heat loss due to increased air flow.

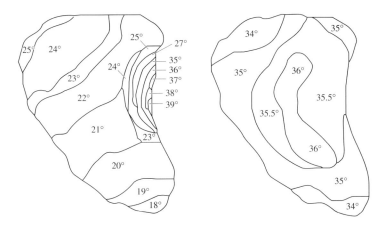

Figure 4.12: Heat transfer from an elephant pinna. Here the pinna acts like a fin for heat transfer purposes. Shown is the right pinna at two different ambient temperatures of 18°C (left) and 32.1°C (right). The change in pattern indicates that a change in blood flow occurs at higher temperatures, Phillips and Heath (1992).

A 4000 kg elephant needs to maintain a heat loss of 4.65 kW or more while moving and feeding. This large amount of heat cannot be dissipated by surface evaporation since elephants do not have sweat glands. Thus, the pinna plays a great role in the heat dissipation and by some estimates (Phillips and Heath, 1992), up to 100% of an African elephant's heat loss can be met by movement of its pinna and by vasodialation. Use of pinna for heat loss by convection and radiation is not unique to elephants. New Zealand white rabbits are known to do the same. Also, the Asiatic elephant, with the same size and metabolic rate but with a much smaller pinna, loses about a third as much heat through the ears.

4.5 Chapter Summary—Steady-State Heat Conduction with and without Heat Generation

- **Steady-State (page 45)**

    1. A heat transfer process is at steady state if the temperature does not change with time.

- **Steady-State Heat Conduction in a Slab (page 47)**

  1. Steady heat conduction in a slab results in a linear temperature profile given by Eqn. 4.5.
  2. A composite slab can be treated as series or parallel combinations of the thermal resistances of individual slabs.

- **Thermal Resistance (page 48)**

  1. Thermal resistance is the resistance to heat flow. It is the analog of electrical resistance to current flow. Thermal resistances for conduction and convection are defined by the terms shown in Eqn. 4.11.
  2. Conductive thermal resistance decreases with increase in thermal conductivity and area of heat flow and decrease in the path length for heat flow.

- **Steady-State Heat Conduction in a Cylinder (page 50)**

  1. The steady state radial temperature profile for heat transfer through a hollow cylinder is given by Eqn. 4.22. Temperature profile is not linear as it has to adjust with the varying cross-sectional area to keep the total heat flow a constant.

- **Steady-State Heat Conduction in a Slab with Internal Heat Generation (page 55)**

  1. Presence of constant internal heat generation with cooling from the surfaces leads to a parabolic temperature profile (Eqn. 4.34).
  2. Thermoregulation (page 56) in mammals to maintain a constant internal body temperature can be seen as a steady-state achieved through a balance between internal heat generation and heat loss at the surface. However, passive heat conduction is not generally sufficient for achieving this steady state and active thermoregulatory effectors in the body are necessary.

- **Steady State Heat Conduction from Extended Surfaces (page 58)**

  1. One way to increase the rate of heat transfer or decrease the thermal resistance is to increase the surface area for heat transfer. This goal is achieved in fins.
  2. The temperature profile in a long fin is given by Eqn. 4.44.

4.6 Concept and Review Questions

1. Explain the difference between steady-state and equilibrium.
2. What are the components (parameters) in thermal resistance?

3. Can you think of a heat transfer situation where the thermal resistances are in parallel instead of being in a series, as in Eqn. 4.11?

4. Does the core temperature in a cold blooded organism, such as a fish, depend on heat conduction?

5. Why is passive heat conduction not sufficient for thermoregulation in warm blooded animals?

6. Why is a straight line profile in case of steady heat transfer through an annulus not physically realistic?

7. What is the purpose of using fins?

8. Can you actually reduce the heat loss by adding fins? Explain under what circumstances this can happen.

4.7 Further Reading

Albright, L. D. 1990. *Environment Control for Animals and Plants.* ASAE, St. Joseph, Michigan.

Bakken, G. S. 1991. Wind speed dependence of the overall thermal conductance of fur and feather insulation. J. Therm. Biol. 16(2):121-126.

Bligh, J. 1983. Temperature regulation. In: *Stress Physiology in Livestock: Volume 1, Basic Principles*, edited by M. K. Yousef. CRC Press, Boca Raton, Florida.

Bligh, J. 1985. Regulation of body temperature in man and other mammals. In: *Heat Transfer in Medicine and Biology*, edited by A. Shitzer and R. C. Eberhart. Plenum Press, New York.

Bruce, J. M. and J. J. Clark. 1979. Models of heat production and critical temperature for growing pigs. Animal Production. 28:353-369.

Cena, K. and J. A. Clark. 1979. Transfer of heat through animal coats and clothing. Environmental Physiology III 20:1-42.

Dewar, H., J. B. Graham, and R. W. Brill. 1994. Studies of tropical tuna swimming performance in a large water tunnel. II. Thermoregulation. Journal of Experimental Biology, 192:33-44.

Gebremedhin, K. G. 1987. A model of sensible heat transfer across the boundary layer of animal hair coat. J. Therm. Biol. 12(1):5-10.

Gilbert, R. D., H. Schröder, T. Kawamura, P. S. Dale, and G. G. Power. 1985. Heat transfer pathways between fetal lamb and ewe. Journal of Applied Physiology, 59:634-638.

Hartwig, M. K. and K. Kiessl. 1997. Calculation of heat and moisture transfer in exposed building components. Int. J. Heat and Mass Transfer 40(1):159-167.

Haslam, R. A. and K. C. Parsons. 1988. Quantifying the effects of clothing for models of human response to the thermal environment. Ergonomics 31(12):1787-1806.

Iberall, A. S. and A. M. Schindler. 1973. On the physical basis of a theory of human thermoregulation. J. of Dynamic Systems, Measurement and Control, Trans. of ASME, March 1973: 68-75.

Kawashima, Y. 1993. Characteristics of the temperature regulation system in the human body. J. Therm. Biol. 18(5/6):307-323.

Lynch, N. J. and R. S. Cherry. 1996. Winter composting using the passively aerated windrow system. Compost Science and Utilization, 4(3):44-52.

Oosterhout, G. R. and G. A. Spolek. 1989. Transient heat and mass transfer in layered walls. In Collected Papers in Heat Transfer, HTD-Vol. 123, ASME, New York.

Pennes, H. H. 1948. Analysis of tissue and arterial blood temperatures in the resting human forearm. J. of Applied Physiology, 1(2):93-122.

Phillips, P. K. and J. E. Heath. 1992. Heat exchange by the pinna of the African elephant (Loxodonta Africana). Comparative Biochemistry and Physiology. 101A (4):693-699.

Robertshaw, D. and V. A. Finch. 1984. Heat loss and gain in artificial and natural environments. *Thermal Physiology*, J. R. S. Hales, Ed., Raven Press, New York.

Shitzer A. and R. C. Eberhart. 1985. *Heat Transfer in Medicine and Biology*. Plenum Press, New York.

van Waversveld, J., A. D. F. Addink, G. van den Thillart, and H. Smit. 1989. Heat production of fish: A literature review. Comparative Biochemistry and Physiology Part A: Physiology, 92(2):159-162.

4.8 Problems

4.8.1 Reduction of Heat Loss Using a Stagnant Air Film

Compare the heat loss in W from two windows of a room, one having a single sheet of glass 10mm thick while the other has two sheets of glass, 5mm each, separated by 5mm of air. The second window is called a thermopane. The window sizes are $1m \times 1m$. The air temperature in the room is $20°C$ and the outside air temperature is $-5°C$. The average value of the thermal conductivity of glass is 0.85 W/m·K and that of air is 0.025 W/m·K over this temperature range of interest. The heat transfer coefficient corresponding to slower moving air inside the room is 20 W/m^2·K and that of the faster moving air outside is 200 W/m^2·K. Consider the thin layer of air between the glass sheets to be stagnant.

4.8.2 Clothing and Conduction Heat Transfer

Compare the total heat loss from a person dressed in summer clothes to the heat loss from a person dressed in winter clothing under the same ambient conditions. The insulating effect of the layer or layers of air trapped between the layers of clothing and between the clothing and the body is reflected in the resistance values of the clothing ensembles. Average temperature of skin is 33°C, ambient temperature is 20°C, total surface area of body is 1.7 m², area of body covered by summer clothing is 1 m², area of body covered by winter clothing is 1.6 m², heat transfer coefficient of bare skin is 27.3 W/m²·K, overall heat transfer coefficient of summer clothing is 18.4 W/m²·K and overall heat transfer coefficient of winter clothing is 4.3 W/m²·K (Shitzer and Eberhart, 1985).

4.8.3 Insulation in a Home Freezer

Consider a home freezer compartment with dimensions of height 60 cm, width 40 cm and depth 40 cm. Assume heat loss to be from the five sides exposed to outside air, i.e., do not consider any heat loss through the side which the freezer sits on. For a styrofoam insulation of thermal conductivity 0.03 W/m·K, what should be the minimum thickness of this insulation so that the total heat loss from the freezer compartment is less than 120 W? Assume the freezer air is at $-20°C$ and the inside surface heat transfer coefficient is 5 W/m²·K, while the outside air temperature is 30°C and the outside surface heat transfer coefficient is 20 W/m²·K.

4.8.4 Steady State Heat Transfer in a Slab with Convection on Surface

Calculate the temperature profile in a slab of thickness ΔL and thermal conductivity k whose one side is maintained at temperature T_2 while the other side exchanges energy by convection with a fluid at a heat transfer coefficient of h. The fluid temperature is T_∞.

4.8.5 Reduction of Heat Loss in a Pipe

A circular sheet metal duct carries refrigerated air to a cold storage room for apples. The duct itself is 250 mm in outer diameter. The duct wall thickness is 1 mm. To reduce the heat gain from the surrounding air, we need to wrap the duct with insulation. The flowing air maintains the inner surface of the duct at 0°C. The outer surface temperature of insulation would be maintained at 25°C by the room air. Thermal conductivity of the sheet metal is 60 W/m·K and that of the insulation is 0.04 W/m·K. Assuming the heat transfer to be at steady state, what thickness of insulation should be put on the duct to keep the rate of heat gain by the refrigerated air per meter length of the duct to 30 W (Albright, 1990).

4.8.6 Convective and Conductive Heat Loss in a Radial Geometry

Consider steady state heat transfer from a hollow cylindrical pipe (Section 4.2) covered with insulation, as discussed in Section 4.2. Only the thermal resistance of the insulation and the external convective resistance are important. The inside wall of insulation is at T_i while the outside air temperature is T_0. 1) Write the expression for steady state heat flow in terms of the temperature difference and the total resistance. 2) Show that the thermal resistance has a minimum value with respect to the thickness of insulation (i.e., below and above this thickness, thermal resistance increases and heat loss decreases) when the pipe outer diameter is kept fixed. 3) Discuss the physical meaning of such a minimum value. 4) For a heat transfer coefficient of 50 W/m^2·K and a thermal conductivity of insulation of 0.2 W/m·K, calculate the radius for minimum thermal resistance and discuss the practical utility of such a result.

4.8.7 Thickness of Insulation in a Radial Geometry

Refrigerant flows in a copper tube of outer diameter of 4.8 cm. The inside surface temperature is $-15°C$ and the room temperature is $20°C$. The surface heat transfer coefficient between the tube and the room air is 25 W/m^2·K. Ignore the thermal resistance of the copper tube, so that the outside wall temperature of the tube is the same as the inside wall temperature. 1) What is the heat gained by the refrigerant per unit length? 2) To reduce the heat gained from the room air, someone decided to put 4 mm of insulation (thermal conductivity 0.75 W/m·K) around this tube. Show that the heat loss per unit length actually increases after adding this insulation. 3) Explain why you can expect the heat loss to increase.

4.8.8 Heat Transfer in the Body

At rest, the human body is producing heat at a constant rate as a byproduct of basal metabolism. This heat is dissipated to the surrounding and a steady-state temperature profile is reached in the body. Here we will approximate the body as a slab with uniform heat generation throughout. Assuming that the heat is transferred to the surface primarily by conduction along the smaller dimension (thickness) of the slab, find and plot the temperature distribution in the body. Assume the body to be symmetrical about its vertical axis. The average rate of metabolic heat generation in the body is 1.4 kW/m^3, the average half thickness of the slab approximating the body is 7.5 cm, body surface temperature is $33°C$ and the average thermal conductivity of the material is 1.05 W/m·K.

4.8.9 Temperature Equilibration in a Fish

Fish are poikilothermic, animals whose body temperature depends heavily on the temperature of their environment (some fish can have somewhat higher level of control of body temperature—see, for example, tropical tuna study by Dewar et al., 1994). Metabolic heat is produced by the fish, and blood flowing though the tissues picks up this heat and carries it to the gills, where the blood exchanges oxygen and carbon dioxide as well as heat with the surrounding water. Large volumes of water move over the gills, so the blood quickly equilibrates to the temperature of the water. Because of this temperature dependence, it is difficult to measure the metabolic heat production of the fish and determine the heat transfer coefficient across the gills.

Using a lumped parameter analysis, find the heat transfer coefficient across the gills of a fish. Assume the process is at steady-state. Mass of fish is 2 kg, total surface area of the gills is 300 cm^2, temperature differential between the gills and the water flowing over the gills is 0.5°C, and the total metabolic heat generation in the fish is given by $0.194 m^{0.85}$ Watts where m is the mass of fish in kg. Assume heat is lost only at the gills.

4.8.10 Heat Transfer in Composting

Composting can be used to treat various organic wastes such as grass clippings, leaves, paper, animal manure, and the organic fraction of municipal solid waste. These wastes are converted by microbial degradation to a stabilized compost. The biochemical conversion process generates heat and consideration of this generated heat is critical in the design of composting systems. Compost managers strive to keep the compost below about 65°C because hotter temperatures cause the beneficial microbes to die off. If the pile gets too hot, turning or aerating will help to dissipate the heat.

Treat a compost pile on the ground as one-dimensional system of height 2 m in the vertical direction. The top of the pile has wind (at temperature 40°C) blowing on it, leading to a heat transfer coefficient of 50 W/m^2·K. The bottom of the pile can be approximated at the ground temperature of 20°C. Consider only conductive heat transfer inside the pile and neglect any convection inside this porous pile. The volumetric biochemical heat generation is 7 W/m^3. The thermal conductivity of the compost material is 0.1 W/m·K. 1) Setup the appropriate governing equation and boundary conditions. 2) Solve for temperature as a function of height from the ground. 3) Calculate the maximum temperature in the pile. 4) Calculate the top surface temperature of the pile.

4.8.11 Heat Transfer Prior to a Surgery

Heart disease is one of the leading causes of death, and cholesterol contributes heavily to this problem. However this same agent can also cause poor circulation and peripheral artery disease in the legs. In some cases the damage to the leg arteries is so severe that the only option is amputation. In that case the leg is often cooled in a refrigerant prior to surgery to reduce the blood flow even further and numb the nerve endings to make surgery and recovery easier for the patient.

Find and plot the leg's temperature profile in the radial direction after the cooling process has reached steady-state. Assume that the skin temperature is equal to the refrigerant temperature, which is unchanged by the cooling process. The leg can be approximated as a cylinder with constant thermal properties in the radial direction. The metabolic heat generation rate is slowed by the cooling, but is still present uniformly throughout the leg, at a value of 1.1 kW/m^3. The thermal conductivity of the leg muscle is 0.6 W/m·K, the radius of leg is 5.0 cm and the refrigerant temperature is 10°C.

4.8.12 Steady State and Metabolic Heat Generation

The steady state radial temperature distribution inside a cylindrical limb being cooled with a refrigerant at 2°C is

$$T = 2 + 0.65\left[1 - \left(\frac{r}{R}\right)^2\right]$$

where T is in °C. Assume constant thermal properties of the limb and the metabolic heat generation being uniform throughout the limb. The radius, R, of the limb is 5.0 cm and its effective thermal conductivity is 0.6 W/m·K. 1) Write the appropriate governing equation assuming only 1D heat transfer. 2) Without solving this governing equation, use it to calculate the rate of metabolic heat generation in W. 3) What is the rate of heat loss from the surface of the limb?

4.8.13 Heat Dissipated During Exercise

Calculate the energy dissipated at steady state per unit length at the surface of a working cylindrical muscle. The heat generated in the muscle is 5.8 kW/m³, the thermal conductivity of the muscle is 0.419 W/m·K and the radius of the muscle is 1 cm. What is the maximum temperature rise ($T_{max} - T_{surface}$) in the muscle?

Chapter 5

CONDUCTION HEAT TRANSFER: UNSTEADY STATE

CHAPTER OBJECTIVES

After you have studied this chapter, you should be able to formulate and solve for a time-varying (unsteady) heat conduction process for the following situations:

1. Where temperatures do not change with position.

2. In a simple slab geometry where temperatures vary also with position.

3. Near the surface of a large body (semi-infinite region).

KEY TERMS

- **internal resistance**
- **external resistance**
- **Biot number**
- **lumped parameter analysis**
- **1D and multi-dimensional heat conduction**
- **Heisler charts**
- **semi-infinite region**

In this chapter, we will consider heat transfer situations where temperature is a function of both position and time. Variations in temperature with time is referred to as unsteady, as opposed to steady state studied in Chapter 4 where temperature did not vary with time. The relationship of this chapter to other chapters in energy transfer is shown in Figure 5.1.

Rate of energy transfer in a solid depends on the internal thermal resistance of the solid and the external thermal resistance (usually the convective thermal resistance of the fluid flowing over the surface). The limiting cases are:

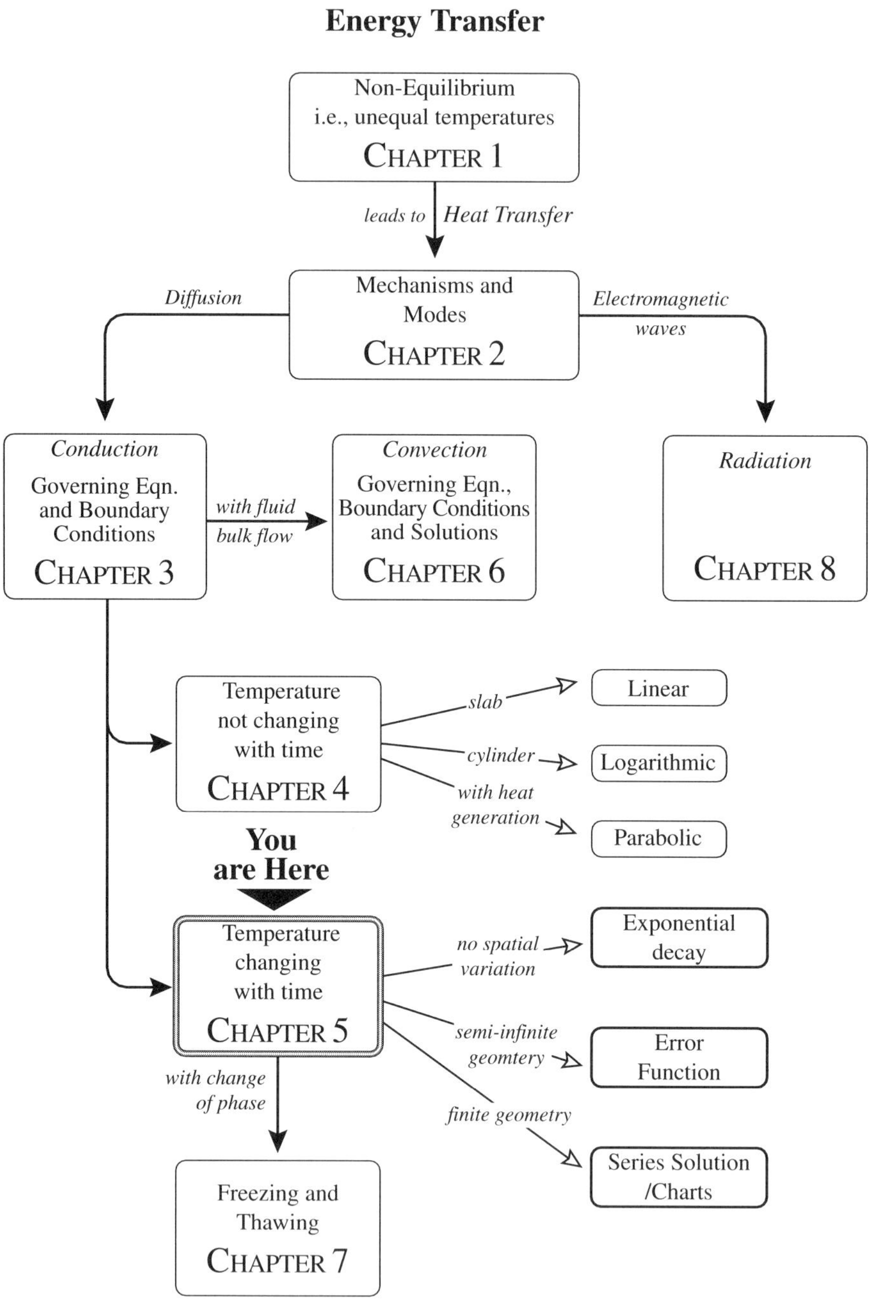

Figure 5.1: Concept map of energy transfer showing how the contents of this chapter relates to other chapters on energy transfer.

1. Negligible internal resistance (Lumped parameter)
2. Negligible external resistance
3. Neither resistance is negligible

5.1 Transient Heat Transfer When Internal Conductive Resistance Is Negligible: Lumped Parameter Analysis

In some special cases, temperature variation in all three spatial directions can be ignored. When the temperature variations are ignored, the situation is considered lumped. Temperature would then vary only with time. Note that this can only be an approximation and it is physically impossible for this to happen exactly. If temperature does not vary spatially, there can be no heat flow and, therefore, no change in temperature with respect to time. The transient heating or cooling of the solid is possible only if there are temperature gradients inside it. In reality, there will be temperature variations inside the solid, except these variations will be small compared to the temperature variation outside the solid (in the fluid). In other words, this lumped parameter condition is possible when the internal resistance in the solid is small compared to the external resistance in the fluid.

In Chapter 3, we derived the governing equations for heat transfer and mentioned that it was general enough to consider all kinds of situations, i.e., whether large thermal conductivity or small. It is in fact possible to start from one of these governing equations and derive the governing equation for this special case of lumped parameter or no internal resistance (see page 371 in Appendix). However, since spatial variations are ignored, there is an alternative and much easier way to derive the governing equation for this situation. Consider the schematic in Figure 5.2. If the temperature change in the solid is ΔT during time Δt, the total energy gained by the solid should be equal to the energy transfer due to the convection at the surface over this time. In other words,

$$\underbrace{-mc_p \Delta T}_{\substack{\text{energy}\\\text{lost}}} = \underbrace{hA(T - T_\infty)\Delta t}_{\substack{\text{energy}\\\text{convected away}}} \qquad (5.1)$$

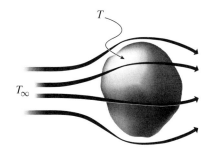

Figure 5.2: A solid with convection over its surface.

where m is the mass of the solid, c_p is its specific heat, h is the convective heat transfer coefficient at the surface, A is the surface area, and T_∞ is the bulk fluid temperature. Note that the right hand side of Eqn. 5.1 uses T as the surface temperature in the formula for convective heat loss since the entire solid is assumed to be at one temperature, T. Rearranging the terms,

$$-\frac{\Delta T}{\Delta t} = \frac{hA}{mc_p}(T - T_\infty)$$

Taking the limit as $\Delta t \to 0$,

$$\frac{dT}{dt} = -\frac{hA}{mc_p}(T - T_\infty) \qquad (5.2)$$

This is the governing equation describing temperature vs. time for a lumped parameter equation. Being a first order equation, it needs one condition which in this case is the initial condition

$$T(t=0) = T_i \tag{5.3}$$

To solve equation 5.2, define transformed temperatures as

$$\theta = T - T_\infty$$
$$\theta_i = \theta(t=0) = T_i - T_\infty$$

Using the new variable θ, the governing equation 5.2 is transformed as

$$\frac{d\theta}{dt} = -\frac{hA}{mc_p}\theta$$

This equation is integrated over time and the initial condition $t = 0$ is set to $\theta = \theta_i$ to obtain

$$\int_{\theta_i}^{\theta}\frac{d\theta}{\theta} = \int_0^t -\frac{hA}{mc_p}dt \tag{5.4}$$

$$\ln\frac{\theta}{\theta_i} = -\frac{hA}{mc_p}t$$

$$\frac{\theta}{\theta_i} = e^{-\frac{hA}{mc_p}t}$$

Changing the variable from θ to T we get temperature as a function of time in lumped parameter heat transfer as

$$\frac{T - T_\infty}{T_i - T_\infty} = \exp\left(-\frac{hA}{mc_p}t\right) = \exp\left(-\frac{t}{\frac{mc_p}{hA}}\right) \tag{5.5}$$

Equation 5.5 shows it takes an infinite time to reach the steady state or the final temperature of T_∞. As the temperature of the solid becomes close to fluid temperature, the rate of heat transfer drops so the solid can never quite reach the fluid temperature T_∞.

Figure 5.3 is a graphical representation of the relationship between time and temperature given by Eqn. 5.5. Note that although much of the total possible change in temperature $T_\infty - T_i$ happens over a short time, the rate of temperature change decreases continuously and the temperature never really reaches T_∞.

5.2 Biot Number: Deciding When to Ignore the Internal Resistance

The lumped parameter solution developed in the previous section is quite simple. It is tempting to use such an approach for many practical situations. We need to decide

5.2. BIOT NUMBER: DECIDING WHEN TO IGNORE THE INTERNAL RESISTANCE

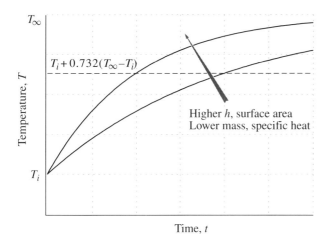

Figure 5.3: Typical temperature change in a lumped parameter analysis.

when is such an approach a valid one, i.e., what is the criterion for ignoring the internal resistance as compared to the external resistance? The parameter that compares the internal and external resistance is

$$Bi \text{ (Biot number)} = \frac{hL}{k} = \frac{\frac{L}{kA}}{\frac{1}{hA}} = \frac{\text{conductive resistance}}{\text{convective resistance}} \quad (5.6)$$

Thus, for large Biot number, the (internal) conductive resistance is higher and therefore the controlling resistance, i.e., we can ignore the (external) convective resistance. Conversely, for small Biot number, the convective resistance is the controlling one and we can ignore the conductive resistance. For a Biot number $Bi < 0.1$, i.e.,

$$\frac{h(V/A)}{k} < 0.1 \quad (5.7)$$

the error in temperature calculation is $< 5\%$ if the lumped parameter approach is used. Qualitatively speaking, lumped parameter is suitable for large surface areas, small volumes, small convective heat transfer coefficients, and large thermal conductivities. Most of the thermal resistance is external, in the fluid surrounding the object.

Characteristic Length The quantity V/A is used in place of L in Eqn. 5.7 has the dimension of length and can be called a *characteristic length* of the system. In heat conduction, characteristic length can be interpreted as the path of least thermal resistance for the geometry of concern. Thus, as surface area increases for the same volume, the effective distance through which heat has to diffuse decreases, i.e., the characteristic length decreases, and heating or cooling the body will require less time. As will be used later, the characteristic length for heat conduction in an infinite slab is the half thickness of the slab. Thus, for an infinite slab, Eqn. 5.7 is written as

$$hL/k < 0.1 \qquad (5.8)$$

Similarly, for a long cylinder or a sphere, characteristic length is the radius. Note that for a very short cylinder, as illustrated in Figure 5.4, the characteristic length can become half thickness since it approaches an infinite slab.

Example 5.2.1 Keeping Eggs Warm in an Incubator

It is desired to find the transient temperature of chicken eggs when they are placed in an incubator. The incubator air temperature is 38°C. The natural convection heat transfer coefficient for air over the eggs in the incubator is 5.2 W/m$^2 \cdot$K. Calculate the temperature of the egg after one hour if its initial temperature is 20°C. Assume the eggs to be spherical with volume of 60 cm^3. The density, specific heat, and thermal conductivity of the egg are 1035 kg/m^3, 3350 J/kg·K, and 0.62 W/m·K, respectively.

Known: Initial temperature of an egg

Find: Temperature of the egg after 60 minutes

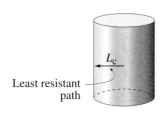

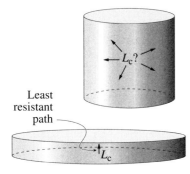

Figure 5.4: Characteristic lengths for heat conduction in various geometries.

Schematic and Given Data: A schematic of the problem is shown in Figure 5.5. The given data are

1. initial temperature of the egg is 20°C
2. ambient air temperature is 38°C
3. surface heat transfer coefficient is 5.2 W/m$^2 \cdot$K
4. density, specific heat, and thermal conductivity of the egg are 1035 kg/m^3, 3350 J/kg·K, and 0.62 W/m·K, respectively

Assumptions:

1. egg is approximately spherical
2. surface heat transfer coefficient provided is an average value and its likely variation over the surface is ignored
3. We hope to use the lumped parameter approach. To do this, we have to check the Biot number. Thus,

$$\text{Bi} = hV/Ak = \left(5.2 \times \frac{4}{3}\pi r^3\right) / \left(4\pi r^2 \times 0.62\right) = .07 < 0.1.$$

Since the Biot number is less than 0.1, the heat transfer problem can be treated in terms of a lumped parameter. What this means is that the heat transfer resistance in this case is mostly external (in the air) as opposed to internal (within the egg). The temperature variation inside the egg can be ignored compared to the temperature variation outside, in the air.

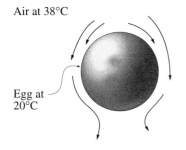

Figure 5.5: Schematic for Example 5.2.1.

5.3. TRANSIENT HEAT TRANSFER WITH INTERNAL RESISTANCE

Analysis: Using the lumped parameter equation (Eqn. 5.5),

$$\frac{T - T_\infty}{T_i - T_\infty} = \exp\left(-\frac{hA}{mc_p}t\right)$$

Substituting the numerical values, we get:

$$\frac{T - 38}{20 - 38} = \exp\left(-\frac{5.2[\text{W/m}^2\cdot\text{K}] \times 0.00785[\text{m}^2]}{1035[\text{kg/m}^3] \times 60 \times 10^{-6}[\text{m}^3] \times 3350[\text{J/Kg}\cdot\text{K}]}3600[\text{s}]\right)$$

where area is $4\pi r^2 = 7.85 \times 10^{-3}\text{m}^2$. Solving, we get temperature after 1 hour as $T = 29.1°\text{C}$.

Comments: To contrast this situation with another, compare the same egg being boiled in water. We would like to know how the temperature increases in the egg. In boiling water, the heat transfer coefficient is much higher, and is 1×10^4 W/m$^2\cdot$K. First, let us consider the Biot number for this situation:

$$\text{Bi} = hV/Ak = \left(1 \times 10^4 \times \frac{4}{3}\pi r^3\right) / \left(4\pi r^2 \times 0.62\right) = 134 >> 0.1$$

Hence the problem cannot be classified as a lumped parameter any more. Note the absolute of the internal resistance of the egg really did not change much (it is the same egg), but this internal resistance is relatively much higher compared to the decreased external resistance in the boiling water. To calculate the temperature inside the egg for this situation is much more involved, and is done using equations developed in the following section.

5.3 Transient Heat Transfer When Internal Resistance Is Not Negligible: Example of Slab Geometry

In the previous section, it was noted that when the Biot number (which is a function of properties of the solid and fluid and the velocity of the fluid) is not small, internal diffusional resistances are significant. This means temperature variation within the solid is now significant, unlike what we dealt with for the lumped parameter analysis. In this section we will learn how to analyze the spatial variation of temperature in a solid. For simplicity, we will consider a one-dimensional slab where temperature will vary along the thickness, as shown in Figure 5.6. To keep things simple, we will consider the situation where the surface is at a specified temperature, i.e., external fluid resistance is negligible. The governing equation for symmetric heating or cooling of an infinite slab without any heat generation can be simplified from:

$$\underbrace{\rho c_p \frac{\partial T}{\partial t}}_{\text{storage}} + \underbrace{\cancel{u\frac{\partial T}{\partial x}}^{0}}_{\text{no bulk flow}} = \underbrace{k\left(\frac{\partial^2 T}{\partial x^2}\right)}_{\text{conduction}} + \underbrace{\cancel{\dot{Q}}^{0}}_{\text{no heat generation}}$$

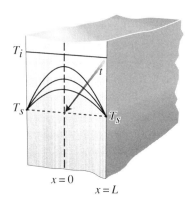

Figure 5.6: Schematic of a slab showing the line of symmetry at $x = 0$ and the two surfaces at $x = L$ and at $x = -L$ maintained at temperature T_s. The material is very large (extends to infinity) in the other two directions.

as

$$\frac{\partial T}{\partial t} = \frac{k}{\rho c_p} \frac{\partial^2 T}{\partial x^2} \tag{5.9}$$

The boundary conditions are

$$\left.\frac{\partial T}{\partial x}\right|_{x=0,t} = 0 \quad \text{(from symmetry)} \tag{5.10}$$

$$T(L, t > 0) = T_s \quad \text{(surface temperature is specified)} \tag{5.11}$$

and the initial condition is

$$T(x, t = 0) = T_i \tag{5.12}$$

where T_i is the constant initial temperature and T_s is the constant temperature at the two surfaces of the slab at time $t > 0$. Due to the same temperature T_s on both sides and the same initial temperature everywhere, the temperature profile will always be symmetric, leading to the symmetry condition in Eqn. 5.10. Note that we needed two boundary conditions in x since Eqn. 5.9 is second order in position x and one (initial) condition in time since it is first order in time. The solution to Eqn. 5.9 is based on separation of variables (see page 365 for details), and is given by

$$\frac{T - T_s}{T_i - T_s} = \sum_{n=0}^{\infty} \frac{4(-1)^n}{(2n+1)\pi} \cos\frac{(2n+1)\pi x}{2L} e^{-\alpha\left(\frac{(2n+1)\pi}{2L}\right)^2 t} \tag{5.13}$$

where $\alpha = k/\rho c_p$ is the thermal diffusivity, defined on page 20. Equation 5.13 provides temperature T as a function of position x and time t. Note that the position appears as

5.3. TRANSIENT HEAT TRANSFER WITH INTERNAL RESISTANCE

a non-dimensional quantity x/L and time as another non-dimensional quantity $\alpha t/L^2$. The non-dimensional time is called the Fourier number Fo so that

$$Fo = \frac{\alpha t}{L^2} \qquad (5.14)$$

Use of these non-dimensional quantities are discussed below.

5.3.1 How Temperature Changes with Time

We now try to visualize the relationship of temperature vs. position and time provided by Eqn. 5.13. The infinite series makes it difficult to visualize this relationship and we will try to simplify it for special situations. As can be seen in Figure G.1 on page 368, we need a large number of terms in the series to satisfy the initial condition. Conversely, if we are interested in times long after $t = 0$, we may not need as many terms. To compare the contribution of the various terms at different times, the relative values of the exponential terms for $n = 1, 2, ...$ in Equation 5.13 are shown in Figure 5.7 for a large and a small value of time. For large value of t (=600 s), the exponential terms in Equation 5.13 decay more rapidly as shown in this figure. Therefore, the contributions of the subsequent terms to the summation can perhaps be ignored for large values of time and we would have less error as compared to ignoring terms for smaller values of time where the terms decay less rapidly (see Figure 5.7 for $t = 30$ s). After long times (Figure 5.7 for $t = 600$ s), often keeping only the first term ($n = 0$) is a good approximation to the exact solution. It turns out that "long" time is defined as $Fo > 0.2$. Substituting $n = 0$ in Equation 5.13 to keep only the first term, we get

$$\frac{T - T_s}{T_i - T_s} = \frac{4}{\pi} \underbrace{\cos \frac{\pi x}{2L}}_{\text{spatial}} \underbrace{e^{-\alpha \left(\frac{\pi}{2L}\right)^2 t}}_{\text{time}} \qquad (5.15)$$

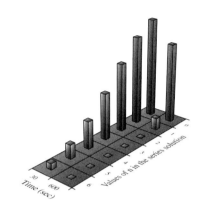

Figure 5.7: The terms in the series ($n = 0, 1, ...$ in Equation 5.13) drop off rapidly for large values of time. Calculations are for $Fo = 0.0048$ at 30 s and $Fo = 0.096$ at 600 s for a thickness of $L = 0.03$ m and a typical $\alpha = 1.44 \times 10^{-7} \text{m}^2/\text{s}$ for biomaterials.

This expression clearly shows that eventually (for times long after $t = 0$), the time-temperature relationship is exponential at a given position (x value). It can never reach the final temperature T_s. As the temperature difference between any location and the surrounding (T_s) decreases, the rate of heat flow decreases. If we take the natural logarithm of both sides of Eqn. 5.15, we get:

$$\ln \frac{T - T_s}{T_i - T_s} = \ln \left(\frac{4}{\pi} \cos \frac{\pi x}{2L}\right) - \alpha \left(\frac{\pi}{2L}\right)^2 t \qquad (5.16)$$

This linear relationship between $\ln(T - T_s)/(T_i - T_s)$ and time t can be seen in Figure B.1 on page 327 for various values of position x/L. Note that at times close to $t = 0$, they are not linear. This is expected since by dropping terms in the series we fail to satisfy the initial condition.

5.3.2 Temperature Change with Position and Spatial Average

It can be easily seen from Equation 5.15 that eventually, at a given time, the temperature varies as a cosine function. Note that it stays as a cosine function for all large values of time, except the amplitude of the cosine wave drops exponentially with time. In practice, a spatial average temperature is often needed. A spatial average temperature would be defined by

$$T_{av} = \frac{1}{L} \int_0^L T\, dx \qquad (5.17)$$

Applying this definition of average to Eqn. 5.16, we obtain an equation for the average temperature as

$$\ln \frac{T_{av} - T_s}{T_i - T_s} = \ln \frac{8}{\pi^2} - \alpha \left(\frac{\pi}{2L}\right)^2 t \qquad (5.18)$$

5.3.3 Temperature Change with Size

Consider the equation for average temperature as a function of time for large values of time. It can be rewritten as:

$$\frac{\alpha t}{L^2} = -\frac{4}{\pi^2} \ln \left[\frac{\pi^2}{8} \left(\frac{T_{av} - T_s}{T_i - T_s} \right) \right] \qquad (5.19)$$

This implies that for a given change of average temperature (measured in terms of fractional change of the total possible change $T_i - T_s$), the time required increases with the square of the thickness, i.e., $t \propto L^2$. This observation can be generalized for other geometries by saying the time required is proportional to the square of the characteristic dimension.

5.3.4 Charts Developed from the Solutions: Their Uses and Limitations

The solution given by Eqn. 5.13 would require much effort in computing every time we need to use it. Fortunately, this series has already been computed and plotted for most practical situations, as shown in Figure B.1 on page 327. We simply need to use these charts. To see how these charts are developed, note that Eqn. 5.13 can be rearranged as below:

$$\begin{aligned}
\frac{T - T_s}{T_i - T_s} &= \sum_{n=0}^{\infty} \frac{4(-1)^n}{(2n+1)\pi} \cos \frac{(2n+1)\pi x}{2L} e^{-\alpha \left(\frac{(2n+1)\pi}{2L}\right)^2 t} \\
&= \sum_{n=0}^{\infty} \frac{4(-1)^n}{(2n+1)\pi} \cos \left[\frac{(2n+1)\pi}{2} \frac{x}{L}\right] e^{-\left(\frac{(2n+1)\pi}{2}\right)^2 \frac{\alpha t}{L^2}} \quad (5.20)
\end{aligned}$$

Equation 5.20 shows that the non-dimensional temperature $(T - T_s)/(T_i - T_s)$ is a function of non-dimensional position x/L and non-dimensional time $\alpha t/L^2$, and therefore these three non-dimensional variables are sufficient to describe the time-temperature relationship at various locations in the slab. Figure B.1 on page 327 has plots of the relationships between these variables and is known as Heisler chart. Note, however,

5.3. TRANSIENT HEAT TRANSFER WITH INTERNAL RESISTANCE

Figure B.1 is not a plot of Eqn. 5.20, but of Eqn. 5.15, which has only the first term in the series and thus the straight lines in this figure are valid at long times. For smaller times, we have to keep more terms in the series. If more terms are kept, the relationship between temperature and time (Eqn. 5.20) would not be a straight line near $t = 0$.

In Figure B.1, the set of lines corresponding to $m = 0$ are plots of Equation 5.15 for the boundary condition of specified surface temperature T_s that is equivalent to $h \to \infty$. Note that $n(= x/L)$ in the charts has different meaning than in Eqn. 5.20. Other sets of lines corresponding to various values of $m(= k/hL)$ are for convective boundary conditions (finite values of h) and are discussed in Section 5.3.5.

Since the charts were developed from the analytical solution, the assumptions that went into the analytical solution are implicit in the charts. It is important to remind ourselves of these conditions that have to be satisfied in order to be able use the charts. These are:

- Uniform initial temperature
- Constant boundary fluid temperature
- Perfect slab, cylinder or sphere
- Far from edges
- No heat generation ($Q = 0$)
- Constant thermal properties (k, α, c_p are constants)
- Typically for times long after initial times, given by $\alpha t/L^2 > 0.2$

Example 5.3.1 Temperatures Reached During Food Sterilization

Sterilization of food and biological material is a very important application of heat transfer to biological systems. Consider sterilization of a tuna can (Figure 5.8) that is filled with tuna meat and water. Considering the rate of bacterial death, it has been decided to heat this container for 30 minutes to reach sterilization. Calculate the temperature after 30 minutes at the location where temperature is the lowest at any time, i.e., the location in the container that stays the coldest during heating. The thermal diffusivity of the tuna meat with water is $2 \times 10^{-7} \mathrm{m^2/s}$. Heating is done in steam that maintains all surface of the can at $121°C$. For simplicity, assume heat transfer at the coldest point to be mostly along the thickness, i.e., one-dimensional. Initial temperature of the tuna is $40°C$. Assume the thermal resistance of the can material to be negligible and ignore any convective movement of water in the can during heating.

Known: A slab of tuna whose surface temperature is suddenly increased

Find: The coldest point in the can is the geometric center. Thus, we need to find the temperature at the center of the 25 mm slab after 30 minutes of heating

Figure 5.8: A cylindrical can containing food to be sterilized.

Schematic and given data: 1. thickness of slab = 25 mm
2. thermal diffusivity of the slab, $\alpha = 2 \times 10^{-7} \text{m}^2/\text{s}$
3. Initial temperature = 40°C
4. Surface temperature = 121°C
5. Time of heating = 1800 s

Assumptions: 1. Since thickness is half the size of the diameter, we assume heating from the side may be ignored.
2. The thermal diffusivity value does not change with temperature during the heating process

Analysis: For transient heat conduction in a slab, the Heisler charts are to be used. The parameters needed to read the chart are the non-dimensional distance n at the center given by

$$n = \frac{x}{x_1} = \frac{0}{.0125} = 0$$

and parameter m signifying the convective condition at the surface given by

$$m = \frac{k}{hx_1} = 0$$

where $h \to \infty$ has been used since the surface temperature is specified (see discussion under the third kind of boundary condition in Section 3.2). The non-dimensional time is calculated as

$$Fo = \frac{\alpha t}{L^2} = \frac{2 \times 10^{-7} \, [\text{m}^2/\text{s}] \, 1800 \, [\text{s}]}{(0.0125)^2 \, [\text{m}^2]} = 2.3$$

With these values of Fo, m and n, value of non-dimensional temperature is obtained from the vertical axis of the chart on page 327 as

$$\frac{T - T_\infty}{T_i - T_\infty} = 0.0043$$

So that temperature $T = 120.65$°C after 30 minutes of heating

Comments: The actual temperature will be higher since the can will also heat from the sides. This temperature can be calculated considering heat transfer from both directions using formulas described in Section 5.4.

5.3.5 When Both Internal and External Resistances are Present: Convective Boundary Condition

So far in Section 5.3 we have considered a negligible external fluid resistance to heat transfer ($h \to \infty$) represented by the boundary condition of specified surface temperature. A convective boundary condition on the slab surface (denoted by location s in

5.3. TRANSIENT HEAT TRANSFER WITH INTERNAL RESISTANCE

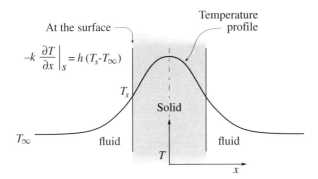

Figure 5.9: In convective boundary condition, surface temperature is not the same as the bulk fluid temperature, T_∞, signifying additional fluid resistance.

Figure 5.9) is written as

$$-k \left.\frac{\partial T}{\partial x}\right|_s = h(T_s - T_\infty)$$

When h is finite (instead of $h \to \infty$), external resistance of the fluid $(1/h)$ needs to be considered. Here T_∞, the external fluid temperature, is known, but the surface temperature, T_s is not. The solution procedure used to obtain Eqn. 5.13 can be generalized for any value of h, and as expected, the right hand side of Eqn. 5.13 would become a function of h as well. In the final solution, T_s is replaced by T_∞, the fluid temperature, i.e., the non-dimensional temperature on the left of Eqn. 5.13 is replaced by $(T - T_\infty)/(T_i - T_\infty)$. Thus, additional sets of lines are added in Figures B.1, B.2, and B.3 on pages 327, 328, and 329 for various values of $m = k/hL$ where m is the inverse of the non-dimensional Biot number hL/k.

5.3.6 Numerical Methods as Alternatives to the Charts

The limitations mentioned in the previous section of the analytical solution can indeed be very serious, particularly when applied to biological materials. Such materials and processes often involve complex natural shapes, variations in properties, initial conditions that vary throughout the material and boundary conditions that vary with time.

Limitations of the analytical solutions can be overcome using numerical, computer-based solutions. In numerical methods, the governing equation developed here would be discretized and essentially solved at finite number of locations in the material and at finite time steps, as opposed to solving exactly for all locations in the material and at all times. Such methods are quite versatile and can often accommodate arbitrary shape, size, initial condition, boundary conditions, properties, etc. These methods are rapidly becoming the standard procedure for solving heat and mass transfer problems. A large number of commercial software are available for this purpose, see for example, the web site: http://icemcfd.com/cfd/CFD_codes.html

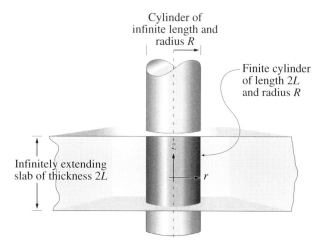

Figure 5.10: A finite cylinder can be considered as an intersection of an infinite cylinder and a slab.

5.4 Transient Heat Transfer in a Finite Geometry— Multi-Dimensional Problems

Frequently, situations are encountered where one-dimensional approximation is inadequate and two- and three-dimensional effects need to be considered. Qualitatively speaking, if the location we are interested in is at a comparable distance from surfaces in more than one direction, heat transfer from both of these directions must be considered. To solve for multi-dimensional heat transfer, a finite geometry is considered as the intersection of two or three infinite geometries. A rectangular box, for example, is considered as intersection of three infinite slabs and the temperature $T_{x,y,z,t}$ can be calculated as

$$\frac{T_{xyz,t} - T_s}{T_i - T_s} = \left(\frac{T_{x,t} - T_s}{T_i - T_s}\right)_{\substack{\text{infinite} \\ \text{x slab}}} \left(\frac{T_{y,t} - T_s}{T_i - T_s}\right)_{\substack{\text{infinite} \\ \text{y slab}}} \left(\frac{T_{z,t} - T_s}{T_i - T_s}\right)_{\substack{\text{infinite} \\ \text{z slab}}} \quad (5.21)$$

Similarly, for a finite cylinder (Figure 5.10), the temperature $T_{r,z,t}$ can be calculated as

$$\frac{T_{r,z,t} - T_s}{T_i - T_s} = \left(\frac{T_{r,t} - T_s}{T_i - T_s}\right)_{\substack{\text{infinite} \\ \text{cylinder}}} \left(\frac{T_{z,t} - T_s}{T_i - T_s}\right)_{\substack{\text{infinite} \\ \text{slab}}} \quad (5.22)$$

5.5 Transient Heat Transfer in a Semi-Infinite Region

A semi-infinite region extends to infinity in two directions and a single identifiable surface in the other direction. The surface shown in Figure 5.11 extends to infinity in the y and z directions and has an identifiable surface at $x = 0$. The semi-infinite region provides a useful idealization for many practical situations where we are interested in heat

5.5. SEMI-INFINITE REGION

transfer for a relatively short time and/or in a relatively thick material. Examples can be heating and cooling near the earth's surface, minor skin burns, etc. The governing equation for conduction heat transfer with constant properties and no heat generation is given by

$$\underbrace{\frac{\partial T}{\partial t}}_{\text{storage}} + \underbrace{u\frac{\cancel{\partial T}^{0}}{\partial x}}_{\text{no bulk flow}} = \underbrace{\frac{k}{\rho c_p}\frac{\partial^2 T}{\partial x^2}}_{\text{conduction}} + \underbrace{\frac{\cancel{Q}^{0}}{\rho c_p}}_{\text{no generation}}$$

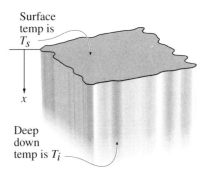

Figure 5.11: Schematic of a semi-infinite region showing only one identifiable surface.

which is simplified to:

$$\frac{\partial T}{\partial t} = \alpha \frac{\partial^2 T}{\partial x^2} \quad (5.23)$$

The boundary conditions are given by:

$$T(x=0) = T_s \quad (5.24)$$
$$T(x \to \infty) = T_i \quad (5.25)$$

and the initial condition is given by:

$$T(t=0) = T_i \quad (5.26)$$

The solution to the above equation, although not difficult, is lengthy (see Appendix G.2 on page 369). The temperature T as a function of position x and time t is given by:

$$\frac{T - T_i}{T_s - T_i} = 1 - \text{erf}\left[\frac{x}{2\sqrt{\alpha t}}\right] \quad (5.27)$$

where the function $\text{erf}(\eta)$ is called the error function and is given by

$$\text{erf}(\eta) = \frac{2}{\sqrt{\pi}} \int_o^\eta e^{-\eta^2} d\eta$$

To get a sense of what the error function looks like qualitatively, consider Figure 5.12 where the error function erf is compared with an exponential. For more accurate values, the error function is also tabulated in Table B.3 on page 325.

Heat flux at the surface of the semi-infinite region is often an important quantity that is of practical interest. The surface heat flux can be calculated from temperatures given by Eqn. 5.27 as:

$$\begin{aligned}
q_s'' = -k\left.\frac{dT}{dx}\right|_{x=0} &= -k\left.\frac{dT}{d\eta}\frac{d\eta}{dx}\right|_{x=0} \\
&= -k(T_s - T_i)\left(-\frac{2}{\sqrt{\pi}}e^{-\eta^2}\right)_{\eta=0}\frac{1}{2\sqrt{\alpha t}} \\
&= \frac{k(T_s - T_i)}{\sqrt{\pi \alpha t}} \quad (5.28)
\end{aligned}$$

Eqn. 5.28 shows that the surface heat flux decreases with time. This is expected since the temperature inside approaches the surface temperature, thus decreasing the temperature gradient at the surface.

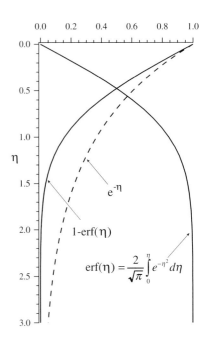

Figure 5.12: Comparison of the complementary error function $(1 - \text{erf}\eta)$ with an exponential $e^{-\eta}$.

When can we use the semi-infinite approximation As mentioned earlier, there is no such thing as a semi-infinite geometry in reality, it is only an idealization. The question is under what conditions can we approximate a real body as a semi-infinite? If the edges of a real body are far enough from the point of interest on the surface *for the time period of interest*, the propagating energy front does not know that the material is actually finite. The process, therefore, works as if the body was infinite. This would be true for thin materials over short times or thick materials over longer times. A more quantitative measure of when such approximations are valid can be derived from Figure 5.12. Since for positions and time combinations given by

$$x/2\sqrt{\alpha t} \geq 2$$
$$x \geq 4\sqrt{\alpha t} \tag{5.29}$$

the temperature change is less than 0.5%, any material whose thickness L is larger than $4\sqrt{\alpha t}$ can be approximated as semi-infinite for this purpose. Note that the appropriate thickness L is defined in terms of the time of interest t. For small times, even a small thickness can be considered semi-infinite. As an illustration, for a thermal diffusivity of $1 \times 10^{-7} m^2/s$, that is the typical order of magnitude for biomaterials containing a lots of water, the relationship given by Eqn. 5.29 is plotted in Figure 5.13.

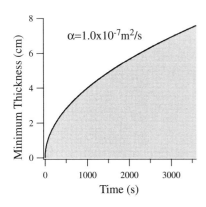

Figure 5.13: Plot of Eqn. 5.29, illustrating the minimum thickness of a material for which error function solution can be used.

Other Boundary Conditions The above solution was for a specified surface temperature, i.e., for no external resistance. Often, there is external resistance in the fluid flowing over the surface. This leads to the convective boundary condition (see Section 3.2)

$$-k \left.\frac{\partial T}{\partial x}\right|_{surface} = h(T_{surface} - T_\infty)$$

The solution for the semi-infinite region for this boundary condition is given by:

$$\frac{T - T_i}{T_\infty - T_i} = 1 - \text{erf}(\frac{x}{2\sqrt{\alpha t}})$$
$$- \exp(\frac{hx}{k} + \frac{h^2 \alpha t}{k^2})(1 - \text{erf}(\frac{x}{2\sqrt{\alpha t}} + \frac{h\sqrt{\alpha t}}{k})) \tag{5.30}$$

Note that Eqn. 5.30 would become Eqn. 5.27 when $h \to \infty$, since the specified surface temperature boundary condition is a special case of convective boundary condition over a surface. For a specified surface heat flux boundary condition (see Section 3.2)

$$q''_{surface} = q''_s \tag{5.31}$$

the solution for the temperature is given by:

$$T - T_i = \frac{2}{k}q''_s \left(\frac{\alpha t}{\pi}\right)^{1/2} \exp\left(-\frac{x^2}{4\alpha t}\right) - \frac{q''_s x}{k}\left(1 - \text{erf}\left(\frac{x}{2\sqrt{\alpha t}}\right)\right) \tag{5.32}$$

Example 5.5.1 Analysis of Skin Burns

A thermal burn occurs as a result of an elevation in tissue temperature above a threshold value for a finite period of time. The values of both the absolute temperature and the exposure time are crucial in determining the extent of injury. Since heat transfer to inside the skin is limited by the thermal conduction, the temperature history in the affected depth is non-uniform, and regions of graded injury develop with the most acute involvement at the surface. Detailed physiological responses such as changes in blood flow, coagulation process, etc. are a very significant part of the total burn injury.

Figure 5.14 shows a cross-section of a normal skin and underlying tissue. Macroscopic characterization of burn injuries (see Krizek et al., 1973) describes a first degree burn as one involving a temporary discomfort, with no permanent scarring or skin discoloration. A superficial second-degree burn involves some but not all of the basal layer. A deep second degree burn involves complete loss of basal layer. In a third-degree burn, all epidermal elements are destroyed. A fourth degree burn can extend all the way to the bone. Ignoring the effects of blood flow and other physiological changes, we may analyze the temperature history and thermal injury for up to a third-degree burn, using semi-infinite region approximation. See problem 5.9.18 for a numerical calculation of the depth of skin layer affected.

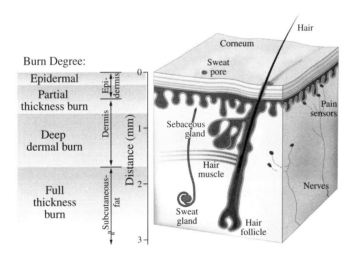

Figure 5.14: Section of a skin with degrees of burn superimposed on it. Data from Lawton (1994).

5.6 Chapter Summary—Transient Heat Conduction

- **No Internal Resistance, Lumped Parameter (page 71)**

  1. The thermal resistance of the solid can be ignored in comparison with the thermal resistance due to convection at the solid surface, if a Biot number (page 72) condition given by $h(V/A)/k < 0.1$ is satisfied.

  2. As thermal resistances are ignored, temperature is not a function of position, is a function of time only and is given by Eqn. 5.5.

- **Internal Resistance Is Significant**

  1. When internal resistance is significant (Biot number > 0.1), temperature is a function of both position and time.

  2. For an infinite slab (page 75), infinite cylinder and spherical geometry, the solutions are given by Figures B.1, B.2, and B.3 on pages 327, 328, and 329, respectively.

  3. For finite slab and finite cylinder (page 82), the solutions are given by Eqns. 5.21 and Eqn. 5.22, respectively.

  4. Materials with thicknesses $L \geq 4\sqrt{\alpha t}$ are considered effectively semi-infinite (page 82) and the solution is given by Eqn. 5.27.

5.7 Concept and Review Questions

1. Contrast the physical meanings of thermal diffusivity and thermal conductivity

2. What is a semi-infinite solid? Give examples.

3. Explain how the one-dimensional Heisler charts can be used to solve two- and three-dimensional problems.

4. Before you begin any calculation for transient temperatures in a solid with fluid flowing over it, what parameter value should you check, to reduce your work? What is the physical meaning of this parameter?

5. Discuss several advantages and disadvantages of the analytical methods of solutions that you are learning. Why are numerical methods of solution often preferrable for practical problems?

5.8 Further Reading

Bird, R. B., W. E. Stewart, and E. N. Lightfoot. 1960. *Transport Phenomena*. John Wiley & Sons, New York.

Cole, G. W. and N. R. Scott. 1977. A mathematical model of the dynamic heat transfer from the respiratory tract of a chicken. Bulletin of Mathematical Biology, 39:415-433.

Diller, K.R. 1985. Analysis of skin burns. In *Heat Transfer in Medicine and Biology* (Vol. 2), edited by A. Shitzer and R.C. Eberhart. New York: Plenum Press, New York.

Geankoplis, C.J. 1983. *Transport Processes and Unit Operations*. Allyn and Bacon, Inc., Boston.

Incropera, F.P. and D.P. Dewitt. 1990. *Introduction to Heat Transfer*. Wiley, New York.

Krizek, T. J., M.C. Robson, and R.C. Wray, Jr. 1973. Care of the burned patient. In: *Management of Trauma*, edited by W.F. Ballinger, R.B. Rutherford, and G.D. Zuidema. W.B. Saunders Co., Philadelphia.

Lawton, B. 1994. Heat dose to produce skin burns in humans. In *Fundamentals of Biomedical Heat Transfer*, edited by M. A. Ebadian and P. H. Oosthuizen, HTD-vol. 295, ASME, New York.

Pages, T., J. F. Fuster, and L. Palacios. 1991. Thermal responses of the fresh water turtle *Mauremys Caspica* to step-function changes in the ambient temperature. J. Thermal Biol., 16(6):337-343.

Pan, J. C. and S. R. Bhowmik. 1991. The finite element analysis of transient heat transfer in fresh tomatoes during cooling. Trans. of ASAE, 34(3):972-976.

Sastry, S.K., G.Q. Shen, and J.L. Blaisdell. 1989. Effect of ultrasonic vibration on fluid-to-particle convective heat transfer coefficients. Journal of Food Science 54(1):229-230.

Sien, H. P. and R. K. Jain. 1979. Temperature distributions in normal and neoplastic tissues during hyperthermia: Lumped parameter analysis. Journal of Thermal Biology, 4:157-164.

5.9 Problems

5.9.1 Thermocouple and the Lumped Parameter Approach

Figure 5.15 shows a thermocouple for measuring temperature. The junction that produces a voltage due to changes in temperature has a spherical lead "bead" of diameter 1 mm, as shown in the enlarged view. For lead, the thermal conductivity is 35.3 W/m·K, density is 11340 kg/m^3 and the specific heat is 129 J/kg·K. The thermocouple is exposed to air flow that leads to a h value of 10 W/m^2·K for the bead. 1) Determine if you can use a lumped parameter approximation for the bead. 2) If the bead is suddenly inserted into the air at 200°C from an inital temperature of 20°C, calculate the time it takes to reach 99.9% of the air temperature? 3) How long does it take for the bead to reach the air temperature exactly?

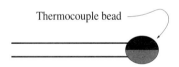

Figure 5.15: Schematic of a thermocouple with a "bead" that measures the temperature.

5.9.2 Changing Heat Flux with Time

Using the temperature solution for a lumped parameter heat transfer situation, show that the heat flux at the surface (i.e., energy leaving or entering the body), drops exponentially. What is the physical explanation for the reduction in heat flux?

5.9.3 Heat Transfer in a Leaf

Consider a typical thin leaf on a tree. When the air temperature surrounding the leaf drops suddenly, we would like to calculate how the center temperature of the leaf changes with time. The surface convective coefficient for a 5mph wind speed is 20W/m^2·K and the thermal conductivity of the leaf is 0.3 W/m·K. 1) What formula would you use and why? 2) Write down all the additional input parameters needed, without making any temperature calculations.

5.9.4 Ultrasound-Aided Food Sterilization

In sterilizing foods such as a chunky soup that has liquid as well as large solid particles, the bottleneck in the rate of heating is the heat transfer from the liquid to the solid. One potential approach to increase the convective heat transfer coefficient, h, between the liquid and the solid particles is the use of ultrasonic agitation, which currently has wide application in industrial cleaning and cell homogenization processes.

To determine this heat transfer coefficient, h, often particles made of aluminum that have a very high thermal conductivity are used (why?). In heating a mixture of aluminum particles and water, compare the time needed for the temperature of aluminum particles to equilibrate to within 3°C of the water temperature with and without the ultrasonic agitation. Heat transfer coefficient between the particles and the water with and without ultrasound are 1323 and 558 W/m^2·K, respectively, in absence of any other mechanical agitation. Assume that the temperature of the water bath is constant at 85°C. Initial uniform temperature of the particles is 20°C. The mass of the particles are 18 g. The density, specific heat, and thermal conductivity of aluminum are 2701.1 kg/m^3, 938.3 J/kg·K and 229 W/m·K, respectively.

5.9.5 Sterilization of a Bacterial Clump

To properly sterilize, all bacteria must be heated for the required duration and time. Although individual bacterial size is small, when present in large clumps, some bacteria located deep inside the clump may take significant time to reach the medium (air or water) temperature. Thus, they cannot be assumed to be at the medium temperature all the time. Consider bacterial clumps of different sizes heated in air. The shape of the clump can be assumed spherical with a radius R which is related to n, the number of bacteria in a clump, and r, the radius of individual bacteria, by the equation $R = r[n/(1-E)]^{1/3}$ where $E = 0.40$ is porosity. The surface heat transfer coefficient for air flow over the clump is given by $h = 4 \times 10^{-4} k_f/R$ where k_f is the air thermal conductivity and is equal to 0.0317 W/m·K. Consider an individual bacterial radius of 1μm and the specific heat, density, and thermal conductivity of the bacterial clump of

3.6 kJ/kg·K, 1080 kg/m^3, and 0.49 W/m·K, respectively. For three bacterial clumps given by $n = 10^2$, 10^5 and 10^6, 1) Determine the clump radius and the heat transfer coefficient, h. 2) Show that the lumped parameter approach is valid. 3) Calculate the time required for the bacterial temperature to increase by 99% of the total possible change in temperature. 4) Comment on if it is always reasonable to assume the bacterial clump to be at air temperature.

5.9.6 Heating of Liquid in a Vessel with Agitation

In many industrial applications, liquid is heated or cooled in a closed vessel and agitated to speed up the heat transfer. The agitation maintains the inside temperature to be fairly uniform, somewhat analogous to the lumped parameter solution described by Eqn. 5.5. In this case, the mass average temperature T_{av} of the liquid will replace the temperature T of the solid in Eqn. 5.5. Also, the heat transfer coefficient will now be the inside heat transfer coefficient between the vessel wall and liquid. 1) Calculate how the mass average temperature (T_{av}) of the liquid would change with time during agitated heating of a containerized liquid where the inside heat transfer coefficient is 120 W/m^2·K. The container wall temperature is at 121°C and the initial temperature of the fluid is 30°C. The container has a height of 10.67 cm and a radius of 4.19 cm. The density and specific heat of the liquid are 950 kg/m^3 and 4100 J/kg, respectively. 2) How long does it take for the average temperature to reach 95% of the maximum possible temperature?

5.9.7 Rate of Heating in a Transient Process

Consider a vertical cylindrical kettle of diameter 1 m filled with liquid that is constantly stirred and the vertical surface of the kettle heated with steam. The steam maintains the kettle wall at 110°C. The initial uniform temperature of the fluid is 40°C and the stirring produces a heat transfer coefficient of 200 W/m^2·K. The density and specific heat of the fluid are 950 kg/m^3 and 4.1 kJ/kg·K, respectively. Calculate the time it will take for the *average* temperature of the fluid to reach 1) 109°C, 2) 109.9°C. 3) Compare the two answers. Hint: Use the lumped parameter approach, replacing the solid temperature by the mass average temperature of the liquid.

5.9.8 Bulk Warming of Soil

The temperature of the topmost layer of a soil surface depends heavily on the relatively small quantity of solar flux that gets absorbed into the soil. Find the *average* rate of temperature increase in the soil, assuming no convective heat transfer or energy generation in the soil. The solar heat is absorbed in the topmost soil layer, so that the soil may be assumed to be well insulated beyond that layer. Solar flux absorbed into the soil is 40 W/m^2, depth of soil layer is 10 cm, bulk density of soil is 1200 kg/m^3 and specific heat of soil is 2500 J/kg·K.

5.9.9 Tissue-Blood Equilibrium

As blood travels through the body, its temperature, T_b, is constantly attempting to equilibrate with the temperature T_s of the surrounding tissue. Assuming that the blood temperature varies only in the direction of blood flow, this variation in T_b with distance can be easily related to the heat exchange between the blood and the tissue by

$$\frac{\pi}{4} d^2 \rho_b c_b u \frac{dT_b}{dx} = \pi d h (T_t - T_b)$$

where x is the distance along the blood vessel, d is the diameter of the blood vessel, u is the velocity of blood, ρ_b and c_b are the density and specific heat of blood, respectively. 1) Solve the above equation for blood temperature, T_b, as a function of x. 2) Calculate the blood temperature at the vessel outlet ($x = L$). 3) Comment on the distance the blood travels in the vessel before reaching equilibrium with the surrounding tissue. The input parameters are: constant tissue temperature of 35°C, temperature of blood at the vessel (large artery) inlet is 37°C, diameter of the large artery is 1.5 mm, $\rho_b c_b = 4.46 \times 10^6$ J/m^3·K, velocity of blood is 40 cm/s, length of artery is 20 cm, heat transfer coefficient between the blood and the artery wall is 200 W/m^2·K.

5.9.10 Experimental Determination of Thermal Property

The solution to the heat equation is also used to determine a thermal property that is needed. Consider a slab 20 cm thick that was initially at a uniform temperature and cooled with surface temperature at 15°C. A thermocouple fixed at a particular location inside the slab measures the temperature to be 4.23°C and 5°C after 15 and 20 minutes, respectively. From these measurements, calculate the thermal diffusivity of the slab.

5.9.11 Average Temperature in a Slab

Starting from the approximate solution for large times for temperature in a symmetrically heated slab as a function of time and position, find the average temperature of the slab.

$$\text{Answer} \quad \frac{T_{av} - T_s}{T_i - T_s} = \frac{8}{\pi^2} \exp\left[-\alpha \left(\frac{\pi}{2L}\right)^2 t\right]$$

5.9.12 Effect of Size on the Rate of Cooling (or Heating)

We would like to know how heating (or cooling) processes speed up as we reduce the size. Consider a slab geometry. The average temperature of a slab is given by problem 5.9.11. Using this equation, comment on how the required time to heat (or cool) would change for reaching the same final average temperature when the slab size is halved.

5.9. PROBLEMS

5.9.13 Changing Heat Flux with Time on a Slab Surface

Consider a slab initially at constant temperature T_0 with the surface temperature maintained at T_1. 1) Using the expression for temperature in the slab as a function of position and time at long times, calculate the heat flux as a function of time on the surface of the slab. 2) Explain in words how does this heat flux vary with time and why?

5.9.14 Heating Time for Food Sterilization

Consider a product to be sterilized if its coldest location during heating reaches 115°C. Consider sterilization of potatoes where metal cans would be filled with water and potatoes (4 cm diameter) and the can is heated in steam at 121°C. Assume symmetrical heating of potatoes and an *overall heat transfer coefficient between the steam and the potatoes* as 20 W/m²·K. The thermal diffusivity for potatoes is 1.5×10^{-7} m²/s and the thermal conductivity is 0.4 W/m·K. The potatoes are initially at 30°C. Calculate the heating time needed for sterilization.

5.9.15 Cooling of a Donor Heart

Human to human heart transplants are becoming more common. Often the heart donor is hundreds or even thousands of miles away from the recipient, so it is very important to quickly preserve the heart to prevent the damage or death of the tissue during its cross-country journey. This is done by submerging the heart in a saline solution and packing it in ice inside a commercial picnic cooler.

The heart can be approximated by a solid, finite cylinder with constant, equal thermal properties in all directions. During the time the heart is between bodies and packed in ice, it can be assumed the tissue activity has slowed enough that the metabolic heat production is negligible. Assume also that there is no heat conduction in the axial direction and that the saline solution is agitated so that it maintains the heart surface temperature at the constant value of the saline solution temperature. The thermal diffusivity of the heart tissue is 1.4×10^{-7} m²/s, initial temperature is 37°C, temperature of the saline solution is 0°C, and the radius and the height of the cylinder approximating the heart are 3.75 cm and 9 cm, respectively. 1) Find the temperature at the center of the heart after 3 hours assuming an infinite cylinder. 2) Recalculate the temperature at the center for a finite cylinder with the dimensions given.

5.9.16 Effect of Heat Transfer Coefficient on Food Sterilization

For the two heat transfer coefficients given in problem 5.9.4, calculate the difference in heating time that would be necessary for the center of a spherical food particle of 5 mm radius, to heat to a presumed sterilizing temperature of 115°C starting from an intial temperature of 30°C and a water temperature of 121°C. The density, specific heat, and thermal conductivity of food are 850 kg/m³, 1680 J/kg·K and 1.42 W/m·K, respectively. Note: Requires interpolation in Heisler's charts.

5.9.17 Effect of Heat Transfer Coefficient on Food Sterilization

For simplicity, consider a product to be sterilized if its coldest location during heating reaches 115°C. Consider sterilization of potatoes of 4 cm diameter using an old and a newer process as follows: In the old process, metal cans would be filled with water and potatoes and the can heated in steam at 121°C. 1) For symmetrical heating of potatoes and an overall heat transfer coefficient between the steam and the potatoes as 20 W/m^2·K, calculate the heating time needed for sterilization. In a newer process, the water and the potatoes are agitated inside a large vessel, to increase the overall heat transfer coefficient to 200 W/m^2·K. The vessel is also heated in steam at 121°C. 2) Calculate the heating time needed for sterilization in the new process. Make approximations in reading the charts. 3) Comment on which potatoes you would prefer (for the same price) and why?

5.9.18 Skin Burn Injury

During a skin burn from an oven, approximate the skin and tissue layer to be infinitely thick compared to the damaged layer. The temperature throughout the entire skin and tissue layer is uniform at 33°C before contact with the oven, and the surface layer of skin increases to the temperature of the oven, 200°C, instantaneously upon contact. Find the depth of the damaged layer of skin after 2 seconds of exposure to the oven temperature. Consider skin becomes damaged when it reaches 62°C. The thermal diffusivity of skin is 2.5×10^{-7} m^2/s.

5.9.19 Induction Thermocoagulation in the Brain

Brain tumors are often treated by creating small thermal lesions known as induction thermocoagulation. In this procedure, a small section of the brain is heated to thermally destroy some of the tissue. Assuming one-dimensional heat transfer, find the thickness of the thermal lesion if the tissue dies when its temperature reaches 55°C. Assume the tissue to be thick and at an initial temperature of 36°C. The surface of the tissue is maintained at 75°C by a probe tip. The time of exposure is 3 minutes. The thermal conductivity, density, and specific heat of the tissue are 0.586 W/m·K, 1050 kg/m^3, and 3763 J/kg·K, respectively.

5.9.20 Heat Transfer Near the Surface

Biological material are often chilled to decrease the biological activity and therefore increase their storage life. A slab of certain biological material 50 mm thick and at an initial temperature of 30°C is to be chilled using an air blast that maintains both the surfaces of the biomaterial at 0°C. The thermal diffusivity of the biomaterial is 0.625×10^{-4} m^2/s. 1) Using the Heisler charts, calculate the temperature at a depth of 5 mm from either surface at 10 seconds after it is put in the air blast. 2) Using the semi-infinite model, calculate the same temperature (at a depth of 5 mm from the surface at 10 seconds). 3) Compare the answers to questions 1 and 2 and explain why the answers are within a few degrees.

5.9. PROBLEMS

5.9.21 Cooling of the Ground

The depth to which freezing temperature reaches inside the soil is important for plant growth and construction of buildings. We will postpone the inclusion of latent heat from freezing until the chapter "Heat Transfer with Phase Change." Thus, only cooling is considered in this problem, i.e., no phase change. Approximate the soil temperature to be uniform at 15°C toward the end of fall, when the air temperature suddenly drops to −20°C. The wind over the soil surface produces a convective heat transfer coefficient of 15 W/m²·K. The thermal conductivity and thermal diffusivity of the soil are 0.85 W/m·K and 4.5×10^{-7} m²/s, respectively. Calculate the following at $t = 3$ hours: 1) The depth to which soil temperature is at 0°C or lower. 2) The surface temperature of the soil. 3) The heat flux from the surface of soil.

5.9.22 Cooling of the Ground with Varying Surface Temperature

The daily periodic variations (diurnal variations) in air temperature and solar radiation leads to periodic variations in the soil temperature. Consider the soil temperature deep inside the soil to be T_{av} and the surface temperature of soil to be $T_{av} + (T_{max} - T_{av})\cos\omega t$, where T_{max} is the maximum temperature that occurs at the soil surface, and time t is measured from the time from the occurence of the maximum temperature. The frequency of diurnal variations is given by ω. The solution to the problem is

$$\frac{T - T_{av}}{T_{max} - T_{av}} = \exp\left(-\sqrt{\frac{\omega}{2\alpha}}z\right)\cos\left(\omega t - \sqrt{\frac{\omega}{2\alpha}}z\right)$$

1) Assuming the soil surface to be a semi-infinite region, write the governing equation and the boundary and initial conditions for this problem. 2) Plot approximately the solution given above and the surface temperature. 3) Explain the physical meaning of the exponential term and the cosine term in the solution. For a hint on the solution technique, see Bird et al., 1960.

5.9.23 Heating in a Microwave Oven

In microwave heating of foods and biomaterials, polarized molecules of water oscillate in the electromagnetic field of the microwaves. The rate of heating decays from the surface into the material, since the wave loses its electromagnetic energy (which appears as thermal energy) as it travels. This decay in the rate of heating is often approximated as

$$Q = Q_0 e^{-\frac{x}{\delta}}$$

where Q is the rate of heat generation in W/cc at a distance x from the surface, Q_0 is the rate of heat generation in W/cc at the surface, and δ is the penetration depth, a material property that is a measure of how easily the material absorbs energy. The rate of heating in microwaves is often so fast that diffusion is not significant for the first few seconds of heating. For these initial times, 1) Find the temperature profile $T(x)$, in a thick slab heated with microwaves from both faces and sketch approximately this temperature profile. 2) Describe qualitatively how you would expect the profile to change with time. Assume any other property or parameter that you need.

Chapter 6

CONVECTION HEAT TRANSFER

CHAPTER OBJECTIVES

After you have studied this chapter, you should be able to:

1. Explain the process of convection heat transfer in a fluid over a solid surface in terms of the relatively stagnant fluid layer over the surface called the boundary layer.
2. Calculate the convective heat transfer coefficient h knowing the flow situation.

KEY TERMS

- convective heat transfer coefficient
- velocity boundary layer
- thermal boundary layer
- Reynolds number
- Nusselt number
- Prandtl number
- Grashof number
- natural and forced convection
- laminar and turbulent flow

In this chapter we will develop a more general version of the governing equation that we derived in Chapter 3. It is more general since it will now include fluids that can flow and carry energy by virtue of their motion. The relationship of this chapter to other chapters in energy transfer is shown in Figure 6.1.

6.1 Governing Equation for Convection

As mentioned in Chapter 2, convective heat transfer is the movement of heat through a medium as a result of the net motion of a material in the medium. Convection takes

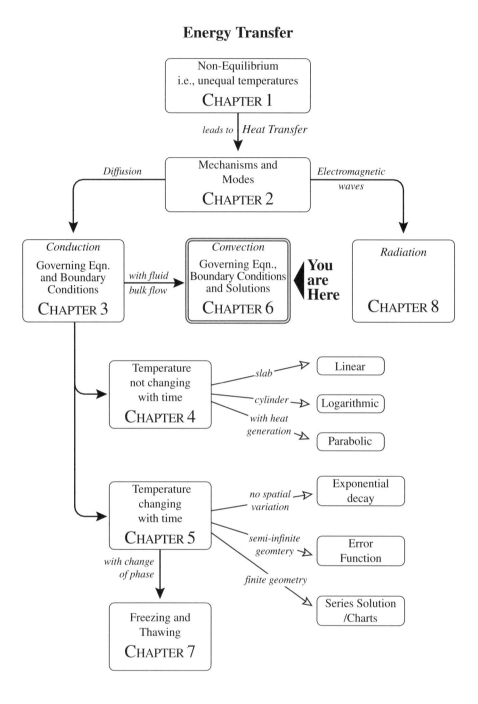

Figure 6.1: Concept map of energy transfer showing how the contents of this chapter relates to other chapters on energy transfer.

place in most heat transfer applications involving a liquid or gas, *in addition to conduction*. In the previous chapters this convection was considered only as a boundary condition at the surface of a conducting solid. The parameter h, the convective heat transfer coefficient describing the convecting process was assumed to be known. In this chapter we look at the details of the convection process with one of the goals being to investigate ways to predict the convective heat transfer coefficient h.

The general equation (Eqn. 3.5) describing energy transfer is repeated here from page 31 as:

$$\underbrace{\frac{\partial T}{\partial t}}_{\text{storage}} + \underbrace{u\frac{\partial T}{\partial x}}_{\text{bulk flow}} = \underbrace{\frac{k}{\rho c_p}\frac{\partial^2 T}{\partial x^2}}_{\text{conduction}} + \underbrace{\frac{Q}{\rho c_p}}_{\text{generation}} \quad (6.1)$$

In this chapter we will be applying the equation to fluids (as opposed to solids in the previous chapters). Thus, for this chapter, all the properties in the above equation are for the fluid. We will need to retain the convection term, i.e., cannot drop it as in the previous chapters. If the convection term is kept, knowledge of fluid velocity u is needed. Thus, the equations governing fluid flow are also needed in addition, to solve the energy equation given by Eqn. 6.1. This makes the study of convection heat transfer often more complex than that of simple conduction.

Although boundary layer approximations (described below) are made to simplify the energy and flow equations for many practical situations, solution to convective heat transfer problem (Equation 6.1 and the related flow equations) is often involved. We will not pursue such detailed solution to the convection equations, but instead, discuss the expected form of the solution and spend our effort in understanding the implications of the solutions.

6.2 Temperature Profiles and Boundary Layers Over a Surface

Figure 6.2 is a schematic of flow over a flat plate. When fluid of initial uniform velocity flows over a surface, such as a flat plate, the velocity becomes zero at the surface of the plate. The decrease in velocity, from the stream velocity to zero, takes place in a relatively small layer of fluid known as the boundary layer. The fluid velocity approaches the free stream velocity u_∞ as we move further away from the plate. It may take a very large distance from the plate for the velocity to be exactly equal to u_∞, but most of this change in velocity takes place over this small distance known as the boundary layer. The thickness of the boundary layer $\delta_{velocity}$ is defined as the distance where the velocity is 99% of the free stream velocity, i.e.,

$$\delta_{velocity} = y|_{u=0.99u_\infty} \quad (6.2)$$

Thus, the fluid can be considered to have two separate regions: a boundary layer where velocity gradients are large and outside the boundary layer where the velocity gradients are small. In other words, the effect of the flat plate on the flow is essentially restricted

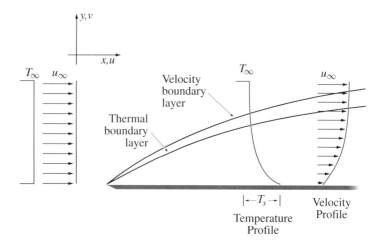

Figure 6.2: Velocity and temperature variation in the boundary layer over a flat plate.

to the boundary layer. The concept of this boundary layer was first introduced by Prandtl (1904).

Like the velocity boundary layer, a thermal boundary layer develops if the surface temperature for the flat plate is different from the uniform initial fluid temperature T_∞. Since the fluid in contact with the surface will be at rest, it will come to the surface temperature. The temperature will vary from the value T_s at the surface to T_∞ in the free stream. Like the velocity variation, this temperature variation is asymptotic, and a thickness of thermal boundary layer $\delta_{thermal}$ is defined as the distance at which temperature T is given by:

$$\frac{T_s - T_{\delta_{thermal}}}{T_s - T_\infty} = 0.99 \tag{6.3}$$

Thus $\delta_{thermal}$ is the distance over which most of the temperature change takes place. Like the velocity boundary layer, the fluid can be considered to have two distinct thermal regions: a thermal boundary layer where thermal gradients are large and outside the thermal boundary layer where the thermal gradients are small and temperatures are uniform. In other words, the thermal effect of the flat plate on the flow is essentially restricted to the thermal boundary layer. While a velocity boundary layer will always exist in a flow situation, thermal boundary layer exists only if there is a difference in temperature between the surface and the bulk fluid.

Thickness of the velocity boundary layer can be shown to relate to the flow parameter Reynolds number, Re, defined as

$$Re_x = \frac{u_\infty x \rho}{\mu} \tag{6.4}$$

where u_∞ is the free stream velocity, as shown in Figure 6.2, x is the distance along the flow from the leading edge of the plate, ρ is the density of the fluid and μ is the viscosity of the fluid. The thermal boundary layer additionally depends (shown in section 6.5) on a characteristic number called the Prandtl number, Pr, for the fluid:

$$Pr = \frac{\mu c_p}{k_{fluid}} = \frac{\mu/\rho}{k_{fluid}/\rho c_p} = \frac{\text{Momentum diffusivity}}{\text{Thermal diffusivity}} \quad (6.5)$$

where c_p is the specific heat of the fluid and k_{fluid} is the thermal conductivity of the fluid. Thus, Prandtl number depends only on the fluid properties. As an example of boundary layer thicknesses, for laminar flow over a flat plate, velocity boundary layer thickness is given by:

$$\delta_{velocity} = \frac{5x}{Re_x^{\frac{1}{2}}} \quad (6.6)$$

Thus, as the velocity (and therefore Re_x) increases, the boundary layer thickness δ decreases. The thermal boundary layer thickness for laminar flow over a flat plate is related to the velocity boundary layer thickness by:

$$\frac{\delta_{velocity}}{\delta_{thermal}} = Pr^{1/3} \quad (6.7)$$

or

$$\delta_{thermal} = \frac{\delta_{velocity}}{Pr^{1/3}}$$

which implies that thermal boundary layer thickness also decreases as velocity increases. Since $Pr \approx 5$ for water, the thermal boundary layer will be much thinner than the velocity boundary layer on a flat plate. This is in contrast with gases, for which $Pr \approx 1$ and the thermal and velocity boundary layers on a flat plate are approximately equal.

6.3 Laminar and Turbulent Flows

In the previous section, we discussed boundary layer over a flat plate. Although fluid motion on such a flat plate starts out to be orderly (streamlined) at the leading edge, as shown in Figure 6.2, as the fluid moves along the plate, the flow changes to a more chaotic situation at some distance from the leading edge, as illustrated in Figure 6.3. This change from a streamlined to a more chaotic condition changes the nature of the flow quite drastically. In particular, the *mixing process* due to the more chaotic motion or turbulence *enhances the rate of heat transfer.* The region of the plate over which the fluid motion is orderly is called the laminar region, whereas the region with a more chaotic fluid motion is called the turbulent region. The transition from laminar to turbulent can occur over a region, which is called the transition region. In determining the type of boundary layer (laminar or turbulent) that exists in a certain flow situation, it is frequently assumed that the transition from laminar to turbulent occurs at a critical Reynolds number. For example, for flow over a flat plate, the critical Reynolds number

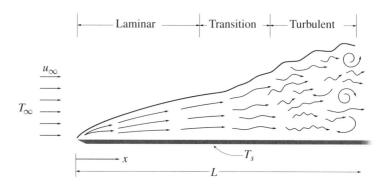

Figure 6.3: Schematic showing laminar, transition, and turbulent flow over a flat plate.

based on distance x along the plate ($Re_x = u_\infty x \rho / \mu$) has been found to vary experimentally between 2×10^5 and 3×10^6, i.e., the following conditions define the laminar, turbulent and transition region over a flat plate:

$$\begin{array}{lrl} \text{laminar region} & Re_x & < 2 \times 10^5 \\ \text{transition region} \quad 2 \times 10^5 < & Re_x & < 3 \times 10^6 \\ \text{turbulent region} \quad 3 \times 10^6 < & Re_x & \end{array}$$

As will be shown later, when calculating rates of convective heat transfer, it is critical to consider whether the flow is in laminar or turbulent region, as characterized by the corresponding Reynolds number.

6.4 Convective Heat Transfer Coefficient Defined

Consider fluid flowing over a surface, as shown in Figure 6.2. Since the layer of fluid in contact with the surface is at rest, heat transfer *in the fluid* at the surface is by conduction only and the flux is given by

$$\begin{array}{c} \text{conductive heat} \\ \text{flux in the fluid} \end{array} = -k_{fluid} \left. \frac{\partial T}{\partial y} \right|_{y=0, \text{in fluid}}$$

where k_{fluid} is the thermal conductivity *of the fluid* and T is the *fluid* temperature at a distance y from the surface. However, by definition of the convective heat transfer coefficient, h, the same heat flux can be written as

$$\begin{array}{rl} \text{convective} & \\ \text{heat flux} & = h(T_{y=0, \text{in fluid}} - T_\infty) \\ & = h(T_s - T_\infty) \end{array}$$

6.5. SIGNIFICANT PARAMETERS IN CONVECTIVE HEAT TRANSFER

Since the two expressions represent the same heat flux:

$$-k_{fluid} \left.\frac{\partial T}{\partial y}\right|_{y=0, \text{ in fluid}} = h(T_s - T_\infty)$$

$$h = \frac{-k_{fluid} \left.\frac{\partial T}{\partial y}\right|_{y=0, \text{ in fluid}}}{T_s - T_\infty} \tag{6.8}$$

The temperature profile $T(y)$ can be obtained by solving the energy equation for the fluid (see next section). Knowing $T(y)$, we can calculate the derivative $\partial T/\partial y$ in Eqn. 6.8 and therefore the heat transfer coefficient h. Since $T(y)$ includes the effect of conduction and flow, h would also include *conduction and flow effects*.

6.5 Significant Parameters in Convective Heat Transfer

To obtain the convective heat transfer coefficient h for many practical situations, one would like to obtain the solution to the convection equation. In this section we will discuss the expected form of the solution to the convection equation for the boundary layer approximation. We will not solve the equations, as mentioned earlier, for brevity. Such solutions can be found in most heat transfer texts such as Incropera and Dewitt (1996). As an example, consider steady convection over a flat plate (Figure 6.2); the flow is two-dimensional (along and perpendicular to the plate) and the governing energy equation for the fluid without any heat generation and with constant property values is (generalized from Eqn. 3.5 in one dimension):

$$\underbrace{u\frac{\partial T}{\partial x} + v\frac{\partial T}{\partial y}}_{\text{bulk flow}} = \underbrace{\alpha\left(\frac{\partial^2 T}{\partial x^2} + \frac{\partial^2 T}{\partial y^2}\right)}_{\text{conduction}} \tag{6.9}$$

Note that the conduction term is included in the above equation, signifying convection heat transfer solutions that will be developed includes the conduction effects. The additional governing equation to be solved for fluid flow (refer to your fluid mechanics book) is:

$$u\frac{\partial u}{\partial x} + v\frac{\partial u}{\partial y} = -\frac{1}{\rho}\frac{\partial p}{\partial x} + \nu\left(\frac{\partial^2 u}{\partial x^2} + \frac{\partial^2 u}{\partial y^2}\right) \tag{6.10}$$

$$u\frac{\partial v}{\partial x} + v\frac{\partial v}{\partial y} = -\frac{1}{\rho}\frac{\partial p}{\partial y} + \nu\left(\frac{\partial^2 v}{\partial x^2} + \frac{\partial^2 v}{\partial y^2}\right) \tag{6.11}$$

with the continuity equation:

$$\frac{\partial u}{\partial x} + \frac{\partial v}{\partial y} = 0 \tag{6.12}$$

where u, v, p, ν are velocity in x direction, velocity in the y direction, pressure, and kinematic viscosity, respectively. The solution to this set of equations will be considerably simplified by using the boundary layer approximations discussed earlier. The assumptions we will use for the boundary layer are:

$$u \gg v \tag{6.13}$$

which signifies the velocity along the plate is much larger than that perpendicular to the plate. The gradient is also much larger in the y direction for u,

$$\frac{\partial u}{\partial y} \gg \frac{\partial u}{\partial x}, \frac{\partial v}{\partial x}, \frac{\partial v}{\partial y} \tag{6.14}$$

For the thermal boundary layer, the thermal gradient in the y direction is much larger than the gradient in the x direction, i.e.,

$$\frac{\partial T}{\partial y} \gg \frac{\partial T}{\partial x} \tag{6.15}$$

With these approximations (Equations 6.13, 6.14, 6.15), the governing equations for the boundary layer (Equations 6.9, 6.10) for convection over a flat plate can be approximated as:

$$u\frac{\partial T}{\partial x} + v\frac{\partial T}{\partial y} = \alpha \frac{\partial^2 T}{\partial y^2} \tag{6.16}$$

$$u\frac{\partial u}{\partial x} + v\frac{\partial u}{\partial y} = -\frac{1}{\rho}\frac{\partial p}{\partial x} + \nu \frac{\partial^2 u}{\partial y^2} \tag{6.17}$$

Equation 6.11, together with Eqn. 6.14, leads to the condition $\partial p/\partial y = 0$ which means the pressure is a function of x. This $p(x)$ can be obtained from other conditions and so we can treat p as known in Equation 6.17.

We would now change Equations 6.16 and 6.17 to a non-dimensional form. For this purpose, we define non-dimensional quantities or similarity parameters as:

$$\begin{aligned} x^* &= x/L \\ y^* &= y/L \\ u^* &= u/u_\infty \\ v^* &= v/u_\infty \\ T^* &= (T - T_s)/(T_\infty - T_s) \end{aligned}$$

where L is the length of the plate and u_∞ is the free stream velocity as shown in Figure 6.2. Using these non-dimensional quantities, Equations 6.16 and 6.17 can be written as:

$$u^*\frac{\partial T^*}{\partial x^*} + v^*\frac{\partial T^*}{\partial y^*} = \frac{\alpha}{u_\infty L} \frac{\partial^2 T^*}{\partial y^{*2}} \tag{6.18}$$

6.5. SIGNIFICANT PARAMETERS IN CONVECTIVE HEAT TRANSFER

$$u^* \frac{\partial u^*}{\partial x^*} + v^* \frac{\partial u^*}{\partial y^*} = -\frac{dp^*}{dx^*} + \frac{\nu}{u_\infty L} \frac{\partial^2 u^*}{\partial y^{*2}} \qquad (6.19)$$

where $p^* = p/\rho u_\infty^2$ is a non-dimensional pressure and $p^*(x^*)$ depends on geometry in general. Defining Reynolds number Re_L as:

$$Re_L = \frac{u_\infty L}{\nu} \qquad (6.20)$$

and Prandtl number

$$Pr = \frac{\mu c_p}{k} \qquad (6.21)$$

Equations 6.18, 6.19 can be written as:

$$u^* \frac{\partial T^*}{\partial x^*} + v^* \frac{\partial T^*}{\partial y^*} = \frac{1}{Re_L \, Pr} \frac{\partial^2 T^*}{\partial y^{*2}} \qquad (6.22)$$

$$u^* \frac{\partial u^*}{\partial x^*} + v^* \frac{\partial u^*}{\partial y^*} = -\frac{dp^*}{dx^*} + \frac{1}{Re_L} \frac{\partial^2 u^*}{\partial y^{*2}} \qquad (6.23)$$

From these equations, we can write symbolically:

$$T^* = f(x^*, y^*, Re_L, Pr, dp^*/dx^*) \qquad (6.24)$$

For a given geometry, contribution from the pressure gradient term is fixed and Equation 6.24 can be written as:

$$T^* = f(x^*, y^*, Re_L, Pr) \qquad (6.25)$$

Let us now consider the definition of the convective heat transfer coefficient (Equation 6.8) which can be written in terms of the non-dimensional temperature and distance as:

$$h = \frac{-k_{fluid} \frac{\partial T}{\partial y}\bigg|_{y=0, \, in \, fluid}}{T_s - T_\infty} = \frac{k_{fluid}}{L} \frac{\partial \left(\frac{T-T_s}{T_\infty - T_s}\right)}{\partial \left(\frac{y}{L}\right)}\bigg|_{y=0} = \frac{k_{fluid}}{L} \frac{\partial T^*}{\partial y^*}\bigg|_{y^*=0} \qquad (6.26)$$

which can be rewritten as:

$$\frac{hL}{k_{fluid}} = \frac{\partial T^*}{\partial y^*}\bigg|_{y^*=0} \qquad (6.27)$$

Using this definition of h and using the functional form of the temperature solution (Equation 6.25), we can write a functional form for the dependence of h as:

$$\frac{hL}{k_{fluid}} = f(x^*, Re_L, Pr) \qquad (6.28)$$

Thus, h should depend on position, in general. An average h can be obtained by integrating over the length of the plate as:

$$h_L = \frac{\int_0^L h\,dx}{L} \qquad (6.29)$$

This average is independent of x. Therefore,

$$\frac{h_L L}{k_{fluid}} = f(Re_L, Pr) \qquad (6.30)$$

This is the final equation showing the functional relationship of the average heat transfer coefficient h_L as related to the two non-dimensional parameters Re_L and Pr. The non-dimensional quantity on the left of Equation 6.30 is defined as the Nusselt number:

$$Nu_L = \frac{h_L L}{k_{fluid}} \qquad (6.31)$$

From Equation 6.27, we can see that the Nusselt number is a non-dimensional temperature gradient. Note the apparent similarity (Table 6.1) between the Nusselt number and the Biot number defined in earlier chapters. Both the numbers involve heat transfer coefficient, length, and thermal conductivity. The Biot number relates to thermal conduction in the solid relative to the convection in the fluid in contact with the solid. It, therefore, involves solid thermal conductivity. The Nusselt number, on the other hand, compares thermal conduction in the fluid relative to the convection in the fluid. It relates entirely to the fluid and involves fluid thermal conductivity. The various non-dimensional parameters defined in this chapter are summarized in Table 6.1.

6.6 Calculation and Physical Implications of Convective Heat Transfer Coefficient Values

Conceptually, for every physical situation, the governing equation for convection can be solved to obtain an explicit form of the function f. This has been performed for many flow situations. Often, for complex situations, h is found experimentally. Such formulas (correlations) for h for various physical situations are now discussed. For definition of Nu, Re, Gr, and Pr, see Table 6.1. In this table, the length L refers to a *characteristic length* that depends on the geometry of the surface over which flow takes place. The characteristic lengths for various geometries are introduced at appropriate places in this section. The correlations provided in this section are used in practice to find h for a given situation. For more details on these equations and to find correlations for situations not discussed below, see standard heat transfer texts such Incropera and Dewitt (1996).

Before presenting the specific convection formulas, some general comments are made about them. As Re increases, the velocity and the thermal boundary layer thickness decreases (see, for example, Eqn. 6.6 for velocity boundary layer thickness over

6.6. CONVECTIVE HEAT TRANSFER COEFFICIENT CALCULATIONS

Table 6.1: Non-dimensional parameters and their physical significance

Dimensionless number	Definition and physical significance
Reynolds number	$Re = \dfrac{u_\infty L \rho}{\mu} = \dfrac{\text{Inertia force}}{\text{Viscous force}}$
Nusselt number	$Nu = \dfrac{hL}{k_{fluid}} = \dfrac{\text{Diffusive resistance}}{\text{Convective resistance}}$
Prandtl number	$Pr = \dfrac{\mu c_p}{k} = \dfrac{\mu/\rho}{\alpha} = \dfrac{\text{Viscous effect}}{\text{Thermal diffusion effect}}$
Biot number	$Bi = \dfrac{hL}{k_{solid}} = \dfrac{\text{Internal diffusive resistance}}{\text{Surface convective resistance}}$
Grashof number	$Gr = \dfrac{\beta g L^3 \Delta T}{\nu^2} = \dfrac{\text{Buoyancy force}}{\text{Viscous force}}$
Rayleigh number	$Ra = Gr \times Pr$

a flat plate). Since the slower moving fluid in the boundary layer provides the thermal 'insulation,' reduced thickness will mean higher heat transfer and therefore larger value of h.

The heat transfer coefficient h varies considerably with location. Typical variation of h in a flat plate is shown in Figure 6.4. At the entrance ($x = 0$), h is very large due to the jump in temperature from T_s for the fluid on surface to T_∞ in the bulk fluid, such that

$$h = \dfrac{-k_{fluid} \left.\dfrac{\partial T}{\partial y}\right|_{y=0,\text{ in fluid}}}{T_s - T_\infty} \rightarrow \infty \quad \text{at } y = 0 \qquad (6.32)$$

since $\left.\dfrac{\partial T}{\partial y}\right|_{y=0,\text{ in fluid}}$ goes to ∞ at the entrance. Thus, when using the correlations for h, one needs to be aware whether the correlations are for local h at a location or for average h over a certain length. In the equations described below, Nu_x is calculated from local h at a location x, whereas Nu_L or Nu_D are averaged over length or diameter, respectively.

Since the temperature varies across a thermal boundary layer, fluid properties such as viscosity, density and thermal conductivity will also vary across the thermal boundary layer, influencing the heat transfer rate. This influence of temperature variation is generally treated in a simple way by evaluating the properties at an average tempera-

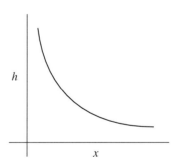

Figure 6.4: Variation of heat transfer coefficient on a flat plate. The edge of the plate is at $x = 0$.

ture, called the *film temperature*, defined as

$$T_f = \frac{T_s + T_\infty}{2} \quad (6.33)$$

where T_s is the surface temperature and T_∞ is the bulk fluid temperature. Thus, for all the correlations presented below, the properties are to be evaluated at this film temperature, T_f.

6.6.1 Flat plate, Forced Convection

The characteristic length for forced convection over a flat plate is the distance along the flow on the plate. As mentioned earlier, the plate surface heat transfer coefficient varies with distance along the flow. This variation is captured in a local Nusselt number, Nu_x ($= u_\infty x/\nu$), at a location x along the flow, and an average Nusselt number, Nu_L ($= u_\infty L/\nu$), over a distance L along the flow. Two sets of formulas are provided for each laminar and turbulent flow:

$$Nu_x = 0.332 Re_x^{\frac{1}{2}} Pr^{\frac{1}{3}} \quad \text{for laminar } (Re_x < 2 \times 10^5) \quad (6.34)$$

$$Nu_L = 0.664 Re_L^{\frac{1}{2}} Pr^{\frac{1}{3}} \quad \text{for laminar } (Re_L < 2 \times 10^5) \quad (6.35)$$

$$Nu_x = 0.0288 Re_x^{\frac{4}{5}} Pr^{\frac{1}{3}} \quad \text{for turbulent } (Re_x > 3 \times 10^6) \quad (6.36)$$

$$Nu_L = 0.0360 Re_L^{\frac{4}{5}} Pr^{\frac{1}{3}} \quad \text{for turbulent } (Re_L > 3 \times 10^6) \quad (6.37)$$

Example 6.6.1 Heat Transfer Coefficient Over a Building Wall

Wind is blowing at 5 km/hour over a building wall of size 5 m × 10 m, as shown in Figure 6.5. Air temperature is 0°C and the wall surface temperature is 10°C. 1) Calculate the average heat transfer coefficient along the 10 m width of the wall. 2) Calculate the local heat transfer coefficient at a location 10 cm from the leading edge of the wall.

Known: Speed of wind that is blowing over a building wall

Find: The average heat transfer coefficient over the wall surface and heat transfer coefficient at a given location

Schematic and Given Data: Schematic of this problem is shown in Figure 6.5. The given data are

1. The wind speed is 5 km/hour
2. The wall dimensions are 5 m × 10 m
3. The air temperature is 0°C and the wall surface temperature is 10°C

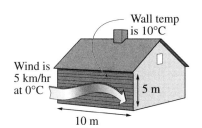

Figure 6.5: Schematic of wind blowing over a building wall.

6.6. CONVECTIVE HEAT TRANSFER COEFFICIENT CALCULATIONS

Assumptions:

1. Heat transfer coefficient is assumed to vary only along the direction of the air flow
2. The wall can be treated as a perfect flat plate

Analysis:

This heat transfer situation is treated as a flat plate under forced convection. We have to then determine if the flow is laminar or turbulent. For this, we need to find the Reynolds number $Re = u_\infty L \rho / \mu$ as follows:

$$
\begin{aligned}
u_\infty &= \frac{5 \times 1000}{3600} \\
&= 1.39 \text{ m/s} \\
\rho &= \text{density of air at the film temperature} \\
&\quad \text{of } 1/2(0+10) = 5°C \\
&= 1.2708 \text{ kg/m}^3 \quad \text{(from Appendix C.8)} \\
L &= \text{characteristic length along the flow} \\
&= 10 \text{ m} \\
\mu &= \text{viscosity of air at the film temperature of } 5°C \\
&= 1.7404 \times 10^{-5} [\text{kg/m·s}] \quad \text{(from Appendix C.8)}
\end{aligned}
$$

Therefore

$$
\begin{aligned}
Re_L &= \frac{u_\infty L \rho}{\mu} \\
&= \frac{1.39 [\text{m/s}] 10 [\text{m}] 1.2708 [\text{kg/m}^3]}{1.7404 \times 10^{-5} [\text{kg/m·s}]} \\
&= 1.015 \times 10^6
\end{aligned}
$$

Based on this Reynolds number, the flow is closer to being turbulent. We assume turbulent flow and, therefore, the appropriate equation for average heat transfer coefficient is Eqn. 6.37. To use it, we need the Prandtl number. From Appendix C.8, $Pr = 0.714$ at the film temperature of $5°C$. Thus, using Eqn. 6.37,

$$
\begin{aligned}
Nu_L &= 0.036 Re_L^{\frac{4}{5}} Pr^{\frac{1}{3}} \\
\frac{h[\text{W/m}^2 \cdot \text{K}] 10[\text{m}]}{0.02451 [\text{W/m·K}]} &= 0.036 (1.015 \times 10^6)^{4/5} (0.714)^{1/3} \\
h &= 5.04 \text{W/m}^2 \cdot \text{K}
\end{aligned}
$$

For heat transfer coefficient at a distance 10 cm from the leading edge, we need the local Reynolds number

$$Re_x = \frac{u_\infty x \rho}{\mu}$$
$$= \frac{1.39[\text{m/s}] 0.1[\text{m}] 1.2708[\text{kg/m}^3]}{1.7404 \times 10^{-5}[\text{kg/m·s}]}$$
$$= 1.015 \times 10^4$$

Thus the local Reynolds number at $x = 10$ cm shows the flow is laminar. Therefore, Eqn. 6.34 needs to be used

$$Nu_x = 0.332 Re_x^{\frac{1}{2}} Pr^{\frac{1}{3}}$$
$$\frac{h[\text{W/m}^2\text{·K}]\, 0.1[\text{m}]}{0.02451[\text{W/m·K}]} = 0.332(1.015 \times 10^4)^{1/2}(0.714)^{1/3}$$
$$h = 7.32 \text{ W/m}^2\text{·K}$$

Comments: Note that the local Reynolds number (at $x = 10$ cm) is higher than the average Reynolds number, as would be expected closer to the leading edge (see Figure 6.4).

6.6.2 Flat plate, Natural Convection

During natural convection, fluid rises or sinks depending on its change in density compared to the surroundings, due to changes in temperature. The coefficient of thermal expansion, β, characterizes the change in density that causes the flow, and is given by

$$\beta = -\frac{1}{\rho} \left.\frac{\partial \rho}{\partial T}\right|_{P=\text{constant}} \quad (6.38)$$

where ρ is the density, T is the absolute temperature, and P is the pressure. The quantity β is a thermodynamic property of the fluid. The manner in which β depends on temperature T varies with fluids. For an ideal gas (see Section 9.1.1 on page 177), $\rho = P/R_g T$ and

$$\beta = -\frac{1}{\rho} \left.\frac{\partial \rho}{\partial T}\right|_{P=\text{constant}} = \frac{1}{\rho}\frac{P}{R_g T^2} = \frac{1}{T} \quad (6.39)$$

Eqn. 6.39 can be used to calculate β for ideal gases. For liquids, β is obtained experimentally, as shown in Table C.12 on page 344.

The velocity of the fluid arises from the density differences. Since there is no forced velocity, the concept of the Reynolds number is not that useful. Instead, we define another dimensionless parameter, the Grashof number, as:

$$Gr = \frac{\beta g \rho^2 L^3 \Delta T}{\mu^2} \quad (6.40)$$

6.6. CONVECTIVE HEAT TRANSFER COEFFICIENT CALCULATIONS

where g is the acceleration due to gravity, L is the characteristic dimension, ΔT is the temperature difference (typically between a surface and the bulk fluid), and μ is the viscosity of the fluid. Another non-dimensional number Ra, called the Rayleigh number, is defined in terms of the Grashof number Gr and the Prandtl number Pr as:

$$Ra = Gr \times Pr \tag{6.41}$$

Since Grashof number characterizes natural convection, Rayleigh number also relates to natural convection.

Formulas for natural convection depend on the orientation of the surface since buoyancy drives the flow. In the vertical position (Figure 6.6), the air with difference in density can flow more easily and therefore the heat transfer coefficient is higher. In the extreme case of two heated horizontal plates (Figure 6.7), one facing upward and the other downward, the one with heated face downward gives a much lower value. This results from the absence of flow for the latter situation due to the warmer and lighter fluid being at the top, which is its stable position.

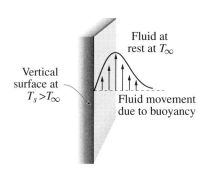

Figure 6.6: Natural convection over a heated vertical surface.

Vertical surface For a heated vertical surface (Figure 6.6), the heat transfer coefficient is calculated from

$$Nu_L = \left(0.825 + \frac{0.387 Ra_L^{1/6}}{\left[1 + (0.492/Pr)^{9/16}\right]^{8/27}}\right)^2 \tag{6.42}$$

Here the characteristic length L is the height of the vertical surface.

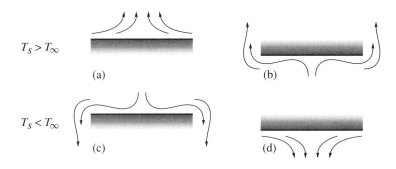

Figure 6.7: Natural convection over (a) hot surface facing up, (b) hot surface facing down, (c) cold surface facing up, and (d) cold surface facing down.

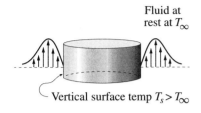

Figure 6.8: Natural convection over a heated vertical cylinder.

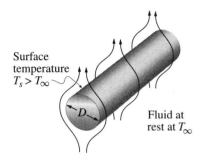

Figure 6.9: Natural convection over a heated horizontal cylinder.

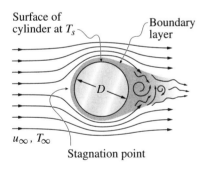

Figure 6.10: Forced convection over a heated cylinder.

Horizontal surface For a horizontal surface (Figure 6.7), with hot side facing up or cold side facing down, the heat transfer coefficients are calculated from

$$Nu_L = 0.54 Ra_L^{\frac{1}{4}} \quad 10^5 < Ra_L < 2 \times 10^7 \quad (6.43)$$

$$Nu_L = 0.14 Ra_L^{\frac{1}{3}} \quad 2 \times 10^7 < Ra_L < 3 \times 10^{10} \quad (6.44)$$

For a horizontal surface, with cold side facing up or hot side facing down, the heat transfer coefficients are calculated from

$$Nu_L = 0.27 Ra_L^{\frac{1}{4}} \quad 3 \times 10^5 < Ra_L < 10^{10} \quad (6.45)$$

The characteristic length in the equations for horizontal plates is calculated as $L = A/P$, where A is the surface area and P is the perimeter of the plate.

6.6.3 Flow Over Cylinder, Natural Convection

For a vertical cylinder (Figure 6.8), expressions for plane surface can be used provided the diameter D is large enough compared to the length L, specifically, the following relationship has to be satisfied:

$$\frac{D}{L} \geq \frac{35}{Gr_L^{\frac{1}{4}}} \quad (6.46)$$

Thus, the characteristic length would be the height of the cylinder, L. For natural convection over a horizontal circular cylinder (Figure 6.9), the characteristic length is the outside diameter of the cylinder. One correlation that is valid for this situation for a wide range of Rayleigh number is

$$Nu_D = \left(0.60 + \frac{0.387 Ra_D^{1/6}}{\left[1 + (0.559/Pr)^{9/16}\right]^{8/27}}\right)^2 \quad \text{for } 10^{-5} < Ra_D < 10^{12} \quad (6.47)$$

6.6.4 Flow Over Cylinder, Forced Convection

For forced convection over a cylinder (Figure 6.10), the characteristic length is the outside diameter. For both laminar and turbulent flow over such a cylinder, the following correlation is widely used:

$$Nu_D = B Re_D^n Pr^{1/3} \quad (6.48)$$

where $Re_D = u_\infty D/\nu$ is Reynolds number based on the outside diameter D of the cylinder, and the values of B and n are given by

Re_D	B	n
0.4-4	0.989	0.330
4-40	0.911	0.385
40-4000	0.683	0.366
4000-40,000	0.193	0.618
40,000-400,000	0.027	0.805

6.6.5 Flow Through Cylinder, Forced Convection

For internal flow through a cylindrical tube (Figure 6.11), the characteristic length is the inner diameter of the tube such that $Nu_D = hD/k$ and $Re_D = u_{av}D/\nu$, u_{av} being the mass average velocity of the fluid. The following correlations are provided for laminar and turbulent flow when the velocity profile is fully developed:

$$Nu_D = 3.66 \text{ for } Re_D \leq 2300 \qquad (6.49)$$
$$Nu_D = 0.023 Re_D^{0.8} Pr^n \text{ for } Re_D \geq 10,000 \text{ and } L/D \geq 10 \qquad (6.50)$$
$$0.6 \leq Pr \leq 160$$
$$n = 0.3 \text{ for fluid being cooled and}$$
$$n = 0.4 \text{ for fluid being heated}$$

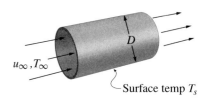

Figure 6.11: Forced convection through a cylindrical tube.

6.6.6 Flow Over Sphere, Natural Convection

For natural convection over a sphere (Figure 6.12), the following correlation is used:

$$Nu_D = 2 + 0.43 Ra_D^{\frac{1}{4}} \text{ for } 1 < Ra_D < 10^5, \ Pr \cong 1 \qquad (6.51)$$

Here the characteristic length is the diameter of the sphere such that $Nu_D = hD/k$ and $Gr_D = \beta g D^3 \Delta T / \nu^2$.

6.6.7 Flow Over Sphere, Forced Convection

Flow over a sphere is similar to that over a cylinder discussed earlier. For forced convection over a sphere, the following correlation developed by Whitaker (1972) is widely recommended in various textbooks:

$$Nu_D = 2 + \left(0.4 Re_D^{1/2} + 0.06 Re_D^{2/3}\right) Pr^{0.4} \qquad (6.52)$$
$$\text{for } 3.5 < Re_D < 7.6 \times 10^4 \text{ and } 0.71 < Pr < 380$$

Here the characteristic length is also the diameter of the sphere such that $Nu_D = hD/k$ and $Re_D = u_\infty D / \nu$.

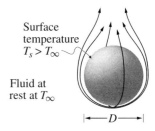

Figure 6.12: Natural convection over a heated sphere.

6.6.8 Laminar vs. Turbulent Flow

As mentioned in Section 6.3, turbulence increases the bulk mixing and is more effective for heat transfer. This is shown schematically in Figure 6.13 and can be seen from the equations of h for laminar and turbulent flow. For example, for flow over a flat plate, Eqns. 6.35 and 6.37 can be used to show that the ratio

$$\frac{Nu_{L,\text{turbulent}}}{Nu_{L,\text{laminar}}} = 0.0542 \, Re_L^{0.3} \qquad (6.53)$$

is more than one in the high Reynolds number range. Thus for turbulent flow, Nu or h values will be higher.

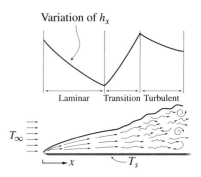

Figure 6.13: Variation of local heat transfer coefficient for flow over a flat plate as flow changes from laminar to turbulent.

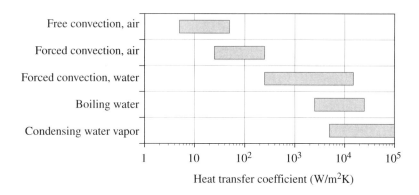

Figure 6.14: Order of magnitude estimates of the heat transfer coefficient h for various situations.

6.6.9 Orders of Magnitude for Heat Transfer Coefficient Values

An order of magnitude estimate for the values of h for various physical situations are given in Figure 6.14. Note that, this figure does not replace the equations described above, which are to be used for engineering calculations.

6.6.10 Coefficients for Air Flow Over Human Subjects

As an example of a biological application, consider the convection of air over human body. For this specific heat transfer situation, the convective coefficient for three different conditions are shown in Figure 6.15. Note that this plot is intended only to show the nature of variation. Significant variability is expected for any real situation. Such data are useful for designing indoor environment to achieve thermal comfort.

6.6.11 Wind Chill Factor and Boundary Layer Thickness

During the winter, the weather report often provides us with a "wind chill." This wind chill temperature or wind chill factor is lower for higher wind speed but same ambient temperature. Also, the lower the wind chill, the colder we feel. Thus, how cold we feel depends on the wind speed in addition to the temperature. What is its explanation from a heat transfer standpoint?

At higher wind speed, the boundary layer, which is a layer of relatively stagnant air providing insulation, becomes thinner (Figure 6.16). This increases the heat loss from the skin and tends to lower its temperature. As a result, a person or an animal feels colder. This is equivalent to being exposed to a colder air, but without the wind. As shown in Figure 6.16, still air that has a thicker boundary layer that can produce the same gradient (same heat loss) at the skin surface for a lower temperature.

To develop a relationship between wind chill and the ambient air temperature and velocity, note that they are equivalent, e.g., they lead to the same heat loss from the

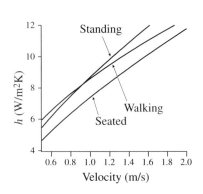

Figure 6.15: Typical values of the convective heat transfer coefficient for humans. For walking, the velocity corresponds to the speed of walking. Prediction equations are taken from Shitzer and Eberhart (1985).

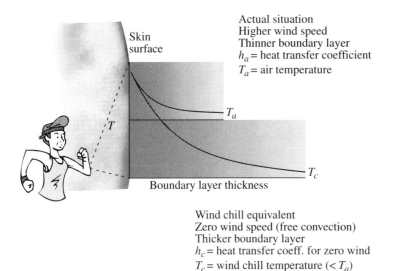

Figure 6.16: Boundary layer explanation of the wind chill factor. The wind chill temperature, T_c, is a fictitious temperature that can produce the same temperature gradient (and therefore the same heat loss) at the skin surface for a thicker boundary layer corresponding to still air.

body. If the skin temperature is assumed constant at T_s, heat loss in terms of the ambient (true) temperature is $h_a(T_s - T_a)$ where h_a is the heat transfer coefficient for the ambient conditions. Heat loss in terms of the wind chill temperature T_c is $h_c(T_s - T_c)$ where h_c is the heat transfer coefficient without any wind blowing (under natural convection). Since these two heat losses are equivalent:

$$h_c(T_s - T_c) = h_a(T_s - T_a) \tag{6.54}$$

Upon simplification, we obtain:

$$T_c = \frac{h_a}{h_c} T_a + \left(1 - \frac{h_a}{h_c}\right) T_s \tag{6.55}$$

Figure 6.17 shows the relationship between the wind chill temperature and the ambient temperature. We can try to explain the data in Figure 6.17 using Equation 6.55. At 0 mph wind, $h_a = h_c$ and therefore $T_c = T_a$. As wind speed increases, the heat transfer coefficient increases. If we use the formula for flat plate in laminar flow (Eqn. 6.35) at low air speeds, $h_a \propto u^{0.5}$ and the slope of Eqn. 6.55 increases with speed as $u^{0.5}$. Beyond a certain wind speed, the flow is likely to be turbulent and the relationship will change to $h_a \propto u^{0.8}$ as given by Eqn. 6.37. In the velocity range the flow changes to turbulent, h_a will increase faster with u as seen in the rapidly increasing slope and intercept between 5 and 15 mph. Eventually, the turbulent relationship will continue and the curves become closely spaced.

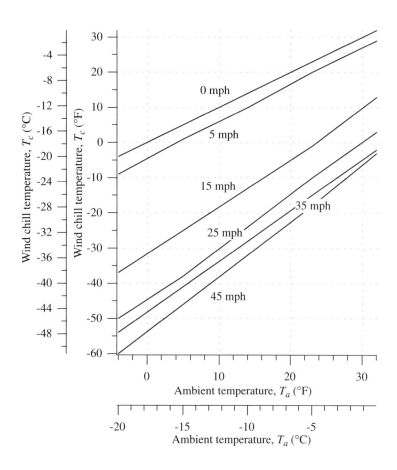

Figure 6.17: The wind-chill chart. Data from Folk (1966).

6.7 Chapter Summary—Convective Heat Transfer

- **Temperature Profiles and Boundary Layers (page 97)**

    1. The convective resistance to heat transfer between a solid surface and a fluid flowing over it is restricted to a thin layer on the surface where the fluid moves relatively slowly. This layer is called a boundary layer.

- **Convective Heat Transfer Coefficient Calculations (page 104)**

    1. Convective heat transfer coefficient, h, represents the inverse of the thermal resistance between a surface and a fluid flowing over it.

    2. h includes the effect of conduction in the fluid and flow. It is therefore also a function of the flow parameter Re in addition to the thermal conductivity k of the fluid. Thus h should not be confused with a material property such as the thermal conductivity, k.

    3. The many significant variables such as the flow velocity that h depends on are grouped into dimensionless parameters as shown in Table 6.1.

    4. Since h varies with the flow situation, its exact value depends on the flow velocity and other variables. The exact value of h for a situation is given by formulas in Section 6.6 (page 104). For forced convection, the flow parameter is Reynolds number, Re. For natural convection, the Re is replaced by the Grashof number, Gr.

6.8 Concept and Review Questions

1. What is meant by a velocity boundary layer? How is its thickness defined?

2. What is meant by a thermal boundary layer? How is its thickness defined?

3. Is the velocity boundary layer always thicker than the thermal boundary layer?

4. Distinguish between laminar and turbulent flow from a physical standpoint.

5. Define the film temperature and describe how it is used.

6. Define the Reynolds number and describe in words its physical significance.

7. Define the Grashof number and describe in words its physical significance.

8. Define the Prandtl number and describe in words its physical significance.

9. What are some of the advantages and disadvantages in using the heat transfer coefficient, h, in the analysis of convection heat transfer.

10. How does the Archimedes principle relate to natural convection?

11. What dimensionless group in natural convection substitutes for the Reynolds number in forced convection?

12. Describe the flow pattern for natural convection a) in a room having people with one wall exposed to cold outside, b) inside a stationary vertical cylindrical can of liquid being heated from all sides for sterilization.

13. Give examples of various kinds of convective heat transfer in a) plant system, b) mammalian system, c) the environment.

6.9 Further Reading

Anantheswaran, R. C. and M. A. Rao. 1985. Heat transfer to model Newtonian foods in cans during end-over-end rotation. Journal of Food Engineering, 4:1-19.

Chandra, S., P. Lindsey, and N. Bassuk. 1992. Measurement of the mass flow rate of water in trees. Presented at the National Heat Transfer Conference, Aug. 9-12, San Diego, CA.

Folk, G. E. Jr. 1966. *Introduction to Environmental Physiology.* Lea and Febiger, Philadelphia.

Fox, R.W. and A.T. McDonald. 1985. *Introduction to Fluid Mechanics.* John Wiley & Sons, New York.

Geankoplis, C.J. 1983. *Transport Processes and Unit Operations.* Allyn and Bacon, Inc., Boston.

Griffin, O.M. 1973. Heat, mass and momentum transfer during the melting of glacial ice in seawater. Transactions of the ASME, J. Heat Transfer, 95:317-323.

Incropera, F. P. and D. P. Dewitt. 1996. *Fundamentals of Heat and Mass Transfer.* John Wiley & Sons, New York.

Phillips, P.K. and J.E. Heath. 1992. Heat exchange by the pinna of the African elephant (Loxodonta africana). Comparative Biochemistry and Physiology. 101A (4):693- 699.

Prandtl, L. 1904. Fluid motion with very small friction (in German). Proceedings of the Third International Congress on Mathematics, Heidelberg. English translation available as NACA TM 452, March 1928.

Shitzer, A. and R.C. Eberhart. 1985. Heat generation, storage, and transport processes. 1:137-151. In *Heat Transfer in Medicine and Biology* edited by A. Shitzer and R.C. Eberhart, Plenum Press, New York.

Whitaker, S. 1972. Forced convection heat transfer correlations for flow in pipes, past flat plates, single cylinders, single spheres, and for flow in packed beds and tube bundles. AIChE Journal, 18:361-371.

6.10 Problems

6.10.1 Thickness of Boundary Layer Over Face

Consider cold wintry breeze parallel to your face, with an average facial dimension along the flow of 5 cm. The air velocity is 1 m/s (corresponding to 2.2 miles per hour). Assume laminar flow. Air propeties can be found in Appendix C.8. 1) Calculate the thickness of the velocity boundary layer at the end of 5 cm length. 2) Calculate the thickness of the thermal boundary layer at the end of 5 cm length. 3) Provide the physical meanings of the two thicknesses in items 1) and 2).

6.10.2 Average and Point Values of h

Starting from Eqn. 6.34 for heat transfer coefficient at a location for laminar flow over a flat plate, derive Eqn. 6.35, the expression for the average heat transfer coefficient for the same plate.

6.10.3 Understanding Convective Heat Transfer Coefficient

Consider a cold fluid is flowing over a warmer flat surface, at a velocity low enough for the flow to be laminar. 1) Is the rate of heat removal the same at any location x measured from the edge of the plate? Illustrate this by writing the equation for h as a function of x. 2) Plot (roughly) the h vs. x relationship and explain (qualitatively) the physical reasoning behind such variation. 3) For the same bulk velocity of the fluids and same film temperature (300 K), how many times faster would water cool (remove heat from) the plate as compared to air? Assume plate temperature does not change. 4) Keeping all other parameters the same, if the bulk velocity (flow rate) doubles (still laminar flow), what happens to *average h*? The properties of air and water at 300 K are given below.

Property	Air	Water
density [kg/m^3]	1.1421	995.8
viscosity [kg/m·s]	17.84×10^{-6}	8.6×10^{-4}
specific heat [J/kg·K]	1.0408×10^3	4.179×10^3
thermal conductivity [W/m·K]	0.0262	0.614

6.10.4 Heat Transfer Coefficient in a Motionless Fluid

Heat transfer coefficient, h, involves both heat conduction and flow. Thus, in the limit, when there is little flow, this coefficient would measure only the effect of conduction. Calculate the heat transfer coefficient for a heated sphere in a large, *motionless* fluid in terms of its diameter, D, and the thermal conductivity of the fluid, k. The sphere surface can be assumed to be at a temperature T_s and far from the sphere surface, the fluid temperature is T_∞. (Hint: Set up the heat conduction problem (without bulk flow or convection) for the fluid in spherical coordinates with temperature T_s at a radius $D/2$ and temperature T_∞ at infinity.)

6.10.5 Convective Heat Transfer Coefficient for Various Fluids

Consider a cold fluid is flowing over a warmer flat surface, at a velocity low enough for the flow to be laminar. Assume plate temperature does not change. 1) For the same bulk velocity of the fluids, how many times faster would water cool (i.e., remove heat from) the plate as compared to air? 2) Keeping all other parameters the same, if the bulk velocity (flow rate) doubles (still laminar flow), what happens to *average h*? Water and air properties are given in the Appendix. Consider a film temperature of 350 K.

6.10.6 Measuring Air Velocity

Hot wire anemometry depends on the convective heat transfer from forced flow across a small diameter wire, as shown in Figure 6.18. A known electric energy input of Q watts per meter length of wire is needed to maintain a known wire surface temperature of T_s when the known temperature of the flowing air is T_∞. For a length L of the wire and diameter D, 1) Write the total energy input to the wire in terms of Q. 2) The total energy lost from the wire by convection. You need to include the appropriate equations for convective heat transfer coefficient, h, and area in terms of the diameter, D. 3) Under steady state conditions, use the information in the previous two steps to find an expression for the air velocity, u_∞. 4) For an air temperature of 30°C and heat input to a 0.15 mm wire at 15 W/m, the wire temperature was measured to be 80°C. What is the air velocity, u_∞, in m/s?

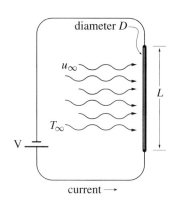

Figure 6.18: Schematic of hot wire anemometry.

6.10.7 Convection Heat Transfer and Exercise

Although it is well-known that exercise increases the overall rate of heat generation in the body, it is also important to realize that the rate of heat loss must increase (by sweating and increased convection) during exercise to avoid large increases in the temperature of the body. Here we are interested in the convective heat loss from an individual walking on a treadmill. For simplicity assume the subject is not clothed, and that the mean skin temperature increases 1°C during exercise. Compare the heat loss due to convection when the subject is seated to the heat loss when the subject is walking on the treadmill at 6.4 kilometers per hour. The surface area of the body is 1.7 m², the ambient air temperature is 20°C, and the skin temperature at rest is 33°C. The heat transfer coefficient at rest is 2.1 W/m²·K and the heat transfer coefficient walking on a treadmill is $6.5(u)^{0.36}$ W/m²·K, where u is the speed of treadmill in m/s.

6.10.8 Heat Loss from Elephant Pinna

The African elephant has developed the pinna or external ear that functions as a radiator-convector (see page 60 for more discussion). The pinna can be modeled as a flat plate with air flow over it. 1) Find and plot the convective heat loss from the pinna as the surface temperature of the pinna ranges from 21°C to 36°C. 2) If the rate of metabolic heat production for a 2000 kg elephant is 1650 W, what is the maximum percentage of the metabolic heat that can be lost by convection through the pinna? Remember that an elephant has two ears, and heat is lost equally from both sides of the ear. The ambient air temperature is 20°C, the air velocity is 2 m/s, the surface area of one ear is 0.84 m², and the average length of one side of the ear is 1 m. For properties of air, see Table C.8, page 340.

6.10.9 Iceberg as a Source of Clean Water

Scientists have been interested in transferring glacier ice from the Antarctic to arid regions of the world as an alternate source of clean water. The technical and economic feasibility hinged to a significant extent on knowledge of the melting or ablation of ice that would occur while it is being towed in warmer sea water.

Consider a very simple ablation process where as warmer sea water flows over the submerged surface of a flat iceberg, heat is transferred by convection to the surface of the iceberg. Consider the ablation to be uniform perpendicular to the direction of flow. The flat surface of ice (parallel to flow) will cease to be flat as more ablation occurs from the front edge, but we will consider it flat for purposes of our calculation. Consider average ice surface temperature to be -10°C and sea water temperature to be 10°C. The average length of Antarctic icebergs is 1000 m. Density of ice is 850 kg/m³. Density of sea water is 1025 kg/m³ and viscosity of sea water is approximately 2×10^3 Pa s. For a towing speed of 0.5 m/sec, 1) Calculate the heat transfer coefficient as a function of position from the edge of the iceberg. 2) Calculate the rate of heat gain and rate of melting (m thickness/day) as a function of position from the edge of the iceberg. 3) Plot the results from (2). 4) Calculate the average ablation (melting) rate in m/day and comment on the technical feasibility of towing it from Antarctica to Southern California. 5) Would towing at a slower velocity make it more feasible? Explain qualitatively.

6.10.10 Wind Chill and Heat Loss

Wind chill charts used in practice often come from an empirical formula developed by Paul Siple in 1945 (Folk, 1966). He was an Eagle Scout who accompanied Richard E. Byrd on his second Antarctic expedition in 1933. His empirical formula was

$$q'' = \left(\overbrace{\sqrt{100u} + 10.45 - u}^{h} \right) (33 - T_A) \ \frac{\text{kcal}}{\text{cm}^2 \text{min}}$$

where u is wind speed in m/s and T_A is the air temperature, in °C. A skin temperature of 33°C is assumed in the above equation. The units in the above equation are definitely wrong. 1) Compare this equation to the equation for heat loss as a function of velocity for a flat plate (in average temperature of $-50°C$) for the velocities of 5, 10, 20, and 30 km/h and estimate the multiplying factor to make the above formula approximately same as the flat plate formula over this range of velocities. 2) What is the ratio of the heat loss per unit area from the skin for the calm day (assume zero wind speed) to that for a windy day with 40 km/h wind? 3) What air temperature on a calm day would produce the same heat loss occurring with the air temperature at $-20°C$ on the windy day?

6.10.11 Steady State Temperature and Natural Convection

You left one of the heaters in your cooking range on by mistake as you hurried for your lectures. For simplicity, consider the heater as a flat disk of 15 cm diameter that is continually producing 500 W of heat in a room whose air temperature stays at 25°C. 1) Discuss why the temperature of the heater would reach a steady-state. 2) What would be the steady-state temperature? Additionally, assume that the heater does not lose any heat from the bottom or side, and unobstructed air surrounds the heater. Hint: You need to assume a temperature of the disk to get started in the iterative process.

6.10.12 Forced Water Cooling of an Apple After Picking

Apples of diameter 12 cm are picked from a tree and cooled in a stream of cold water to reduce its metabolism so that it stores longer. Consider a stream of water at 5°C and a velocity of 0.04 m/s flowing over an individual apple. If the initial uniform temperature of the apple was 25°C, calculate the time it takes for the center of the apple to reach 8°C. The thermal properties of apple and water are given in Appendix C. Hint: Calculate Biot number to confirm that both external convection and internal conduction are important in this problem.

6.10.13 Heat Loss from a Free Water Surface

A cattle watering fountain has a free water surface with a size equal to 0.6 m × 0.3 m, with the longer side being parallel to the prevailing wind. Wind at −23°C is blowing over the surface at 1 m/s. A electrical heating element maintains the water surface and interior temperature at 10°C. Find the power needed for the heater. Assume that the bottom and the sides of the container are perfectly insulated, the evaporative cooling is small, and water surface is level with the container top edge.

6.10.14 Natural Convection Heating of Containerized Liquids

In food and biochemical processes, sometimes liquid to be sterilized is heated in a container without any kind of agitation, thus heating by natural convection. Such heating sets up flow patterns, as shown in Figure 6.19. Assume the average heat transfer coefficient between the fluid and the wall reduces with time as

$$h = 200 \exp(-t/600) \quad (6.56)$$

where t is the time of heating, in seconds and h is in W/m^2·K. For simplicity, we will treat this problem as a lumped parameter situation, and use Eqn. 5.4. The mass average temperature of the fluid, T_{av} will replace the temperature T in this equation. 1) Explain why you would expect the heat transfer coefficient to reduce with time. 2) Are there changes in the fluid that would tend to increase the heat transfer coefficient? 3) Integrate Eqn. 5.4 with a variable h given by Eqn. 6.56 to obtain an expression for temperature as a function of time. 4) Find the time to heat a fluid to reach an average temperature of 115°C from an initial uniform temperature of 30°C when the surface of the container is kept at 121°C. The fluid in the container has a mass of 0.2 kg, specific heat of 4190 J/kg·K and the container has a surface area of 0.017 cm².

6.10.15 Agitation to Improve Heat Transfer in Containerized Liquids

The rate of heating of a canned liquid food is increased substantially by agitating, an example of which is shown in Figure 6.20. For this type of agitation, the heat transfer coefficient was given by (Anantheswaran and Rao, 1985) as

$$Nu = 2.9 Re^{0.436} Pr^{0.287} \tag{6.57}$$

where the characteristic diameter used was not the diameter of the can but the diameter of rotation ($= 2R_r$ in Figure 6.20) and is given by $Re = D_r^2 N \rho / \mu$ where N is the speed of rotation, in revolutions per second. By what factor does the rate of heat transfer increase, when the speed of rotation is doubled?

6.10.16 Heat Transfer Between Fluid and Particles in Continuous Flow

Continuous sterilization (as opposed to canning) of foods containing particulates such as chunky soups is of major interest to the food industry as it provides much improved quality. Consider a spherical particle (such as a piece of potato) flowing with a carrier liquid (such as a soup) in a tube. Whereas the liquid can be agitated to heat faster, heating of a solid particle is limited by heat conduction inside the solid and the heat transfer coefficient at the surface of the particle that depends on the relative velocity between the fluid and the particle. A small relative velocity between the fluid and a particle would lead to lower surface heat transfer coefficient, i.e., slower heating of the particle. Since sterilization processes are to be designed so that all particles are sterilized, some researchers have suggested that the smallest possible heat transfer coefficient between the fluid and a particle, i.e., worst heating scenario, be used to decide how long to heat. 1) For a pea of 5 mm diameter that is in a carrier fluid of water (thermal conductivity of 0.5 W/m·K), what is this smallest value of the heat transfer coefficient? 2) What would be the disadvantage of deciding the heating time based on the smallest value of the heat transfer coefficient?

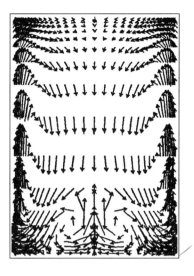

Figure 6.19: Natural convection patterns in an unagitated fluid.

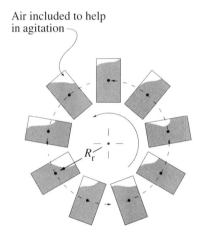

Figure 6.20: End-over-end agitation of a canned liquid to increase rate of heat transfer.

Chapter 7

HEAT TRANSFER WITH CHANGE OF PHASE

CHAPTER OBJECTIVES

After you have studied this chapter, you should be able to:

1. Explain the process of gradual freezing of a biomaterial over a temperature range, as opposed to one fixed temperature.

2. Explain the process of freezing in a cellular tissue and how the success of freezing depends on the rate of freezing.

3. Describe the changes in thermal and physical properties of biomaterials during freezing.

4. Calculate the time to freeze a plant or an animal tissue of given size.

KEY TERMS

- **heat of fusion or latent heat**
- **supercooling**
- **nucleation**
- **crystal growth**
- **freezing point depression**
- **semi-permeable membrane**
- **cryopreservation**
- **"mushy" zone in frozen biomaterials**
- **vapor pressure**
- **boiling**
- **boiling point elevation**
- **evapotranspiration**

This chapter deals with energy transport when a change of phase is involved. The first part of this chapter deals with freezing, the change of state from liquid to solid. Figure 7.1 shows how this chapter relates to other chapters in energy transfer.

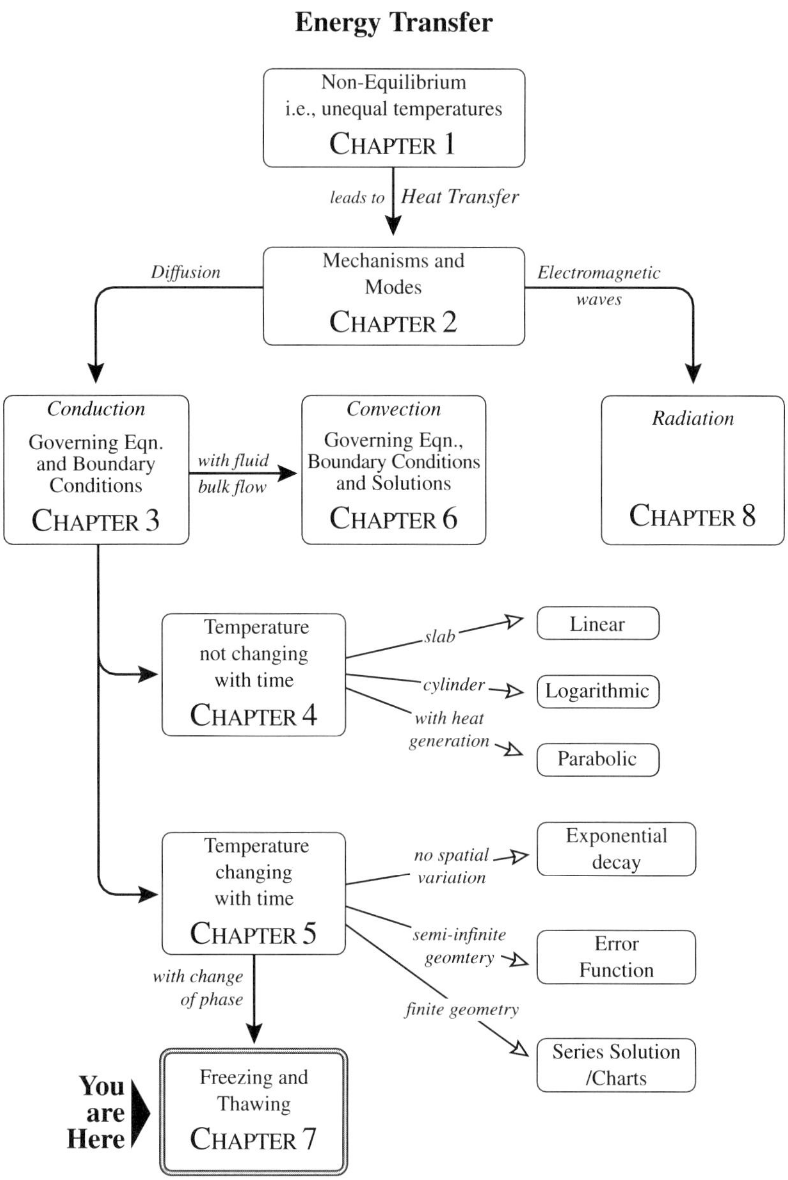

Figure 7.1: Concept map of energy transfer showing how the contents of this chapter relates to other chapters in the part on energy transfer.

Phase change brings complications in studying heat transfer, particularly in biomaterials, due to drastic changes in property. One primary comment that can be made about phase change processes is that they are energy intensive; i.e., a high energy ex-

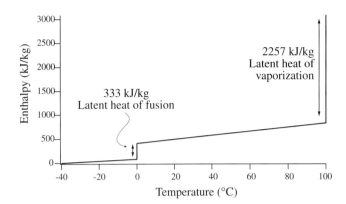

Figure 7.2: Enthalpy changes in water showing the latent heats involved for phase change from ice to water and from water to steam.

change occurs in the change from one state to another, as shown in Figure 7.2. The enthalpy difference due to change of phase from solid to liquid, occurring at the melting temperature, is termed the *latent heat of fusion*. Similarly, the enthalpy difference due to phase change from liquid to vapor, occurring at the boiling temperature, is termed the *latent heat of vaporization*. Figure 7.2 shows that for water the latent heat of vaporization is almost seven times the latent heat of fusion. Also, sensible heat or enthalpy change due to temperature difference is typically smaller than the latent heats.

7.1 Freezing and Thawing

Freezing and thawing is one of the most promising methods of preserving biological material. The term *cryopreservation* generally refers to the preservation of a living system at low temperatures in such a way that the system can be brought back to life. One of the ultimate goals, to preserve a whole human body reversibly by freezing and storage at low temperatures, is still at the science fiction stage. At present success in cryopreservation is limited to smaller mammalian systems, including blood, embryos, spermatozoa, culture cells, hepatocytes, bone marrow cells, pancreatic tissue and corneas. Other successfully frozen systems include protozoa and other microorganisms, insects, fish, plants, and algae.

Freezing is one of the most common methods of preserving large quantities of food materials. Biochemical spoilage reactions are reduced due to lower temperatures and the unavailability of water in liquid form. Frozen foods are considered the best alternative to fresh foods, as demonstrated by the $15 billion of frozen foods sold in the U.S. in 1990, which represents 7-8% of all foods sold in supermarkets. Intact food tissues, such as in fresh whole vegetables or large cuts of meat, have many similarities to living specimens, and the goal in food freezing is to preserve a material as closely as possible to its original state so that it is close to the fresh state upon thawing.

As well as preserving materials, freezing is sometimes used to purposely destroy living material. In cryosurgery a localized necrosis in diseased tissues or organs is produced by freezing. A cryosurgical probe is inserted into a tissue and suddenly brought to a cryogenic temperature. The diseased tissue in the vicinity of the probe undergoes a phase change and is destroyed.

Quantitative studies of the thermodynamic and heat transfer processes during freezing and thawing have resulted in significant progress in cryobiology. In industrial food processing, these studies have led to reduced energy use and better quality frozen food. In cryosurgery they help determine the optimum freezing protocol to minimize unwanted damage to surrounding tissues.

Since biomaterials are mostly water, the freezing process of pure water should provide significant information about the freezing of biomaterials. It makes sense, therefore, to study the basic aspects of the liquid-solid phase change of pure water. After an explanation of the pure water system, more complex systems such as solutions and tissues will be studied.

7.2 Freezing of Pure Water

7.2.1 Freezing Process

Several steps are involved in the freezing of pure water—supercooling, nucleation, crystal growth, and maturation. The temperature at which crystallization is initiated is uncertain. Ice does not always form at precisely 0°C. Water frequently cools to a sub-zero temperature before freezing, a phenomenon called *supercooling* or *undercooling*. Ultrapure water may not freeze until around −40°C. For any kind of crystal to grow, a stable seed that can act as a foundation is required. Such seeds are termed nuclei. At around −40°C, homogeneous nucleation takes place in ultrapure water— nuclei spontaneously generate, and the water freezes. However, homogeneous nucleation happens rarely in unpurified water because extraneous (heterogeneous) nuclei are usually present as impurities and ice crystals can form easily around such nuclei. Thus, heterogeneous nucleation takes place at temperatures much higher than −40°C, typically −2 to −3°C.

Crystal growth is possible once nucleation has taken place. As long as a stable ice crystal is present, further growth is possible. The rate of this growth is controlled in part by the rate of heat removal from the system. Rapid freezing tends to form small ice crystals, whereas slow freezing tends to produce large ice crystals. During storage, even at constant temperatures, a *maturation* process occurs in which smaller ice crystals decrease in size, while larger ice crystals enlarge. With time the number of ice crystals decreases, and their average size increases, in part reversing the initial effects produced by rapid freezing. This increase in crystal size with time can be important. For example, ice cream stored for a long time in the freezer grows large ice crystals, which diminish its appeal.

7.2.2 Property Changes During Freezing

As the temperature of water decreases, the water's specific volume and density change, as shown in Figure 7.3. As it freezes, the water expands greatly, around 9%. This high degree of expansion has many consequences. Water pipes can rupture. In plant and animal cells expansion can rupture the cell membranes. Freezing in order to preserve food material can lead to cracks in the food. Stresses caused by expansion during a phase change are also a major problem in cryosurgery.

Soils or pavement in colder climates often heave from the expansion of water in the soil due to freezing. However, the heaves are much bigger than what the normal amount of water in soil can account for. As temperature decreases near the surface, the surface tension of the water increases. Water moves up from the warmer interior soil and groundwater due to capillary action. Freezing and expansion of this accumulated water lead to the larger heaves. The total water movement is partly due to a difference in vapor pressure from the warmer groundwater (higher vapor pressure) to the colder surface soil (lower vapor pressure). During the spring, when a substantial amount of ice melts and greatly increases the water content of the soil, the strength of the soil greatly decreases, and pavements can suffer serious structural damage. Measures to avoid such damage include lowering of the groundwater table, replacement of frost-susceptible soils, and use of impervious membranes.

The thermal conductivity of ice is more than three times that of water, as shown in Figure 7.4. Such a large change in thermal conductivity has implications, for example, in a thawing process. For a given material, thawing is slower than freezing. The layer near the surface of the material, which provides the thermal resistance to heat flow from the inside, has a higher resistance when it is thawed, due to lower thermal conductivity.

The specific heat of water drops considerably during freezing and continues to drop as the temperature is lowered beyond the freezing point, as shown in Figure 7.5. An average value of the specific heat of water is 4.179 kJ/kg·K, and that of ice, is 2.04 kJ/kg·K.

Dielectric properties measure the interaction of a material with electromagnetic waves such as microwaves. One of the two dielectric properties, dielectric loss, is a measure of the ability of a material to absorb electromagnetic energy and turn it into heat. The dielectric loss of water changes significantly during freezing. At the microwave frequency of 2450 MHz, the frequency used in most domestic microwave ovens in the U.S., the dielectric loss of water changes as shown in Figure 7.6. Since the dielectric loss is a measure of microwave absorption, the figure shows that ice cannot absorb microwaves readily.

The lower absorption of microwaves by ice has many practical implications. For example, in microwave thawing of food, initially very few of the microwaves are absorbed. However, due to differences in heating rates and other factors, one region of the food will thaw first, and that region will selectively absorb more of the microwaves and may start to boil while other areas remain frozen! This non-uniformity of thawing is undesirable, and to minimize it, the thawing cycle in microwave ovens typically does not heat continuously. Instead it cycles between heating and rest periods, so that temperatures at warmer and colder areas can equilibrate by diffusion.

Mechanical properties such as the modulus of elasticity also change drastically

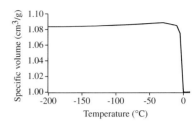

Figure 7.3: Increase in specific volume (decrease in density) of water during freezing.

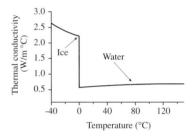

Figure 7.4: Changes in thermal conductivity with temperature showing large increase in thermal conductivity for ice over water.

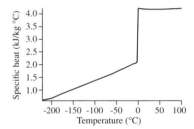

Figure 7.5: Specific heat of water and ice as a function of temperature.

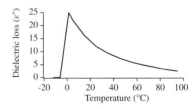

Figure 7.6: Dielectric loss of water and ice at a microwave frequency of 2450 MHz.

during freezing and continue to change as temperature drops, as in cryogenic freezing. The material becomes brittle at very low temperatures and may crack, depending on the rate of freezing, as explained later (see Figure 7.11).

7.3 Freezing of Solutions and Biomaterials

7.3.1 Solutions

Dissolved solutes in water depress the freezing point of the solution below the freezing point of pure water. Solutions experience less supercooling and earlier nucleation than pure liquids. The depressed freezing point T_A of a solution is related to its concentration x_A by

$$\frac{\Delta H_f}{R_g}\left[\frac{1}{T_{A0}} - \frac{1}{T_A}\right] = \ln x_A \qquad (7.1)$$

where ΔH_f is the latent heat of fusion of pure liquid A in J/mol, T_{A0} is the freezing point of pure liquid A in K, T_A is the freezing point of the solution in A in K; R_g is the gas constant equal to 8.314 kJ/kmol, and x_A is the mole fraction of A in the solution. This relationship is plotted in Figure 7.7 for water. Since nucleation occurs easily in solutions, they undergo little supercooling; i.e., solutions in fact freeze at their expected freezing point of T_A, given by Equation 7.1. For dilute solutions, the above equation can be simplified to

$$\frac{\Delta H_f}{R_g}\left[\frac{T_A - T_{A0}}{T_{A0} T_A}\right] = \ln(1 - x_B) = -x_B \quad for\ x_B << 1 \qquad (7.2)$$

where x_B is the mole fraction of solute. Defining

$$\Delta T_f = T_{A0} - T_A$$

as the drop in freezing point or freezing point depression and simplifying for small changes in T_A as $T_A \cdot T_{A0} \cong T_{A0}^2$, we get

$$\frac{\Delta H_f}{R_g} \cdot \frac{\Delta T_f}{T_{A0}^2} = x_B \qquad (7.3)$$

Note that Equation 7.3 provides a linear relationship between the drop in freezing point and mole fraction of solute for dilute solutions, as plotted in Figure 7.7. For dilute solutions the mole fraction can be approximated as below:

$$x_B = \frac{m_B/M_B}{\frac{m_A}{M_A} + \frac{m_B}{M_B}} = \frac{m_B}{M_B} \cdot \frac{M_A}{m_A} = \frac{m_B/M_B}{m_A} \cdot M_A$$

where m_A is mass of water, m_B is mass of solute, M_A is molecular weight of water, and M_B is the molecular weight of solute. Defining $\mathcal{M}$ as the molality or moles of solute per unit mass of water (or other solvent), and substituting in Equation 7.3, we get the freezing point depression equation for a dilute solution

$$\Delta T_f = \frac{R_g T_{A0}^2 M_A \mathcal{M}}{\Delta H_f} \qquad (7.4)$$

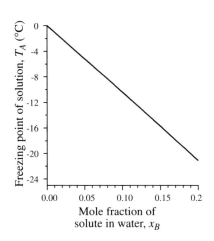

Figure 7.7: Freezing point depression of an aqueous solution as a function of concentration.

7.3. FREEZING OF SOLUTIONS AND BIOMATERIALS

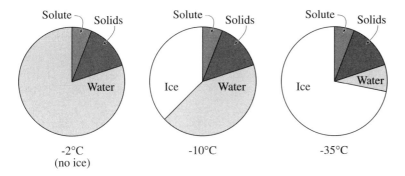

Figure 7.8: Schematic showing various components in gradual freezing of a solution.

which is sometimes written as

$$\Delta T_f = K_f \mathcal{M} \tag{7.5}$$

where K_f is known as the *cryoscopic constant* and depends only on the solvent.

This process of gradual freezing of a solution is shown as a schematic for a hypothetical system in Figure 7.8. The various regions in the pie chart represent the relative amounts of various components at a location in a solution. As the temperature is lowered, the amount of ice is zero until the initial freezing point of the solution is reached, depending on the initial concentration of the solution. As ice begins to form, the rest of the solution concentrates since the ice formed reduces available liquid water while the amount of solute remains the same. As temperature is continually lowered, this process involving freezing of pure water and concentration of the remaining solution continues until all the water that can be crystallized as ice freezes, leaving only the solute and its water of hydration. At this solute concentration, the solution is termed an eutectic. Further reduction in temperature results in the solidification of the remaining solution as a whole, i.e., without formation of pure ice. The lowest temperature at which the solution remains liquid is referred to as the eutectic temperature.

7.3.2 Cellular Tissues

The process of freezing in plant tissues is considerably more complicated than in a solution, primarily due to the presence of cellular structures (Figure 7.9). Cells possess semi-permeable membranes that selectively control the passage of molecules in and out of the cells. During tissue freezing, the most fundamental physical change taking place is the transport of water across a cell's semi-permeable membrane. The cell walls are generally considered freely permeable to water. As cooling proceeds, the intracellular as well as the extracellular fluid reaches the freezing temperature. Depending on the extent of supercooling, the random process of ice nucleation can be initiated preferentially in the intracellular or extracellular fluid. The cell has been considered devoid of nucleating agents, and thus it is normally assumed that ice forms in the extracellular

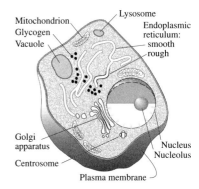

Figure 7.9: Hypothetical composite cell.

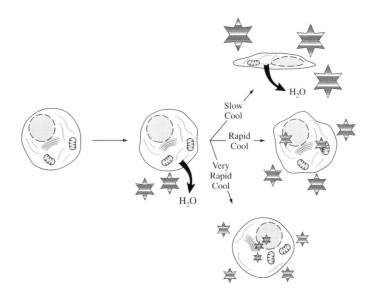

Figure 7.10: Schematic showing possibilities during cell freezing. Faster freezing reduces the transport of water out of the cell.

fluid first. With the formation of ice, the remaining extracellular fluid becomes more concentrated. The cell contents remain unfrozen and supercooled, presumably because the plasma membrane blocks the growth of ice crystals into the cytoplasm. This difference in concentration between the inside and outside develops osmotic pressure, and water flows out of the cell. The expelled water freezes in the extracellular space. The extent of cellular dehydration is important, because the rehydration process depends on the amount and rate of water loss. In addition the presence of too much intracellular ice may rupture the cell membrane.

7.3.3 Cooling Rates and Success of Freezing

The rate of cooling has microscopic as well as macroscopic consequences. At the cellular level, the osmotic water migration mentioned above is higher for a slower cooling process, leading to dehydration and a large change in the cell size, as shown in Figure 7.10. Fast cooling, on the other hand, will cause the least amount of migration and will leave the cell closer to its original shape and size. For this and other reasons, fast freezing is often practiced in cryopreservation and is the recommended procedure for industrial food-freezing applications.

Fast freezing can lead to damage, however. At the cellular level, formation of intracellular ice can cause membrane damage. At the macroscopic level, for specimens as small as a few centimeters, cracks due to stress from thermal expansion can occur. Cracking is more likely for larger sizes. Such thermal stresses arise because all the material is not at the same temperature. Due to the difference in temperature, all the material is not expanding to the same rate. This differential expansion causes stresses in the material. Cracks have been reported for organs and for food systems such as fish, potatoes, and beans frozen in liquid N_2. As shown in Figure 7.11, a potato material can crack significantly when dipped in liquid N_2.

7.4 Temperature Profiles and Freezing Time

7.4.1 Freezing Time for an Infinite Slab of Pure Liquid

Since the freezing process involves latent heat in addition to sensible heat, analysis of this very simple process is rather complicated. A very simplified solution (known as the Plank's solution) that is used in practice and that preserves some of the essential physics of the process is described below.

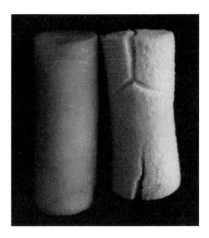

Figure 7.11: Cracks in cylindrical potato sample (1.78 cm dia) dipped in liquid nitrogen.

Consider symmetric (both surfaces at T_∞) freezing of a slab of pure liquid, as illustrated in Figure 7.12. Due to the symmetry, one needs to consider only the half thickness of the slab. The latent heat evolved at the interface of frozen and unfrozen regions is removed through the frozen layer. Although the thermal conductivity of the frozen layer is higher than that of the unfrozen layer, it is still small in absolute terms. The rate of heat transfer in the frozen layer is slow enough to be regarded as in pseudo-steady-state condition. Thus, although the temperature profile changes with time, this happens slowly enough that it approaches close to a steady-state profile at any time. In addition, the following assumptions are made:

1. Initially all material is at freezing temperature T_m but unfrozen.

2. All material freezes at one freezing point.

3. Thermal conductivity of the frozen part is constant.

With the above assumptions, heat loss through a frozen layer whose thickness is x at any time t is

$$q = kA \frac{T_m - T_s}{x}$$

This energy comes from the latent heat given off at the freezing front. The rate at which latent heat is given off is related to the rate at which the freezing front advances. If ΔH_f is the latent heat of fusion of the material per unit mass in J/kg,

$$q = \Delta H_f A \rho \frac{dx}{dt}$$

Equating the two

$$\Delta H_f A \rho \frac{dx}{dt} = kA \frac{T_m - T_s}{x} \qquad (7.6)$$

and solving for t, time to freeze as a function of x, the depth frozen,

$$\frac{k(T_m - T_s)}{\Delta H_f \rho} \int_0^t dt = \int_0^x x\, dx$$

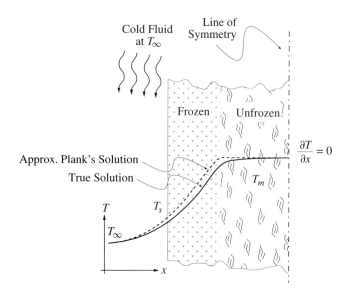

Figure 7.12: Freezing of a pure (stationary) liquid showing two distinct regions, frozen and unfrozen. Plank's assumption of a linear temperature profile is compared with the "true" profile.

$$t = \frac{\Delta H_f \rho}{k(T_m - T_s)} \frac{x^2}{2} \tag{7.7}$$

Freezing of the slab is complete when the freezing front reaches the midpoint, or the line of symmetry, for the slab. If the slab has a total thickness of $2L$, the freezing time for the slab is obtained by substituting $x = L$ in the above equation:

$$t_{slab} = \frac{\Delta H_f \rho}{k(T_m - T_s)} \frac{L^2}{2} \tag{7.8}$$

Two observations can be made from Equation 7.8. First, it shows that the depth frozen is related to time as

$$x \propto \sqrt{t}$$

Let's try to provide some physical explanations for this result. At any given time, a certain thickness of material is frozen. For another layer to freeze, the latent heat given off by that layer must be conducted through the frozen layer. The speed of freezing is determined by the rate at which heat can be conducted out. But as freezing progresses, the rate at which heat can be conducted out drops because the temperature gradient drops (same temperature difference over an increasing distance). Thus, the rate of advancement of the freezing layer slows.

Second, in the thawing process, the outer layer will be the unfrozen layer. The unfrozen material has lower thermal conductivity, and Equation 7.8 clearly demonstrates

7.4. TEMPERATURE PROFILES AND FREEZING TIME

that thawing of the same slab will take longer (although not by the same factor, in reality).

A more common boundary condition is convection at the surface instead of a specified temperature. For this case additional convective resistance $1/hA$ must be added to the conductive resistance x/kA in Equation 7.6 with the total temperature difference becoming $T_m - T_\infty$ instead of $T_m - T_s$, leading to

$$\frac{T_m - T_\infty}{\frac{x}{kA} + \frac{1}{hA}} = \Delta H_f \rho A \frac{dx}{dt} \tag{7.9}$$

$$\frac{T_m - T_\infty}{\frac{x}{K} + \frac{1}{h}} = \Delta H_f \rho \frac{dx}{dt}$$

$$\frac{T_m - T_\infty}{\Delta H_f \rho} \int_0^{t_F} dt = \int_0^L \left(\frac{x}{k} + \frac{1}{h}\right) dx$$

$$\frac{T_m - T_\infty}{\Delta H_f \rho} t_F = \frac{L^2}{2k} + \frac{L}{h}$$

$$t_F = \frac{\Delta H_f \rho}{T_m - T_\infty}\left[\frac{L^2}{2k} + \frac{L}{h}\right] \tag{7.10}$$

Note that Equation 7.8 becomes identical to Equation 7.10 for large values of h, which is equivalent to a surface-temperature-specified boundary condition.

7.4.2 Freezing Time for Biomaterials

The previous calculations of freezing time for a pure liquid will not work for even a simple solution where just one solute is dissolved. For one dissolved solute there will be two freezing fronts or interfaces (why?), and our calculations (Equation 7.10) will not work. For every solute added, there will be an added interface to keep track of. The water in a biomaterial has many components dissolved in it whose many interfaces must be tracked, and this procedure would be hopeless for such materials. The effect of one or more dissolved components is a "mushy" zone in the material, where the material is only partially frozen, as shown in Figure 7.13.

Like spatial variation, time variation of temperature during freezing of a biomaterial would also be substantially different from what was assumed for the simple solution in the previous section. To have a qualitative understanding of the time variation of temperature in a biomaterial freezing process, consider the schematic given by Figure 7.14. Initially, the solid is at 10°C. Cold fluid is flowing over the surface to maintain the surface temperature at −10°C. The temperatures at the two locations A and B would

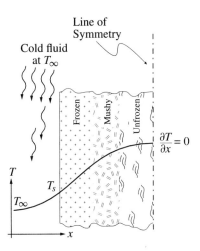

Figure 7.13: Freezing of a biological material showing a partially frozen or mushy zone.

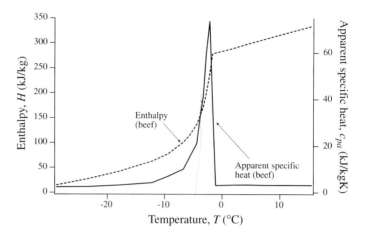

Figure 7.15: Experimental enthalpy (H) vs. temperature (T) relationship for beef muscle tissue around freezing temperature. The apparent specific heat (c_{pa}) is calculated from this data using Eqn. 7.12.

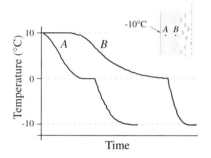

Figure 7.14: Schematic showing typical time-temperature history during a freezing process.

follow approximately the curves shown. The plateau near 0°C is characteristic of a freezing process and occurs due to the need to remove latent heat for the entire region from the surface up to the location. The plateau would also occur near the temperature where most of the freezing takes place (which is 0°C here) and therefore most of the contribution from latent heat happens. The plateau would not be present in a simple cooling process where the material does not go through change of phase.

The gradual freezing of biomaterial due the effect of the "mushy" region can be successfully captured by following an alternative procedure for calculating freezing time. To develop this, we rewrite the heat equation as

$$\frac{\partial}{\partial x}\left(k\frac{\partial T}{\partial x}\right) = \rho c_{pa}\frac{\partial T}{\partial t} \qquad (7.11)$$

where c_{pa} is an apparent specific heat that includes the latent heat. Thus, c_{pa} is defined as:

$$c_{pa} = \frac{\partial H}{\partial T} \qquad (7.12)$$

where H is the heat content, or enthalpy (i.e., the sum of sensible and latent heat) of a material per unit mass. The relationship between H and T

$$H = f(T) \qquad (7.13)$$

is available from experimental data on the material. An example of such experimental data for beef muscle tissue is shown in Figure 7.15. There are also simple empirical equations to estimate H for a material, as shown in this figure, although such equations have considerable error in them. Equations 7.11 and 7.13 together can provide temperature as a function of time as needed for the freezing time calculation. Note that Equation 7.11 automatically includes the phase change effects due to the use of the apparent specific heat, c_{pa}. Thus, even if the cellular structure and other complications are ignored for a biomaterial, its freezing process is more complex than that of a pure liquid.

It is apparent that the calculation of freezing time using Equations 7.11 and 7.13 can be involved. For practical use often the simplified formula (Equation 7.10) derived in the previous section for pure materials is used to obtain an approximate value for freezing time. Note the assumptions mentioned in deriving the simple formula. Depending on the situation, the estimate can have considerable error. Also the latent heat in Equation 7.10 should be that of the biomaterial which is related to its water content by

$$\lambda_{f,bio\ material} = \lambda_{f,water} * \begin{pmatrix} \text{mass fraction} \\ \text{of water in biomaterial} \end{pmatrix} \quad (7.14)$$

7.5 Evaporation

At any temperature above absolute zero, a molecule has kinetic energy. Consider the molecules of a liquid near its surface. At any temperature, some molecules have more than the average kinetic energy, while some molecules have less. While they are moving randomly, some molecules happen to be moving upward and may have enough energy to break out of the liquid. They can leave the liquid and join molecules of the gas above it. This change of state from the surface of a liquid to gas is called evaporation. When the change of state is from a solid to a vapor, the process is termed sublimation. Evaporation or sublimation are spontaneous processes driven by the energy of the molecules, i.e., by their temperature. Since the movement of the vapor molecules is random, some vapor molecules come back to the liquid (in case of evaporation). Initially, more molecules leave the liquid for the vapor phase. However, when the vapor phase has enough vapor, the number of vapor molecules reentering the liquid phase is the same as the number leaving it. An equilibrium is reached at this point. The pressure of the vapor in equilibrium with the liquid (called *vapor pressure*) is a measure of the relative ease of evaporation. Vapor pressure increases with temperature, because the average molecule has more energy to leave the liquid at higher temperature. Vapor pressure of water as a function of temperature is shown in Figure 7.16.

As an example, at a temperature of 60°C, the vapor above water exerts a pressure of 0.02 MPa ($\approx$ 0.2 atm). If the total pressure above the water is one atmosphere, 0.8 atmospheres will be exerted by air. From Figure 7.16 we can see that vapor pressure increases almost exponentially with temperature. At higher temperatures, more vapor can be formed. The evaporation when the vapor pressure at a given temperature is equal to the external pressure, the evaporation that occurs is termed *boiling*. Because the vapor pressure above water at 100°C is one atmosphere, water will boil when it reaches 100°C if the external pressure is one atmosphere. At slightly less than 100°C and under one atmosphere, water will evaporate vigorously but will not be boiling, strictly speaking.

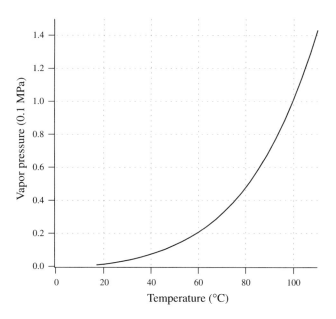

Figure 7.16: Vapor pressure of water at different temperatures.

Both heat and mass are transferred when a liquid evaporates. The rate at which the liquid evaporates depends on the rate at which energy can be supplied to, and the rate at which vapor (mass) can be removed from, the interfacial region. Continued evaporation is possible only if vapor can be removed from the interfacial region. As a liquid evaporates, the more energetic molecules leave the liquid. Thus, the energy of the remaining liquid, i.e., its temperature, is lowered. This cooling process is termed evaporative cooling.

Evaporation is a very important process in biological systems. Evaporation from plant leaves, skin (sweating), soil surfaces, and free water surfaces are all important biological processes.

7.5.1 Evaporation from Wet Surfaces

Evaporation from wet surfaces is frequent: from wet soils, food and biological materials, lakes, or falling droplets. Here, the surface vapor concentration is assumed to be the saturation value at the surface temperature. Thus, the rate of transfer of water vapor (*rate of evaporation*) per unit area at surface temperature T is given by

$$\text{Flux of water vapor} = h_m(c_{\text{sat at } T} - c_{\text{vapor at } \infty}) \tag{7.15}$$

where h_m stands for the mass transfer coefficient and c stands for concentration of water vapor. (Use of the mass transfer coefficient is discussed in detail on page 299 under convective mass transfer.) At any given temperature evaporation is limited by the rate at which vapor can be convected away, as given by Eqn. 7.15. Thus, evaporation is a special case of convective mass transfer from a surface. This mass flux also removes energy and lowers the temperature (the evaporative cooling process). The energy flux due to evaporation is given by

$$\begin{array}{c}\text{Flux of energy}\\ \text{from evaporation}\end{array} = \Delta H_{vap} \left(\begin{array}{c} \text{Flux of}\\ \text{water vapor} \end{array} \right) \quad (7.16)$$

where ΔH_{vap} is the latent heat of vaporization.

Evaporation in the context of various biological systems is now discussed qualitatively.

Plant systems (plant and soil surfaces) *Transpiration* in a leaf is the passage outward of water vapor through stomatal pores. Transpiration is generated by photosynthesis (see Example 12.2.1 on page 247). *Evapotranspiration* is the flux of water vapor just above a plant canopy, which includes transpiration from the leaves and evaporation from the soil surface.

$$\text{Evapotranspiration} = \text{Evaporation} + \text{Transpiration} \quad (7.17)$$

Evaporation from soil and the surface of small water bodies, and transpiration from vegetation, are difficult to separate and are therefore referred to in the single term called evapotranspiration. For fairly dense vegetation, evapotranspiration is appreciable, amounting to 60-90% of the flux of water vapor from a layer of open water at the same ambient air temperature (Nobel, 1974). Evapotranspiration is calculated using empirical formulas.

Penman's equation, a specialized equation for evaporation from a sufficiently wet soil surface, where the contacting air is fully saturated, can be seen in soil science textbooks such as Jury et al. (1991). As expected, Penman's equation to estimate potential evaporation is a function of the velocity and other conditions of the wind, just as it should be for any mass-transfer coefficient. Thus, appropriate mass transfer coefficients as described in section 14.5.3 can be used to find the potential evaporation. Note that the estimate will be for potential evaporation. Even when it is well supplied with water near the soil surface or at the roots, an evaporating vegetation cannot, in general, be considered wet, except after rainfall or dew formation. Thus, the humidity at the surface of the foliage elements is likely to be somewhat smaller than the saturation value at the corresponding temperature, and equations such as Penman's need to be modified as described in Brutsaert (1982).

Evaporation in animal systems (evaporative cooling) As discussed under thermoregulation (page 56), animals routinely use panting and cutaneous evaporation, which are evaporative cooling processes. Evaporation accounts for about 20% of the body heat loss for humans, as discussed on page 9 under human and animal comfort.

Evaporative cooling is used to reduce building air temperature in the U.S. in poultry houses and sometimes in swine and dairy housing. During evaporative cooling, the evaporation of water removes energy from the surrounding air, thus cooling it. Evaporative cooling works best in areas of low relative humidity. In a pad system air is blown through a wet porous pad. High pressure fogging systems that create a fine aerosol are preferred in broiler houses. Direct spray systems where nozzles produce mist or spray droplets that wet animals directly are used during hot weather in swine or dairy confinement facilities with solid concrete or slotted floors.

7.5.2 Evaporation Inside a Solid Matrix

Inside an unsaturated porous matrix, the rate of evaporation is related to temperature, amount of pore space available for evaporation, and permeability (ease with which vapor can escape). As liquid evaporates and more vapor forms, pressure rises. The extent to which it rises depends on how much pressure can be released (measured by permeability). This vapor generation and flow is part of all drying (soil, foods, biomaterials) and similar processes (baking and frying of food, for example). Intensive heating such as the use of microwaves tend to produce higher pressures. A common assumption is that liquid and vapor are at equilibrium in a pore (at any point). Thus, pressure and temperature are related as shown in Figure 7.16. In a typical drying process, moisture is evaporated internally and the vapor either diffuses or is blown to the surface and convected away. Additionally, any liquid on the surface can evaporate.

7.5.3 Evaporation of Solutions

Evaporation is a very common method of concentrating food and other biological solutions. It is an important operation in the processing of liquid food materials, to improve their storability or taste and to reduce bulk in transit. In this context evaporation is really boiling; i.e., it occurs at the boiling point of the solution. A product is concentrated by boiling off the solvent, generally water. Because products such as food materials are often heat-sensitive, i.e., because they degrade at higher temperatures, they are boiled at lower temperatures. A lower boiling point is achieved by lowering the pressure of the enclosure where boiling takes place. Figure 7.16 shows that boiling will take place at any given temperature, if the pressure over the water is equal to the vapor pressure at that temperature. Details of the evaporation process for concentrating food and biological solutions can be found in texts on food process engineering or in Minton (1987).

As mentioned earlier, the evaporation of water is an energy-intensive process, and therefore other means of concentrating food products such as freeze concentration are often used. In freeze concentration the product is partially frozen and the ice is separated out. Because the product is not heated, quality is not degraded as much as in evaporation.

In heating a solution the boiling point of the solution is higher than that of pure liquid. The reasons are the same as those giving rise to freezing point depression in the freezing of solutions, discussed in Section 7.3. Dissolved solutes elevate the boiling point above that of pure liquid. The elevated boiling point, T_b, of a solution is related to its concentration x_A by:

$$\frac{\Delta H_{vap}}{R_g}\left[\frac{1}{T_b} - \frac{1}{T_{b0}}\right] = \ln x_A \qquad (7.18)$$

where ΔH_{vap} is the latent heat of vaporization of pure liquid A in J/mol, T_{b0} is the boiling point of pure liquid A in K, R_g is the gas constant, and x_A is the mole fraction of A in solution. For a small increase in concentration x_A, the increase in T_b is linear.

7.6 Chapter Summary—Heat Transfer with Change of Phase

• Freezing of Water (page 126)

1. As water freezes, its density and specific heat decrease while its thermal conductivity increases. Microwave absorption of ice is much lower than that of water.

• Freezing of Solutions and Biomaterials (page 128)

1. As water freezes in a solution, the rest of the solution is more concentrated and its freezing point lowers. This decrease in freezing point is given by Eqn. 7.1.

2. Water in biomaterials has dissolved solutes. As pure water freezes, it concentrates the rest of the solution, continuously lowering its freezing point. This makes biomaterials freeze over a range of temperatures.

3. In a slow freezing process of a tissue, the solution outside a cell (extracellular fluid) freezes first and concentrates, leading to the transport of intracellular water to the outside.

4. Higher cooling rates minimize the transport of intracellular water and are often preferred for biomaterial freezing.

• Freezing Time Calculations (page 131)

1. Approximate formulas for the freezing time for a slab and cylinder are given by Eqn. 7.10 and problem 7.9.3, respectively.

2. A more accurate procedure for calculating the freezing time of biomaterials is provided in section 7.4.2

• Evaporation (page 136)

1. The spontaneous change of molecules from liquid to vapor state is termed evaporation.

2. Vapor pressure is the partial pressure of water vapor in equilibrium with the liquid, at any given temperature. Vapor pressure increases with temperature.

3. At any temperature, if the external or surrounding pressure is equal to the vapor pressure, the evaporation at that temperature is termed boiling.

4. The evaporation rate from a very wet surface is often limited by the rate at which vapor can be taken away. The rate of vapor removal is given by the mass transfer Equation 7.15.

7.7 Concept and Review Questions

1. Why do biomaterials freeze over a temperature range?

2. Would the freezing of pure water lead to a "mushy" zone?

3. Does Plank's formula to calculate freezing time consider this "mushy" zone?

4. Do biomaterials start to freeze at 0°C? Why?

5. When a tissue is frozen slowly, cells often shrink. Why?

6. Is faster freezing always preferable? Explain.

7. What are the ways to minimize cracking from faster freezing?

8. What would be the problems in cryopreservation of a whole human body?

9. In thawing a frozen food, the microwave oven is programmed to cycle on and off, instead of staying on continuously. Explain why.

10. A given solution can be concentrated by partially freezing it and removing the ice. However, the same solution can also be concentrated by boiling it. Which process would you choose, and why?

7.8 Further Reading

Freezing

Bischof, J., N. Merry, and J. Hulbert. 1997. Rectal protection during prostrate cryosurgery: Design and characterization of an insulating probe. Cryobiology, 34:80-92.

Cleland, A.C. and R.L. Earle. 1984. Assessment of freezing time prediction methods. Journal of Food Science, 49:1034-42.

Fahy, G.M., J. Saur, and R.J. Williams. 1990. Physical problems with the vitrification of large biological systems. Cryobiology, 27:492-510.

Fennema, O., W.D. Powrie, and E. Marth. 1973. *The Low Temperature Preservation of Foods and Living Matter*. Marcel Dekker, New York.

Laverty, J. 1991. Physio-chemical problems associated with fish freezing. In *Food Freezing: Today and Tomorrow*. W.B. Bald, ed. Springer-Verlag, London.

McGrath, J.J. 1985. Preservation of biological material by freezing and thawing. In *Heat Transfer in Medicine and Biology* (Vol. 1), edited by A. Shitzer and R.C. Eberhart. Plenum Press, New York.

Meryman, H.T. 1966. *Cryobiology*. Academic Press, New York.

NEMC. 2000. Cryosurgical Ablation of the Prostate (CSAP) for Patients with Prostate Carcinoma. WWW document at http://www.nemc.org/urology/cryo.htm accessed on June 29.

Shi, X., A.K. Datta, and Y.X. Mukherjee. 1998. Thermal stresses from large volumetric expansion during freezing of biomaterials. Journal of Biomechanical Engineering, Transactions of the ASME, 120:720-726.

Shi, X., A.K. Datta, and J. Throop. 1998. Mechanical property changes during freezing of a biomaterial. Transactions of the ASAE, 41(5):1407-1414.

Evaporation

Brutsaert, W. 1982. *Evaporation into the Atmosphere: Theory, History, and Applications*. D. Reidel Publishing Co., Dordrecht, Holland.

Dowdy, J.A., R.L. Reid, and E.T. Handy. 1986. Experimental determination of heat- and mass-transfer coefficients in aspen pads. ASHRAE Transactions, 92(2A):60-69.

Hagiwara, T. and R.W. Hartel. 1996. Effect of sweetener, stabilizer, and storage temperature on ice recrystallization in ice cream. J. Dairy Sci., 79:735-744.

Jones, F.E. 1991. *Evaporation of Water*. Lewis Publishers, Inc., Chelsea, Michigan.

Jury, W.A., W.R. Gardner, and W.H. Gardner. 1991. *Soil Physics*. John Wiley & Sons, New York.

Kung, S.K.J. and T.S. Steenhuis. 1986. Heat and moisture transfer in partly frozen nonheaving soil. Soil Science Soc. Am. J., 50(5):1114-1122.

Minton, P.E. 1987. *Handbook of Evaporation Technology*. Noyes Publications, Park Ridge, New Jersey.

Nobel, P.S. 1974. *Biophysical Plant Physiology*. W.H. Freeman and Company, San Francisco.

7.9 Problems

7.9.1 Freezing of a Solution

Consider freezing of grape juice. Assume that the initial weight percentage of water in grape juice is 85% and the effective molecular weight of solute is 183.61. 1) Predict the percent of initial water frozen in grape juice when the temperature has been reduced to $-6°C$. 2) Repeat the calculations in 1 when the temperature has been reduced to $-15°C$. 3) Would your answer change if we started from a more concentrated grape juice of, let's say, 50% water? Why?

7.9.2 Storing at Subzero Temperature Without Freezing

A large biological tissue has to be kept at a temperature as low as possible, but without freezing (ice formation). The water in the tissue has dissolved in it several materials, with an effective molecular weight of 50. The concentration of all dissolved materials combined is 10% of the weight of water. What is the lowest temperature that this tissue can be stored at, without ice formation? The latent heat of fusion of water is 335 kJ/kg and the molecular weight of water is 18.

7.9.3 Freezing of an Infinitely Long Cylinder

Obtain an expression for the time to freeze a long cylinder (length much greater than the radius) of pure material of radius R. The ambient freezer temperature is T_s (which is also the cylinder surface temperature) with the freezer volume being very large in comparison to the cylinder. Ignore any convection in the unfrozen material. The cylinder is at an initial temperature T_m, which is close to its freezing temperature, so that any sensible heat change may be ignored. Material has density ρ, latent heat ΔH_f and thermal conductivity k. Note $\int r \ln r \, dr = r^2(\ln r - 1/2)/2$

Ans: $t = \frac{\Delta H_f \rho}{k(T_m - T_s)} \frac{R^2}{4}$

7.9.4 Freezing of a Sphere

Obtain an expression for the time to freeze a sphere of pure material of radius R. The ambient freezer temperature is T_s (which is also the sphere surface temperature) with the freezer volume being very large in comparison to the sphere. Ignore any convection in the unfrozen material. The sphere is at an initial temperature T_m, which is close to its freezing temperature, so that any sensible heat change may be ignored. Material has density ρ, latent heat ΔH_f and thermal conductivity k.

7.9.5 Thawing of Blood

Thawing is the change of phase from solid to liquid. Consider two frozen rectangular packets of blood to be thawed, one having a thickness twice that of the other. How would the thawing time change? Assume the thickness is much smaller than the other two dimensions. Consider the blood to be at the freezing temperature so that only the change in latent heat is to be considered.

7.9.6 Freezing vs. Thawing Time

We want to compare the freezing time vs. thawing time for the same blood as in Problem 7.9.5. For the same difference between the cooling or heating medium temperature and the blood temperature, how does the freezing time compare with thawing time? Thermal conductivity of frozen blood is 1.5 W/m·K and that of unfrozen blood is 0.6 W/m·K.

7.9.7 Freezing of Hamburger Patty

Freezing of food materials for preservation is one of the most important applications of biomaterials freezing. Consider freezing of hamburger patties as shown in Figure 7.17. Cold air at −50°C blows at 0.5 m/s over both sides of the patty. The intial temperature of burgers is −5°C and we will consider all freezing takes place at this temperature. The meat contains 75% water. The thermal conductivity of the frozen meat is 1.4 W/m·K and the density of the meat is 900 kg/m³. The latent heat of fusion of pure water is 335 kJ/kg. 1) Calculate the heat transfer coefficient over the top surface. The properties of air at the relevant temperature is $\rho = 1.3947$ kg/m³, $\mu = 1.596 \times 10^{-5}$ Pa · s, $k = 2.23 \times 10^{-2}$ W/m·K and Pr = 0.72. 2) Calculate the time to freeze the patty (i.e., when the center plane just freezes), when the top and the bottom surface have the same value of heat transfer coefficient.

Figure 7.17: Hamburger patty with dimension.

7.9.8 Inadvertent Thawing

During transportation and storage of food, unexpected and undesired variation in the surrounding temperature can take place that can pose health hazards as bacteria can grow if the temperature rose or lowered to the range for optimum bacterial growth. Consider one of the frozen hamburger patty (Problem 7.9.7) left on countertop by mistake. *Assume no heat transfer from the bottom of the patty.* Neglecting the sensible heat change, how long would the hamburger take before it completely thaws from an initial frozen state with uniform temperature of −5°C? The room air temperature is at 30°C and the heat transfer coefficient at the top of the patty from natural convection is 10 W/m²·K. The unfrozen patty has 75% water and a thermal conductivity of 0.4 W/m·K, density of 1040 kg/m³, and a freezing temperature of −5°C. The latent heat of fusion of pure water is 335 kJ/kg. The dimensions of the patty are as in Figure 7.17.

7.9.9 Freezing of Meat (Use of Characteristic Length)

Consider two pieces of sausage being frozen in a cylindrical form in an air blast freezer. The diameter for each cylinder is 7.5 cm. The height of one cylinder is 10 cm, and that of the other is 5 cm. The meat has already been cooled to a uniform initial temperature of −4°C, which is the initial freezing temperature. The air temperature in the freezer is −30°C, and the blowing leads to a high heat transfer coefficient of 200 W/m². The meat has 70% moisture, and when frozen, its physical properties are known to be a density of 950 kg/m³ and a thermal conductivity of 1.4 W/m·K. The latent heat of fusion of pure water is 335 kJ/kg. Calculate the freezing time for the two cylinders. Be careful to select the right formula based on characteristic dimension.

7.9.10 Depth of Freezing of a Lake

1) Determine the depth to which a large stagnant lake surface will be frozen during the winter that lasts for 60 days. The average air temperature for this period is $-10°C$ with an average heat transfer coefficient of 20 W/m²·K over the lake surface. Neglect any sensible heat effect by assuming that the lake temperature is 0°C during the season. 2) Calculate if it would take longer or shorter for the thawing of the frozen layer you just calculated in the springtime. Assume all the frozen layer is at 0°C when thawing begins. Air temperature for spring is 10°C and it has the same surface heat transfer coefficient value. 3) Explain the reasons for thawing time being longer or shorter. The latent heat for freezing of water is 335 kJ/kg, density of water is 1000 kg/m³, thermal conductivity of water is 0.61 W/m·K and thermal conductivity of ice is 2.3 W/m·K.

7.9.11 Depth of Frost Penetration in a Soil

In regions where surface temperatures fall below the freezing point of water, freezing of the pore water in the soil is possible. A cold front which freezes the pore water in the subsoil is generally defined as the frost front or the depth of frost penetration. Consider a uniform homogeneous silty soil, with a water content equal to 20% and a density of 1750 kg/m³. Assume the soil water freezes at 0°C. The thermal conductivity of the frozen soil is 1.4 W/m·K. Calculate the depth of frost penetration for a winter lasting 5 months, if the temperature at the soil surface remains constant at $-13°C$. Latent heat of water is 335 kJ/kg.

Chapter 8

RADIATIVE ENERGY TRANSFER

CHAPTER OBJECTIVES

After you have studied this chapter, you should be able to:

1. Explain the characteristics of electromagnetic radiation as it is incident on a surface or emitted by a surface.

2. Calculate the magnitude and spectral quality of electromagnetic radiation emitted from a material.

3. Explain the energy emitted by biological and environmental entities.

4. Calculate the radiative energy exchange between two bodies.

KEY TERMS

- **electromagnetic spectrum**
- **reflection, absorption, and transmission of waves**
- **blackbody**
- **emission and emissivity**
- **solar radiation**
- **solar constant**
- **greenhouse effect**
- **radiative exchange**
- **view factor**

There are only two fundamental mechanisms of energy transfer—diffusion and electromagnetic radiation (convection refers to the addition of bulk flow to these modes). This chapter deals with transport of energy by the mechanism of electromagnetic radiation. No medium is necessary in transporting energy using electromagnetic radiation, in contrast with diffusion and convection which, by definition, requires a medium to be present, i.e., cannot occur in vacuum. Although radiative energy transfer often occurs in conjunction with diffusion and convection, we will look at simple radiative energy transfer situations (see Figure 8.1) and exclude situations such as radiation combined with convection.

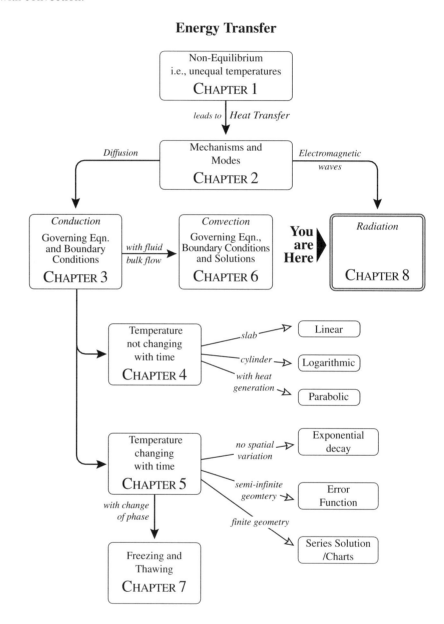

Figure 8.1: Concept map of energy transfer showing how the contents of this chapter relates to other chapters in the part on energy transfer.

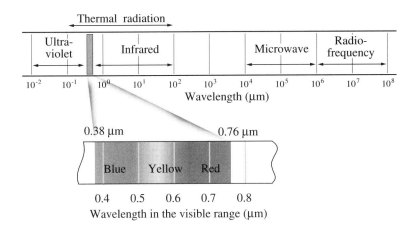

Figure 8.2: The electromagnetic spectrum.

8.1 The Electromagnetic Spectrum

Unlike conductive or convective heat transfer, no medium is required for the radiative exchange between bodies since the exchange is through electromagnetic waves. The magnitude of the radiative exchange depends significantly on temperature level and not just temperature difference. The quality of thermal radiation also depends on the temperature level.

Electromagnetic radiation travels at the speed of light, c, of 3×10^8 m/s. The wavelength and frequency of electromagnetic radiation are related by:

$$\begin{pmatrix} \lambda \\ \text{Wavelength} \end{pmatrix} \begin{pmatrix} \nu \\ \text{Frequency} \end{pmatrix} = \begin{pmatrix} c \\ \text{Speed of Light} \end{pmatrix} \tag{8.1}$$

In the electromagnetic spectrum (Figure 8.2), thermal radiation generally refers to the range 0.1-100 microns (1 micron = 10^{-6} m) and includes some ultraviolet and all infrared radiation, as well as the entire visible spectrum of 0.38-0.76 microns. It is interesting to note that in the visible spectrum, the sensitivity of human eye retina varies with wavelength, as shown in Figure 8.3. Microwaves typically refer to wavelength (in air) of 1-100 cm and radio-frequency waves are 1-100 m. Microwaves and radio frequencies are efficiently absorbed by many materials, including biomaterials. This will be discussed later.

8.2 Reflection, Absorption and Transmission of Waves at a Surface

Being part of the electromagnetic spectrum, infrared, microwave, and radio-frequency waves incident on a surface are reflected, absorbed and transmitted similar to light.

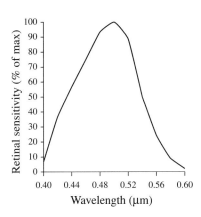

Figure 8.3: Sensitivity of the human eye to various wavelengths in the visible region.

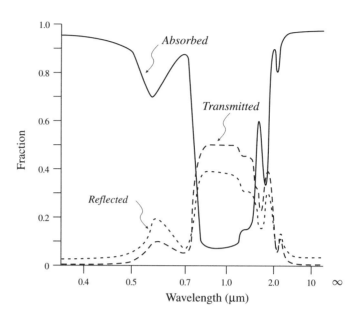

Figure 8.5: Representative spectral values for the reflectivity, absorptivity, and transmissivity of a leaf.

Figure 8.4 shows reflection, absorption, and transmission of electromagnetic waves at a surface. The reflection can be specular or diffuse. For a specular reflection, the angle of incidence is equal to the angle of reflection. For diffuse reflection, which is the case for most materials, waves are reflected in all directions.

Thus, the following relation would apply at the surface:

$$\left(\underset{\text{Reflectivity}}{\rho}\right) + \left(\underset{\text{Absorptivity}}{\alpha}\right) + \left(\underset{\text{Transmissivity}}{\tau}\right) = 1 \qquad (8.2)$$

All three properties reflectivity, absorptivity, and transmissivity are functions of wavelength. In addition, reflectivity and absorptivity are directional, i.e., they vary with the direction of the incident wave. In this text, only directional averages of reflectivity and absorptivity will be used.

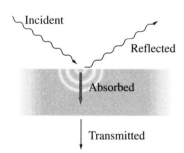

Figure 8.4: Reflection, absorption, and transmission of electromagnetic wave at a surface.

8.2.1 Transmissivity of a Leaf and Photosynthesis

Figure 8.5 shows representative values for the reflectivity, absorptivity, and transmissivity of a leaf. This has implications in plant growth. For example, a shaded leaf in a canopy will get only solar radiation between 0.7 and 2 μm since that is where the leaves have high transmissivity. Solar radiation in this region is not very useful for photosynthesis. As illustrated in Figure 8.6, green chlorophyl (responsible for photosynthesis) absorbs strongly in the red (0.660 μm), blue (0.435 μm) and some of the

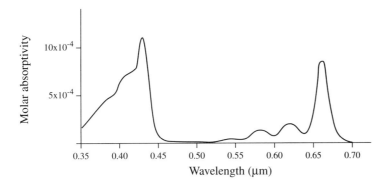

Figure 8.6: Visible spectra of green chlorophyl.

lower wavelength portions of the visible region. Moreover, the stomatal opening in a leaf (see Example 12.2.1 on page 247) is partially closed at the lower illumination levels, which increases the resistance to CO_2 exchange by the leaf and further decreases photosynthesis.

8.2.2 Transmissivity of the Atmosphere: Greenhouse Effect

The atmosphere traps heat. As seen from space, the earth radiates energy at wavelengths and intensities characteristic of a body at $-18°C$. Yet the average temperature at the surface is about $33°C$ higher. Heat is trapped between the surface and the level, high in the atmosphere, from which radiation escapes. The ability of the earth's atmosphere to trap the heat from the sun is known as the greenhouse effect.

The greenhouse effect is a consequence of the spectral dependence of the transmissivity of the atmosphere. Some of the gases in the atmosphere, CO_2, methane, and chlorofluorocarbons are relatively transparent to sunshine (smaller wavelengths) but opaque to the longer-wavelength infrared radiation released by the earth. Thus, the radiation from earth is absorbed by the atmosphere, increasing its temperature. Gases such as CO_2, methane and chlorofluorocarbons are called greenhouse gases. The relationship between CO_2 concentration and the rise in temperature of the earth was discussed in Chapter 1.

8.2.3 Albedo–Reflection from Soil

The overall fraction of incident radiation that is reflected from celestial surfaces such as water, soil, etc., is termed as albedo. Albedo is dependent on, among other factors, position of the sun and the surface composition and roughness. An example of variations in albedo for different surfaces can be seen in Figure 8.7.

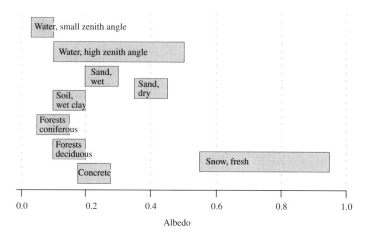

Figure 8.7: Albedo of various surfaces, data from Guyot (1998).

8.2.4 Absorption and Transmission in Biomaterials

As the waves are transmitted in a material, they are also absorbed. The decay in the intensity due to the absorption is often characterized by the simple Beer-Lambert law given by:

$$F = F_0 e^{-x/\delta} \tag{8.3}$$

where F_0 is the flux in W/m^2 entering a material at its surface (incident flux minus the reflected flux) and F is the flux at a location inside the material at a distance x from the surface, as illustrated in Figure 8.8. The penetration depth, δ, defined by Eqn. 8.3, is the distance at which the flux becomes $1/e$ or approximately 37% of its value at the surface, since using $x = \delta$ in Eqn. 8.3 we get

$$F = F_0 e^{-\delta/\delta} = \frac{F_0}{e} \tag{8.4}$$

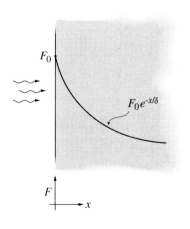

Figure 8.8: Illustration of the decay in the intensity of a wave as it is transmitted through a material, as given by Eqn. 8.3.

A smaller penetration depth means a quicker decay of the flux or good absorptivity and therefore poor transmissivity. Table 8.1 compares the penetration depth for infrared, microwave, and radio-frequency which are the three types of waves normally used in heating applications such as drying, baking, tempering, sterilizing, etc. The low penetration depth in the infrared region signifies that most solids are opaque to infrared radiation. Infrared provides primarily surface heating whereas radio-frequency can provide uniform heating to a much greater depth.

8.3 Thermal Radiation from an Ideal (Black) Body

To describe radiation emitted from a surface, several terms are first defined. Total emissive power of a surface, E [W/m^2], is defined as the total rate of thermal energy

8.3. THERMAL RADIATION FROM AN IDEAL (BLACK) BODY

Table 8.1: Heating characteristics of various types of electromagnetic waves for water or biomaterials.

Type of electromagnetic wave	Frequency (typical) in MHz	Wavelength in air (cm)	Penetration depth, δ in biomaterials
Infrared	70,000,000	0.0004	few microns
Microwave	2,450	12	few centimeters
Radio-frequency	27	1100	hundred centimeters

emitted via radiation in all directions and at all wavelengths per unit surface area. The emissive power of an ideal (black) body, E_b, is the largest possible E for a given temperature.

The monochromatic emissive power E_λ is defined similarly to total emissive power but restricted to one wavelength λ. Thus, the monochromatic emissive power is the energy emitted between wavelengths λ and $\lambda + d\lambda$, such that:

$$dE = E_\lambda d\lambda$$

Energy emitted between two wavelengths λ_1 and λ_2 is given by:

$$\int_{\lambda_1}^{\lambda_2} E_\lambda d\lambda$$

Total energy emitted by a body can be written as:

$$E = \int_0^\infty E_\lambda d\lambda$$

Fraction of total energy emitted between two wavelengths is:

$$F = \frac{\int_{\lambda_1}^{\lambda_2} E_\lambda d\lambda}{\int_0^\infty E_\lambda d\lambda} \qquad (8.5)$$

To describe the radiation characteristics of real surfaces, the concept of radiation from an ideal surface is introduced. This ideal surface neither reflects nor transmits any radiation, i.e., it absorbs all radiation incident upon it. Since no light is considered to be reflected, this body will appear black and hence it is referred to as a blackbody. A blackbody is only an idealization and it is a standard against which radiation from real surfaces are compared. An example of a blackbody can be a cavity with a small hole. Radiation entering this cavity through the small hole is continuously reflected inside

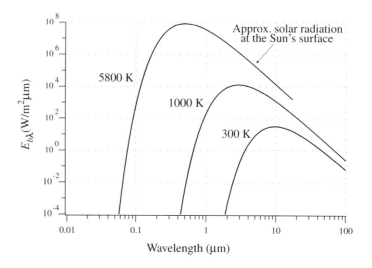

Figure 8.9: Spectral blackbody emissive power.

the cavity and is almost entirely absorbed before it can come out. This approximates a blackbody behavior.

Radiation emitted by a blackbody has a spectral distribution. The amount of radiation emitted by a blackbody at temperature T between two wavelengths λ and $\lambda + d\lambda$ (monochromatic emissive power $E_{b,\lambda}$) is given by Planck's Law of radiation:

$$E_{b,\lambda} = \frac{2\pi c^2 h \lambda^{-5}}{\exp(ch/\kappa\lambda T) - 1} \tag{8.6}$$

where c is the speed of light, h is the Planck's constant and has the value 6.625×10^{-34} J · s, κ is the Boltzmann's constant and has the value 1.380×10^{-23} J/K, and T is the absolute temperature in K. Equation 8.6 is plotted in Figure 8.9 for different temperatures. Note that the vertical axis in the figure is power flux per unit wavelength.

Important features of Planck's law can be seen from Figure 8.9 and Equation 8.6 as:

1. The emitted radiation varies continuously with wavelength.

2. At any wavelength the magnitude of the radiation increases with increasing temperature.

3. The spectral region in which the radiation is concentrated depends on temperature, with comparatively more radiation appearing at shorter wavelengths as the temperature increases. The relationship between the peak of emission $\lambda_{\max}$ and temperature is given by:

$$\lambda_{\max} T = 2897.6 \, \mu\text{mK} \tag{8.7}$$

As a consequence of Equation 8.7, a significant fraction of the radiation emitted by the sun, which may be approximated as a blackbody at 5800 K, is in the visible region of the spectrum. In contrast, for $T \leq 800$ K, emission is predominantly in the infrared region of the spectrum and not visible to the eye.

Using Equation 8.6, the total energy emitted between two wavelengths λ and $\lambda + d\lambda$ is given by $E_{b,\lambda} d\lambda$. To obtain the total energy emitted at all wavelengths, this emitted energy is to be integrated over all wavelengths to yield the Stefan-Boltzmann law:

$$E_b = \int_0^\infty E_{b,\lambda} d\lambda = \frac{2\pi^5 \kappa^4 T^4}{15 c^2 h^3} = \sigma T^4 \tag{8.8}$$

where $\sigma = 5.670 \times 10^{-8}$ W/m$^2 \cdot$K^4 is the Stefan-Boltzmann constant.

8.4 Fraction of Energy Emitted by an Ideal Body in a Given Range of Wavelengths

It is often quite useful to estimate the amount of energy emitted by a body between any two wavelengths or in a certain band. For a black or an ideal body at temperature T, the energy emitted between two wavelengths λ_1 and λ_2 is given by

$$\int_{\lambda_1}^{\lambda_2} E_{b,\lambda} d\lambda$$

as illustrated in Figure 8.10. This emitted energy can be written as a fraction of the total energy emitted by the same blackbody to be:

$$\frac{\int_{\lambda_1}^{\lambda_2} E_{b,\lambda} d\lambda}{\int_0^\infty E_{b,\lambda} d\lambda} \tag{8.9}$$

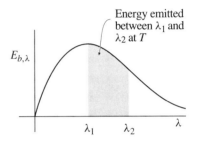

Figure 8.10: Illustration of energy emitted by a blackbody between two wavelengths.

Note that the denominator in Eqn. 8.9 representing the total energy emitted is given by Eqn. 8.8. Substituting this value in the denominator and separating the integral in the numerator,

$$= \frac{\int_0^{\lambda_2} E_{b,\lambda} d\lambda - \int_0^{\lambda_1} E_{b,\lambda} d\lambda}{\sigma T^4}$$

$$= \int_0^{\lambda_2} \frac{E_{b,\lambda}}{\sigma T^4} d\lambda - \int_0^{\lambda_1} \frac{E_{b,\lambda}}{\sigma T^4} d\lambda$$

$$= \int_0^{\lambda_2 T} \frac{E_{b,\lambda}}{\sigma T^5} d(\lambda T) - \int_0^{\lambda_1 T} \frac{E_{b,\lambda}}{\sigma T^5} d(\lambda T) \tag{8.10}$$

Substituting Eqn. 8.6 for the integrand in the two integrals above, we get

$$\frac{E_{b,\lambda}}{\sigma T^5} = \frac{2\pi c^2 h (\lambda T)^{-5}}{\exp(ch/\kappa \lambda T) - 1} \tag{8.11}$$

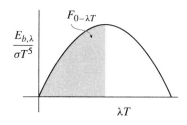

Figure 8.11: Graphical illustration of $F_{0-\lambda T}$.

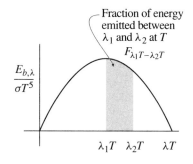

Figure 8.12: Graphical illustration of $F_{0-\lambda_2 T} - F_{0-\lambda_1 T}$.

which shows that the integrand is exclusively a function of λT (the rest are constants). If the integrals are denoted by F such that

$$F_{0-\lambda T} = \int_0^{\lambda T} \frac{E_{b,\lambda}}{\sigma T^5} d(\lambda T) \quad (8.12)$$

The $F_{0-\lambda T}$ values, representing the fraction of total energy emitted by a blackbody at temperature T between wavelengths 0 and λ is illustrated in Figure 8.11. Then the fraction of energy between the two wavelengths λ_1 and λ_2 is given by

$$\begin{array}{c}\text{Fraction of energy between}\\ \lambda_1 \text{ and } \lambda_2 \text{ at temp } T\end{array} = F_{0-\lambda_2 T} - F_{0-\lambda_1 T} \quad (8.13)$$

as illustrated in Figure 8.12. The $F_{0-\lambda T}$ values calculated using Eqn. 8.12 are in Table 8.2.

Example 8.4.1 What Fraction of Solar Energy Is in the Visible Region (0.38-0.76 μm)

As mentioned earlier, solar energy can be assumed to radiate from a blackbody at surface temperature of 5800 K. Thus, fraction of energy between 0.38 and 0.76 μm at 5800 K can be calculated using Eqn. 8.13 as:

$$\begin{aligned}\begin{array}{c}\text{Fraction of energy between}\\ 0.38 \text{ and } 0.76\ \mu\text{m at 5800 K}\end{array} &= F_{0-(0.76)(5800)} - F_{0-(0.38)(5800)}\\ &= F_{0-4408} - F_{0-2204}\\ &= 0.5491 - 0.1013 \quad \text{(from Table 8.2)}\\ &= 0.4478\end{aligned}$$

Thus, 44.78% of the Sun's energy is in the visible region. Note that this value does not include the effect of atmospheric absorption and therefore is true for solar radiation just outside the atmosphere.

8.5 Thermal Radiation from a Real Body: Emissivity

The emissive powers of all real bodies are less than that for the ideal body and depend on surface condition, among other factors. *Total emissivity* ϵ compares the emissive power E of an actual body to the emissive power of the ideal (black) body:

$$\epsilon = \frac{E}{E_b} \quad \text{where } 0 \leq \epsilon \leq 1 \quad (8.14)$$

Figure 8.13 shows the total emissivity values for a number of surfaces. Emissivities of highly polished metallic surfaces are low, those of non-metallic surfaces are higher, and those of biomaterials are in the range 0.9-0.95. For example, low emissivity value of polished aluminum is exploited on the outer surface of space suits to reduce radiative

8.5. THERMAL RADIATION FROM A REAL BODY: EMISSIVITY

Table 8.2: Blackbody radiation functions.

λT [μmK]	$F_{0-\lambda T}$	λT [μmK]	$F_{0-\lambda T}$	λT [μmK]	$F_{0-\lambda T}$
0	0	10200	0.91823	20400	0.986375
200	7.01016e-27	10200	0.91823	20600	0.986733
400	2.20179e-12	10400	0.92197	20800	0.98707
600	9.96926e-08	10600	0.925491	21000	0.987413
800	1.70667e-05	10800	0.928807	21200	0.987735
1000	0.000328258	11000	0.931932	21400	0.988047
1200	0.00216717	11200	0.9344881	21600	0.988349
1400	0.00787475	11400	0.937665	21800	0.98864
1600	0.0198754	11800	0.942781	22000	0.988922
1800	0.0395777	12000	0.945133	22200	0.989195
2000	0.0670396	12200	0.94736	22400	0.989459
2200	0.10126	12400	0.94947	22600	0.989715
2400	0.140672	12600	0.95147	22800	0.989962
2600	0.183564	12800	0.953368	23000	0.990202
2800	0.228346	13000	0.955169	23200	0.990434
3000	0.273688	13200	0.95688	23400	0.990659
3200	0.318549	13400	0.958506	23600	0.990877
3400	0.362168	13600	0.960052	23800	0.991088
3600	0.404021	13800	0.961523	24000	0.991293
3800	0.443775	14000	0.962924	24200	0.991492
4000	0.481246	14200	0.964258	24400	0.991685
4200	0.51636	14400	0.96553	24600	0.991872
4400	0.549119	14600	0.966743	24800	0.992053
4600	0.579582	14800	0.9679	25000	0.99223
4800	0.607839	15000	0.969004	25200	0.992401
5000	0.634007	15200	0.97006	25400	0.992567
5200	0.658212	15400	0.971068	25600	0.992728
5400	0.680584	15600	0.972032	25800	0.992885
5600	0.701254	15800	0.972954	26000	0.993038
5800	0.720351	16000	0.973836	26200	0.993186
6000	0.737997	16200	0.974681	26400	0.99333
6200	0.754307	16400	0.97549	26600	0.99347
6400	0.769389	16600	0.976265	26800	0.993606
6600	0.783343	16800	0.977008	27000	0.993738
6800	0.796263	17000	0.977721	27200	0.993867
7000	0.808234	17200	0.978405	27400	0.993993
7200	0.819335	17400	0.979061	27600	0.994115
7400	0.829637	17600	0.979691	27800	0.994234
7600	0.839206	17800	0.980297	28000	0.994349
7800	0.848101	18000	0.980879	28200	0.994462
8000	0.856379	18200	0.981438	28400	0.994572
8200	0.864089	18400	0.981976	28600	0.994678
8400	0.871276	18600	0.982494	28800	0.994783
8600	0.877981	18800	0.982992	29000	0.994884
8800	0.884244	19000	0.983471	29200	0.994983
9000	0.890098	19200	0.983933	29400	0.995079
9200	0.895575	19400	0.984378	29600	0.995173
9400	0.900703	19600	0.984807	29800	0.995265
9600	0.90551	19800	0.98522	30000	0.995354
9800	0.910019	20000	0.985619		
10000	0.914252	20200	0.986004		

heat loss. Emissivity typically varies with wavelength. The *monochromatic emissivity* is defined as the ratio of the monochromatic emissive power of a body to the monochromatic emissive power of a blackbody at the same wavelength and temperature. Thus

$$\epsilon_\lambda = \frac{E_\lambda}{E_{b,\lambda}} \quad (8.15)$$

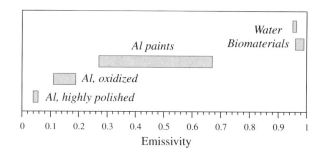

Figure 8.13: Emissivity of biomaterials as compared to other materials such as Aluminum and water.

For example, Figure 8.14 shows the typical spectral dependence of human skin monochromatic emissivity. An average value of 0.98-0.99 for human skin monochromatic emissivity is often used. When the monochromatic emissivity is independent of the wavelength, the body is described as a *gray body*.

Emissivities of soil surfaces range from 0.9 for dry quartz sand to 0.98 or higher depending on organic matter content, mineral composition, and moisture content. Between a wet and dry soil it can vary by almost 0.04. Differences in soil emissivity are not likely to be significant in the energy balance of bare soils (having a negligible effect on surface temperature). The emissivity differences are important for thermal infrared imagery.

Note that the emission discussed here is passive emission, i.e., it happens automatically at a rate dependent on the temperature level. This is to be distinguished from emission studied in spectroscopy. In spectroscopy, an external radiation field is introduced which is reflected, absorbed, or transmitted. The absorption causes molecules to make a transition to an energetically excited state. The transition from the excited state back to the lower state is accompanied by emission of radiation. This emission has information about the molecules, to obtain which is the purpose of emission spectroscopy.

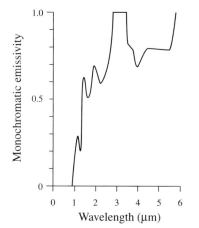

Figure 8.14: The spectral emissivity of human skin at room temperature. The corresponding total emittance is 0.993. Most authors use values between 0.98 and 0.99. Adapted from R. Elam, D. W. Goodwin, and K. L. Williams. 1963. Optical properties of human epidermis. Nature. 198(4884):1001-1002. Reproduced with permission.

8.6 Emission from Human Bodies—Infrared and Microwave Thermography for Clinical Uses

As mentioned earlier, all bodies above a temperature of absolute zero emit radiation naturally over a range of wavelengths. Thus, a human body would also emit radiation. At a temperature of approximately 300 K, which is near the comfortable human skin surface temperature of 306 K(92°F), the spectral curve of a blackbody radiation is shown in Figure 8.15. As shown in this figure, human radiation will cover infrared as well as microwave regions.

Several medical problems, including cancer and vascular occlusions, are accompanied by local changes in temperature of the order of 1°C. Infrared and microwave ther-

8.6. EMISSION FROM HUMAN BODIES

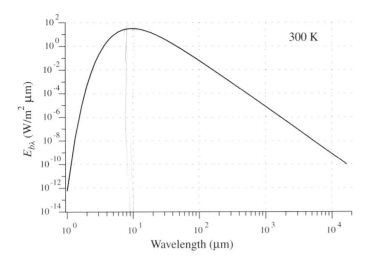

Figure 8.15: Radiation emitted from a blackbody at 300 K showing the approximate radiation emitted from humans.

mography are passive, non-invasive techniques that can detect and diagnose pathologic conditions in which there are disease related temperature differentials (Carr, 1989). These radiometric techniques, for the most part, have been directed at the early detection and diagnosis of breast cancer. Early detection can dramatically increase the survival rate. Present detection techniques other than radiometry require that the tumor have mass and contrast with respect to the surrounding tissue (i.e., physical examination, mammography, ultrasonography, etc.). Average tumor diameter when first detected and diagnosed using these techniques is still quite large and radiometric techniques that depend on thermal activity rather than mass could provide still earlier detection.

At the human body temperature of approximately 300 K, maximum intensity of radiative emission (according to Equation 8.7) occurs at a wavelength of approximately 10 μm. This has led to the operation of infrared thermographs at wavelengths near 10 μm.

At a typical microwave frequency of 3 GHz (approximately 12 cm or $1.2 \times 10^5 \mu$m wavelength), the intensity of radiative emission is reduced considerably. However, microwave radiometers, developed primarily for radio astronomy, can detect radiation with this intensity. These radiometers can detect intensity changes corresponding to emitter temperature changes of less than 1°C. At 3 GHz, microwaves penetrate approximately 5 cm in fat and 0.8 cm in muscle or skin. Thus, it would be possible to get tissue information from a depth of several centimeters. Of course, there are many complications and microwave thermography is still at the research stage (Carr, 1989).

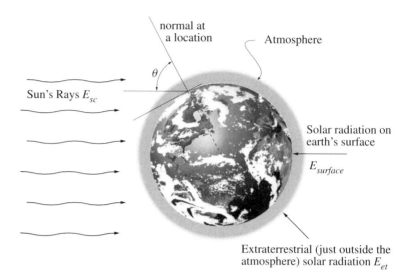

Figure 8.16: Relation of solar radiation on earth surface to the extraterrestrial solar radiation and the solar constant.

8.7 Solar, Atmospheric, and Earth Surface Radiation

8.7.1 Solar Radiation–Magnitude and Spectral Distribution

Outside the atmosphere, the solar radiation reaches an approximate intensity E_{sc} of 1353 W/m². An approximation to the solar constant can be made by considering radiative exchange between two spheres sun and earth at their distance (see Example 8.12.4). Depending on the location on earth (Figure 8.16), the intensity of radiation just outside the earth's atmosphere (extraterrestrial radiation) can be calculated from:

$$E_{et} = E_{sc} \cdot \beta \cdot cos\theta$$

where E_{et} is the extraterrestrial solar irradiation in W/m², E_{sc} is the solar constant of 1353W/m², β is a correction factor to account for the eccentricity of the earth's orbit about the sun with $0.97 \leq \beta \leq 1.03$ and θ is the zenith angle, shown in Figure 8.16. The spectral distribution of solar radiation is compared with that emitted from a blackbody at 5800 K (Figure 8.17). Just outside the atmosphere this solar radiation is incident with an intensity as shown in Figure 8.17. The significant discrepancies which occur in the ultraviolet region (around 0.5 μm) are mainly due to electronic transitions which occur in the overlying gases of the sun. The various atmospheric processes operate to change the spectral distribution as the radiation traverses to the earth surface through the atmosphere to approximately that shown in Figure 8.17. The main absorption is by the water vapor, which is responsible for the several strong bands in the infrared region (most of Figure 8.17), and by high-altitude atmospheric ozone,

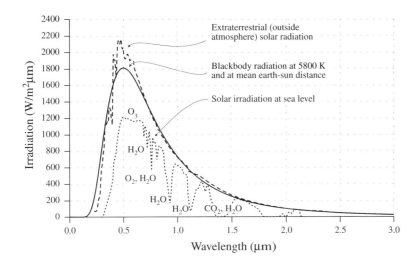

Figure 8.17: Spectral distribution of solar radiation outside the earth's atmosphere and on the earth surface. The absorption of the radiation over various bands of wavelength occurs due to the atmospheric gases, ozone, oxygen, water vapor and carbon dioxide, as noted. Data from Gast (1965).

which effectively limits radiation which reaches the ground in appreciable quantities to wavelengths greater than about 0.30 μm.

8.7.2 Emission from Earth's Surface

The radiation emitted by the earth surface is important since the earth has to lose (and thus balance) the energy it receives from the sun. The energy emitted by the earth is given by:

$$E_{earth} = \epsilon_{earth} \sigma T_{earth}^4 \qquad (8.16)$$

The average temperature of the earth's surface is considered to be 14°C (57°F). The emissivity of the earth surface will be made up of that of water, vegetation, soil, etc. As shown earlier, these emissivities are close to unity.

8.7.3 Atmospheric Emissions

The atmospheric gases such as H_2O and CO_2 emit considerable radiation. Effective concentration of these atmospheric gases can vary from time to time. For example, clouds contain much water in the form of vapor, droplets, and/or crystals. Effective sky temperature can be substantially lower for a cold, clear sky than for a warm, cloudy sky. Atmospheric emissions are the only source of radiation at night. The radiation from these gases is estimated in practice (when measured data is not available) using an apparent sky emissivity as

$$G_{atm} = \epsilon_{sky} \sigma T_a^4 \tag{8.17}$$

where T_a is the air temperature near the ground. The apparent sky emissivity, ϵ_{sky}, is an empirical constant that can be estimated from equations such as (Brunt, 1940):

$$\epsilon_{sky} = 0.55 + 1.8 \left(\frac{p_{H_2O}}{P}\right)^{1/2} \tag{8.18}$$

for a clear sky. Here p_{H_2O} is the partial pressure of water vapor in the atmosphere and P is the total atmospheric pressure.

8.7.4 Global Energy Balance

The approximate energy contributions from various sources (e.g., solar radiation, emission from earth, atmospheric emissions, etc., discussed earlier) averaged over the entire earth's surface for annual mean conditions is shown in Figure 8.18. The solar radiation value, for example, is obtained by dividing the total radiation incident on earth by the total surface area of the earth.

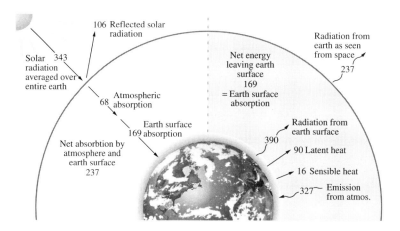

Figure 8.18: The global energy balance for annual mean conditions. Data from Ramanathan (1987).

8.8 Radiative Exchange Between Bodies

When two bodies exchange thermal radiation, the net radiative exchange between them will depend on their size, shape, and the relative orientation of their radiating surfaces.

8.8. RADIATIVE EXCHANGE BETWEEN BODIES

The size, shape and orientation factors are lumped in a parameter called the *configuration factor* or *view factor*, denoted by F_{1-2}, and stands for

$$F_{1-2} = \text{fraction of radiation leaving surface 1 that is intercepted by surface 2} \quad (8.19)$$

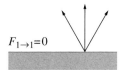
Plane surface

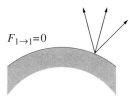
Convex surface

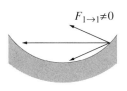
Concave surface

The view factor, F_{1-2}, is dimensionless. Since radiation travels along a straight line in the absence of strong electromagnetic fields, the view factor from a surface to itself will be zero unless the surface "sees" itself. Thus, $F_{1-1} = 0$ for plane or convex surfaces and $F_{1-1} \neq 0$ for concave surfaces, as illustrated in Figure 8.19. The value of the view factor ranges between zero and one. The limiting case of $F_{1-2} = 0$ represents the situation when two surfaces do not have a direct view of each other, and thus radiation leaving surface 1 cannot reach surface 2 directly. The other limiting case of $F_{1-2} = 1$ represents the situation where surface 2 completely surrounds surface 1 so that all radiation leaving surface 1 reaches surface 2.

View factor has been calculated and is available (see, for example, Siegel and Howell, 1981) for a large number of physical situations involving different combinations of size, shape, and orientations. As an illustration, the view factor between two aligned parallel rectangles of same size can be derived as

$$F_{1-2} = \frac{2}{\pi \bar{X} \bar{Y}} \left\{ \ln \left[\frac{(1+\bar{X}^2)(1+\bar{Y}^2)}{1+\bar{X}^2+\bar{Y}^2} \right]^{1/2} \right.$$
$$+ \bar{X}(1+\bar{Y}^2)^{1/2} \tan^{-1} \frac{\bar{X}}{(1+\bar{Y}^2)^{1/2}} + \bar{Y}(1+\bar{X}^2)^{1/2} \tan^{-1} \frac{\bar{Y}}{(1+\bar{X}^2)^{1/2}}$$
$$\left. - \bar{X} \tan^{-1} \bar{X} - \bar{Y} \tan^{-1} \bar{Y} \right\} \quad (8.20)$$

Figure 8.19: Illustration of view factors showing view factor is zero for a surface to itself for a plane or a convex surface, but non-zero for a concave surface.

where X and Y are the sides of the rectangles, D is the distance between them and $\bar{X} = X/D$ and $\bar{Y} = Y/D$. Equation 8.20 is plotted in Figure 8.20 and shows how the view factor F_{1-2} increases as either the size of the rectangles are increased or the distance between the rectangles is decreased so that a greater fraction of radiation leaving surface 1 is intercepted by surface 2. Figure 8.20 can be used, for example, to find the view factor between the two parallel surfaces of a roof and a floor for a structure.

The simplest situation of radiative exchange occurs when two blackbodies are exchanging radiation. We recall that blackbodies absorb all incident radiation and therefore do not reflect any. Energy is only emitted from a blackbody. The net radiative exchange between two such bodies is given by

$$q_{1-2} = \sigma A_1 F_{1-2} \left(T_1^4 - T_2^4 \right) \quad (8.21)$$

or

$$q_{1-2} = \sigma A_2 F_{2-1} \left(T_2^4 - T_1^4 \right) \quad (8.22)$$

where q_{1-2} is the net energy transfer between bodies 1 and 2, in W, A_1 and A_2 are the areas of the bodies in m^2, F_{1-2} is the view factor between bodies 1 and 2, while F_{2-1}

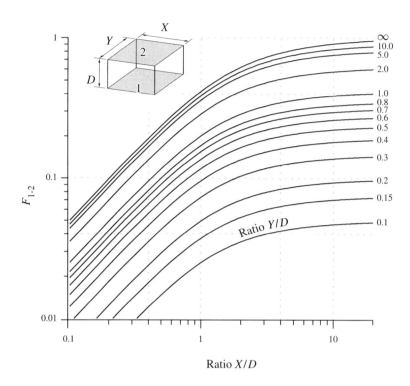

Figure 8.20: View factor between two aligned parallel rectangles of equal size.

is the view factor between bodies 2 and 1, and T_1 and T_2 are the absolute temperature of the respective surfaces, in K. Note that any one of the Eqns. 8.21 or 8.22 can be chosen depending on whether F_{1-2} or F_{2-1} is easier to calculate.

If we have an enclosure made up of several black surfaces all at different temperatures, the net radiative heat transfer between a surface i and all other surfaces is given by

$$q_i = \sum_{k=1, k \neq i}^{n} \sigma A_i F_{i-k} \left(T_i^4 - T_k^4 \right) \tag{8.23}$$

where n is the total number of surfaces.

Instead of blackbodies, when two *gray* bodies are exchanging radiation, the situation becomes more complex since these bodies are also reflecting part of the radiation. For the special case when two gray surfaces form a complete enclosure or close to it, as illustrated in Figure 8.21, the net radiative exchange is given by

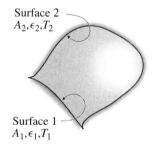

Figure 8.21: Schematic of two gray surfaces that form a complete enclosure.

$$q_{1-2} = \frac{\sigma \left(T_1^4 - T_2^4 \right)}{\dfrac{1 - \epsilon_1}{\epsilon_1 A_1} + \dfrac{1}{A_1 F_{1-2}} + \dfrac{1 - \epsilon_2}{\epsilon_2 A_2}} \tag{8.24}$$

8.8. RADIATIVE EXCHANGE BETWEEN BODIES

where q_{1-2} is the net energy transfer between bodies 1 and 2, in W, ϵ_i is the emissivity of body i, A_i is the area of body i in m², F_{1-2} is the view factor between bodies 1 and 2, dimensionless, and T_i is the absolute temperature of surface i, in K. For example, radiative exchange between two large (infinite) parallel planes, as illustrated in Figure 8.22, would be given by a special case of Eqn. 8.24 for $A_1 = A_2 = A$ and $F_{1-2} = 1$

$$q_{1-2} = \frac{\sigma A \left(T_1^4 - T_2^4\right)}{\frac{1-\epsilon_1}{\epsilon_1} + 1 + \frac{1-\epsilon_2}{\epsilon_2}} \tag{8.25}$$

When a relatively small object 1 is completely enclosed by a much larger surface 2, such as a person inside a large room, $A_1/A_2 \approx 0$ and $F_{1-2} = 1$, and Eqn. 8.24 simplifies to

$$q_{1-2} = \epsilon_1 \sigma A_1 \left(T_1^4 - T_2^4\right) \tag{8.26}$$

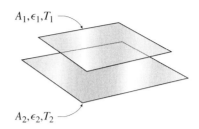

Figure 8.22: Schematic of two gray infinite parallel planes.

8.8.1 Radiative Heat Transfer Coefficient

For a special case of radiative heat exchange, when temperatures T_1 and T_2 are close, Equation 8.21 can be written as:

$$q_{1-2} = \sigma A_1 F_{1-2} 4 T_1^3 (T_1 - T_2) \tag{8.27}$$

$$= F_{1-2} A_1 h_r (T_1 - T_2) \tag{8.28}$$

where:

$$h_r = 4\sigma T_1^3 \tag{8.29}$$

can be termed as a radiative heat transfer coefficient analogous to the convective heat transfer coefficient. Note that the unit for h_r is W/m² · K, same as that for convective heat transfer coefficient. Since Equation 8.29 is an approximation, in some cases, h_r may be determined experimentally. For example, in radiative heat exchange between human and environment (Iberall and Schindler, 1973), the radiative heat transfer coefficient h_r was measured experimentally as:

$$h_r = 3.5[1 + 0.0055(T_s + T_a)] \tag{8.30}$$

where T_s is the average skin temperature and T_a is the environmental temperature, both in °C. For the range of environmental temperatures $0 < T_a < 50°$ C, h_r varies between $3.8 - 5.1$ W/m²·K.

8.8.2 Radiative Exchange Between a Leaf and Surroundings

Radiative heat loss from an unshaded leaf (Figure 8.23) has important consequences. As will be discussed later, in a cloudless night, the effective temperature of the sky can be quite low. If this leaf is radiating to the sky, it can get to very low temperatures and freeze even when the surrounding air is above freezing. Note that the air temperatures in such cases are considerably warmer than the effective sky temperature. Therefore, convection or wind actually warms the plant. In fact, radiative freeze typically occurs when the wind is calm or still. In some orange groves in the state of Florida in the United States, where such radiative freeze frequently causes considerable damage, large fans are employed to blow air when such freezing is imminent. Also, since the uppermost leaves in a tree are more exposed to the sky, such leaves will tend to have the lowest temperatures and frost tends to form there first.

In the daytime, with large amounts of radiative heat gain from the sun, the surrounding air typically cools the leaf. This also has very important consequences, as this lowering of temperature keeps it in the range for optimal photosynthesis. Also, the lower the leaf temperature, the lower is the concentration of water vapor in the pores of the cell walls of the mesophyll cells, and consequently less water would then tend to be lost in transpiration (see mass transfer). This is an important consideration in arid and semi-arid regions.

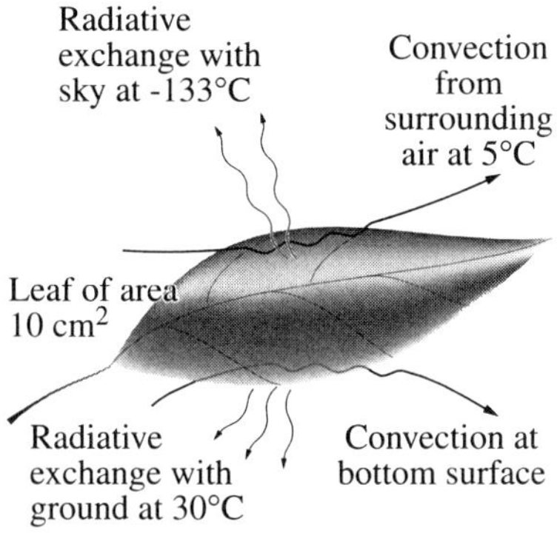

Example 8.8.1 Radiative Freezing of Orange Trees

Consider a flat orange leaf as shown in Figure 8.23 of surface area 10 cm^2 parallel to the ground. The top surface of the leaf radiates to clear sky and the effective radiation temperature of the sky is assumed at $-133°$C. The bottom surface of the leaf radiates to the ground and the ground temperature is 30°C. Both the top and bottom surfaces are also exchanging heat with the air through convection. The air temperature is 5°C. Consider the leaf to be very thin so that the entire leaf (including the top and the bottom surfaces) is at one temperature.

1. By making an energy balance on the leaf involving the convection and the radiation, write an equation from which you can solve for leaf temperature.

2. For a convective heat transfer coefficient of 10 W/m$^2 \cdot$K, calculate the leaf temperature.

3. Suppose that convective heat transfer from the underside of the leaf is much reduced (due to reduced air flow). Ignoring convective heat transfer from the bottom, what would be the leaf temperature?

Solution

This is a steady state problem with no generation of energy. Also, since the leaf has little temperature variation across its thickness (it is very thin), we can treat the heat transfer as lumped (see section 5.1 on page 71). Thus, energy balance on the leaf is written as

$$\begin{matrix} \text{Energy} \\ \text{In} \end{matrix} - \begin{matrix} \text{Energy} \\ \text{Out} \end{matrix} + \underbrace{\begin{matrix} \text{Energy} \\ \text{Generation} \end{matrix}}_{0} = \underbrace{\begin{matrix} \text{Energy} \\ \text{Storage} \end{matrix}}_{0} \qquad (8.31)$$

The various terms are now calculated as

$$\begin{matrix} \text{Radiative heat loss} \\ \text{to sky (energy out)} \end{matrix} = \sigma A (T_l^4 - T_{sky}^4)$$

Figure 8.23: Schematic of radiation exchange of a leaf with the sky and the ground.

$$\begin{aligned}\text{Radiative heat loss} \\ \text{to ground (energy out)}\end{aligned} = \sigma A(T_l^4 - T_{ground}^4)$$

$$\begin{aligned}\text{Convective heat loss to air} \\ \text{from } both \text{ sides of the leaf}\end{aligned} = 2hA(T_l - T_{air})$$

Substituting in Eqn. 8.31, we get

$$(5.676 \times 10^{-8})(10 \times 10^{-4})(T_l^4 - 140^4)$$
$$+ \ (5.676 \times 10^{-8})(10 \times 10^{-4})(T_l^4 - 303^4)$$
$$+ \ 2(10)(10 \times 10^{-4})(T_l - 278) = 0$$

Note that T_l is absolute temperature in K. Also, there is no term corresponding to "Energy In" in Eqn. 8.31. This is due to the formula for *net* heat exchange being used here for the radiative heat loss term. Instead of using the term for heat exchange, the "In" and "Out" can also be considered separately, leading to the same final expression. The leaf temperature T_l can be solved from the above equation by trial and error, and its value is 271.96K or $-1.04°$C.

For the third question, when the underside of the leaf has convective heat exchange from only the top surface, Eqn. 8.31 is modified to

$$(5.676 \times 10^{-8})(10 \times 10^{-4})(T_l^4 - 140^4)$$
$$+ \ (5.676 \times 10^{-8})(10 \times 10^{-4})(T_l^4 - 303^4)$$
$$+ \ (10)(10 \times 10^{-4})(T_l - 278) = 0$$

This leads to a leaf temperature of $-4.22°$C. Note that the leaf now is colder, since it is unable to receive convective heat from the air.

8.8.3 Radiative Exchange Between Human (or Animal) and Its Surroundings

As mentioned in Chapter 4, body heat loss by radiation is an important mechanism for maintaining a thermal balance and comfort. Figure 8.24 shows the (shaded) zone for human thermal comfort as related to surrounding wall temperatures. The mean radiant temperature (MRT) is an average temperature of the walls that produces approximately similar radiative effect. The higher the MRT value, the lower the corresponding air temperature required for comfort. Note that as the MRT increases, a lower air temperature is required to increase the convective heat loss and thus maintain thermal comfort.

8.9 Chapter Summary—Radiative Energy Transfer

- **Electromagnetic Radiation (page 147)**

    1. Electromagnetic waves travel at the speed of light. Waves having a wavelength of 0.1-100 microns are referred to as thermal radiation.

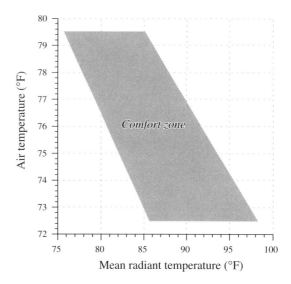

Figure 8.24: Human Thermal Comfort and Mean Radiant Temperature. Data for lightly clothed subjects engaged in sitting activities. Relative humidity is 50% and air velocities range from about 15 to 60 fpm (0.076 to 0.305 m/s) for the data. Adapted from *Concepts in Thermal Comfort* by M. D. Egan, © 1975 by Prentice Hall. Reprinted by permission.

2. The waves are reflected, absorbed and sometimes transmitted at a surface. The reflectivity, absorptivity, and transmissivity are material properties.

- **Radiation Emitted by a Body (pages 150 and 154)**

  1. All bodies above absolute zero emit radiation. The total radiation emitted by an ideal body called a blackbody is given by Eqn. 8.8. The radiation emitted in a given range of wavelength is given by Eqn. 8.13.

  2. Emission from a real body is less than an ideal body. It is calculated by multiplying the emission from an ideal body by a surface property called the emissivity of the real body. A polished body has a lower emissivity. Biomaterials with lots of water have emissivity close to 1.

- **Solar and Atmospheric Radiation (page 158)**

  1. The spectral distribution of solar radiation is given by Figure 8.17. The peak solar radiation is in the visible region. Ozone in the atmosphere attenuates the harmful ultraviolet rays before they reach the earth's surface.

  2. Earth's atmosphere allows shortwave radiations from the sun to reach earth's surface but blocks the longwave radiation emitted by the earth's surface

from leaving the atmosphere. This trapping of energy is termed the greenhouse effect.

- **Radiative Exchange (page 160)**

    1. When two bodies are exchanging radiation, only a fraction of radiation from one body reaches the other and vice-versa. The net exchange of energy is given by equations such as Eqn. 8.21 or 8.24.

    2. Radiation is a major mechanism for losing body heat to maintain thermoregulation in mammals.

8.10 Concept and Review Questions

1. Electromagnetic radiation is characterized by (a) frequency, (b) wavelength, and (c) propagation velocity. Define these characteristics and indicate which ones are independent of the substance through which the radiation is transmitted.

2. What is solar radiation? Does it contain components of microwave and radiowave radiation?

3. List several important biological effects of (a) visible light and (b) ultraviolet radiation.

4. Explain how it is possible for an individual to get a sunburn on a cloudy day, even though the skin actually feels cool.

5. What are microwaves and what role do they play in heat transfer processes that occur in microwave ovens?

6. Examine a microwave oven door and explain how it is possible to see through the window during the cooking process without being harmed by the microwave radiation.

7. Explain why the sky appears blue and the sun appears yellow.

8. What are the main factors responsible for attenuating the irradiation from the sun that reaches the surface of the earth?

9. What is the solar constant?

10. How does thermal radiation differ from other types of electromagnetic radiation?

11. What is the Stefan-Boltzmann law?

12. What is a gray body?

13. What is meant by the radiation shape factor?

14. What is meant by the atmospheric greenhouse effect?

8.11 Further Reading

Albright, L.D. 1990. Environmental Control of Animals and Plants. American Society of Agricultural Engineers, St. Joseph, Michigan.

Bohm, M., A. Browen, and O. Noren. 1991. Thermal environment in cabs. Presented at the International Meeting of ASAE, Dec. 17-20, Chicago, IL.

Brunt, D. 1940. Radiation in the atmosphere. Quarterly Journal of the Royal Meteorological Society, 66(Suppl.):34-40.

Carr, K.L. 1989. Microwave radiometry: Its importance to the detection of cancer. IEEE Transactions on Microwave Theory and Techniques. 37(12):1862-1868.

Coulson, K.L. 1975. *Solar and Terrestrial Radiation*. Academic Press, New York.

Garzoli, K.V. and J. Blackwell. 1987. An analysis of the nocturnal heat loss from a double skin plastic greenhouse. Journal of Agricultural Engineering Research, 36:75-85.

Gast, P.R. 1965. Solar electromagnetic radiation. In *Handbook of Geophysics and Space Environments*. Edited by S. L. Valley. McGraw-Hill, Inc., New York.

Guyot, G. 1998. *Physics of the Environment and Climate*. Praxis Publishing, Chichester, UK.

Hejazi, S., D.C. Wobschall, R.A. Spangler, and M. Anbar. 1992. Scope and limitations of thermal imaging using multiwavelength infrared detection. Optical Engineering. 31(11):2383-2392.

Iberall, A.S. and A.M. Schindler. 1973. On the physical basis of a theory of human thermoregulation. Trans. ASME, J. Dynamic Systems Measurement and Control. 95:68-75.

Incropera, F.P. and D.P. Dewitt. 1996. *Fundamentals of Heat and Mass Transfer*. John Wiley & Sons, New York.

Metaxas, A.C. 1996. *Foundations of Electroheat: A Unified Approach*. John Wiley & Sons, Chichester, UK.

Myers, P., N.L. Sadowsky, and A.H. Barrett. 1979. Microwave thermography: Principles, methods, and clinical applications. Journal of Microwave Power. 14(2): 105-115.

Ramanathan, V. 1987. The role of earth radiation budget studies in climate and general circulation research. Journal of Geophysical Research. 92(D4):4087-4095.

Roussy, G. and J. Pearce. 1995. *Foundations and Industrial Applications of Microwaves and Radio Frequency Fields*. John Wiley & Sons, New York.

Shitzer, A. and R.C. Eberhart (ed). *Heat Transfer in Medicine and Biology* (2 volumes). New York: Plenum Press, 1985.

Siegel, R. and J.R. Howell. 1981. *Thermal Radiation Heat Transfer*. McGraw-Hill, New York.

Steketee, J. 1973. Spectral emissivity of skin and pericardium. Phys. Med. Biol. 18(5):686-694.

ten Berge, H.F.M. 1990. Heat and water transfer in bare topsoil and the lower atmosphere. Center for Agricultural Publishing and Documentation, Wageningen, the Netherlands.

8.12 Problems

8.12.1 Solar Energy Flux

What is the energy flux at the surface of sun? Assume an average surface temperature of sun to be 5800 K. Compare this to average solar flux just outside the earth's atmosphere as 1353 W/m^2 and explain the difference.

8.12.2 Spectral Emission from Sun

The sun can be treated as a blackbody is at temperature 5800 K. What is the wavelength λ_1 below which 10% of the solar emission is concentrated? What is the wavelength λ_2 above which 10% of the solar emission is concentrated? Determine the maximum spectral emission power of the sun and the wavelength at which this emssion occurs?

8.12.3 Fraction of Solar Energy for Photosynthesis

Solar radiation between the wavelengths 0.35 μm and 0.70 μm is useful for photosynthesis. 1) What fraction of total energy radiated from the sun is useful for photosynthesis? 2) Is this the same fraction of the solar energy incident on earth surface that is useful for photosynthesis? Consider sun to be a blackbody at 5800 K.

8.12.4 Extraterrestrial Solar Radiation

Estimate the solar energy flux incident just outside the earth's atmosphere from knowing the sun's surface temperature of 5800 K, sun's radius of 0.695×10^6 km, and the sun to earth distance of 1.5×10^8 km. Assume the sun to be a blackbody.

8.12.5 Simple Energy Balance of Earth

Consider the earth to be a sphere of average radius 6.38×10^6 m. The incident solar radiation *normal to the earth surface* outside atmosphere is 1353 W/m^2. About 30% of this energy is reflected from the atmosphere and the rest of it is transmitted into the atmosphere. The radiation from earth to outer space seems to be from an average earth surface temperature of 15°C. The emissivity of earth surface is approximately 0.97. Calculate the total radiation received by the earth from sun and compare it with the total energy lost by means of radiation from the earth.

8.12.6 Human Radiation

The radiation emitted from the human body has been used for a variety of applications ranging from military night vision devices to diagnostic medical imaging. For thermoregulation, i.e., maintaining body temperature, human being has to lose energy. Consider a standing person (do not consider clothes) in the center of a *large* room. Approximate the person as a vertical cylinder of 1.8 m tall and 0.3 m diameter. Average surface temperature of the person is 33°C, the room surface temperature is 20°C, and the emissivity of skin surface is 0.95. Consult Appendix C.8 (page 340) for properties of air. 1) Calculate the radiative heat transfer from this person. Neglect heat loss from the ends of the cylinder. 2) What is the wavelength λ_{max} where the maximum amount of energy will be radiated? 3) What region (primarily) of electromagnetic spectrum is the radiation calculated in step 2? 4) If a detector is available that can detect only in the wavelength range $\lambda_{max} \pm 5$ μm, what fraction of the total energy from the human being will this detector be sensitive to? 5) Calculate the natural convection heat loss from this person for room air temperature of 20°C. For the vertical surface of the cylinder, formulas for vertical plate can be used. Neglect heat loss from ends of this cylinder. 6) Compare the convective heat loss with radiative heat loss and state if they are close.

8.12.7 Radiative Heating in a Greenhouse

Heat delivery from a steam heating system can be different based on different conditions of the surface that lead to different emissivities. Consider a steam heating pipe in a large greenhouse having a surface temperature of 95°C. The surface has been painted with an aluminized paint having an emissivity of 0.45. 1) Calculate the radiant flux leaving the heated surface. 2) How much would the flux change if the pipe were repainted with an oil-based paint having an emissivity of 0.95? (Albright, 1990)

8.12.8 Heat Loss from Animals in a Barn

Consider a pig alone in a large barn exchanging thermal radiation with the inside walls of the barn. The surface area of the pig can be approximated as 2 m², with its skin temperature of 35°C and a skin emissivity of 0.90. The mean radiant temperature of the inside walls of the barn is 5°C. 1) Calculate the *net* radiative heat exchange (loss) from the pig. 2) Calculate the thermal radiation emitted by the pig and explain the difference between it and what is calculated in step 1 (Albright, 1990).

8.12.9 Radiative Cooling of Orange Groves (Re-solving Example 8.8.1)

Consider a flat orange leaf of surface area 10 cm² parallel to the ground. The top surface of the leaf radiates to clear night sky and the effective radiation temperature of the sky is assumed at 140 K. The leaf absorbs all incident radiation and its emissivity is 1. The bottom surface of the leaf radiates to the ground and the ground temperature is 7°C. Both the top and bottom surfaces are also exchanging heat with the air through convection. The air temperature is 5°C. Consider the leaf to be very thin so that the

entire leaf (including the top and the bottom surfaces) is at one temperature. 1) By making an energy balance on the leaf involving the convection and the radiation, write an equation from which you can solve for leaf temperature. 2) For a convective heat transfer coefficient of 25 W/m²·K, calculate the leaf temperature. 3) Compare your leaf temperature with that obtained in Example 8.8.1 and comment on what would be the simplest or least expensive solution in the orange grove of Example 8.8.1 to prevent the leaf from freezing.

8.12.10 Heat Balance on Greenhouse Glass

Due to the selective absorption by the glass roof and walls of a greenhouse, as shown in Figure 8.26, more of the solar radiation come inside than can leave the greenhouse, effectively trapping the energy and raising the inside temperature. Some gases in the atmosphere around earth have a similar effect as the glass and leads to increase in temperature of earth's surface, thus deriving the name "greenhouse gases." In this problem, we would like to calculate the temperature reached at steady state by the greenhouse glass due to balance of all energy exchanges.

Assume a section of glass receives solar radiation at 1000 W/m², atmospheric radiation at 200 W/m² and radiation emitted from the interior surfaces (plants, walls, etc.) at 500 W/m². The radiation from the atmosphere, the interior surfaces and the glass are concentrated in the far infrared region ($\lambda > 8$ μm) of the spectrum, i.e., consider no energy is emitted below 8 μm. The solar radiation can be assumed to be that from a blackbody at 5800 K temperature. The glass has the property that it absorbs no radiation (transmissivity of 1) for $\lambda < 1$ μm and absorbs all of the radiation (with zero reflectivity) for $\lambda > 1$ μm as shown in Figure 8.25. The glass emissivity is 1 in the wavelength region $\lambda > 1$ μm. The glass emits energy and also exchanges energy with convective air on both sides of it. The heat transfer coefficient and air temperature on the glass outside are 50 W/m²·K and 20°C respectively. The inside glass surface heat transfer coefficient is 15 W/m²·K and greenhouse interior air temperature is 25°C. In order to perform an energy balance on the glass, 1) Write the energy coming into the glass. 2) Now write the energy leaving the glass. 3) Calculate the glass temperature at steady-state.

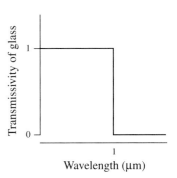

Figure 8.25: Transmissivity of glass for Problem 8.12.10.

8.12.11 Steady State Glass Temperature

Consider solar radiation incident on a glass plate with air on both sides. The solar radiation can be assumed to be that from a blackbody at 5800 K temperature. The glass absorbs all radiation between 0.3 μm and 3 μm. Outside this range of wavelengths, all radiation is transmitted through the glass without any absorption. The radiation incident to the glass is 700 W/m² and the ambient air temperature is 20°C. 1) Calculate the solar energy that is absorbed in the glass, in W/m². 2) Write an energy balance for the glass at steady state, from which you can calculate the steady-state temperature. 3) For a convective heat transfer coefficient of 50 W/m²·K, what is the steady state temperature? Note: Solution requires iteration.

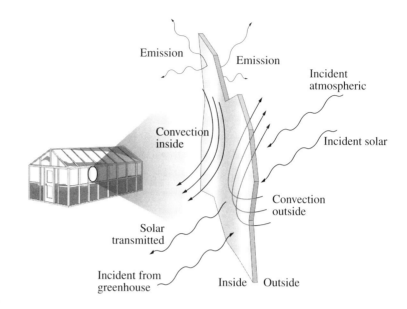

Figure 8.26: Schematic of various energy exchanges on a greenhouse glass for Problem 8.12.10.

8.12.12 Thermocouple Measurement Error Due to Radiation

The thermocouple, a temperature measuring device, can introduce measurement errors due to its radiative exchange with the surroundings. Consider a cylindrical thermocouple sheath that is 4 mm diameter and with a surface emissivity of 0.5. It is kept horizontal during the measurement in a large room where the air can move only by natural convection. The air temperature is 25°C and the mean radiant temperature of the wall surfaces is 35°C. 1) What temperature will the thermocouple indicate? 2) What is the measurement error, i.e., the difference between indicated temperature and true air temperature? Hint: Consider natural convection from the thermocouple and the air properties are provided in Appendix C.8 (page 340).

Part II

Mass Transfer

Chapter 9

EQUILIBRIUM, MASS CONSERVATION, AND KINETICS

CHAPTER OBJECTIVES

After you have studied this chapter, you should be able to:

1. Explain equilibrium between various phases— solid, liquid, and gas.

2. Understand the various components needed for setting up conservation of mass for a single mass species

3. Understand simple rates of reaction in the decay or generation of a mass species.

4. Understand the increase in reaction rate with temperature.

KEY TERMS

- mass species
- mass concentration
- molar concentration
- mass conservation
- equilibrium
- Henry's law
- isotherms
- partition or distribution coefficient
- chemical kinetics
- reaction rate constant
- reaction mechanism
- zeroth and first order reaction
- Arrhenius kinetics
- activation energy

We have concluded the *first part* of the text that dealt with energy transfer. This is the first chapter in the *second part* of the text on mass transfer. Figure 9.1 is a concept map which shows the relationship of this chapter to the other chapters on mass transfer. Like Chapter 1, this first chapter in the part on mass transfer discusses concepts of equilibrium with non-equilibrium in concentration being the driving force.

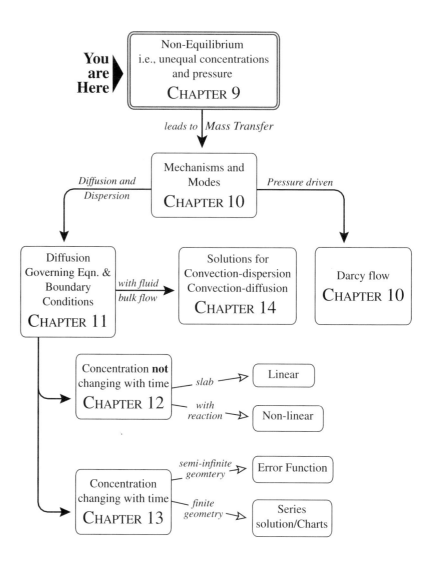

Figure 9.1: Concept map of mass transfer showing how the contents of this chapter relates to other chapters in mass transfer.

9.1 Concentration

The subject of mass transfer involves movement of one mass species through one or more other species. Therefore, we will always be dealing with systems consisting of two or more species, i.e., multi-species systems. The quantity of a certain species is described by its concentration, which is defined as the amount of substance per unit mass or volume. In a multi-species system having many mass species, the concentration of any one species can be expressed in many different ways. We will primarily use two different measures of concentration:

$$\rho_A = \text{mass concentration} = \frac{\text{mass of A}}{\text{unit volume}} = \frac{\text{kg of A}}{\text{m}^3}$$

$$c_A = \text{molar concentration} = \frac{\text{moles of A}}{\text{unit volume}} = \frac{\text{mol of A}}{\text{m}^3}$$

Thus, the two concentrations can be interchanged using the relation

$$\rho_A = c_A M_A$$

where M_A is the molecular weight of component A. For simplicity, however, we will not keep track of two separate symbols of concentration corresponding to two different units (mass and molar). Instead, the symbol c_A will be used to denote concentration of species A in either units, except when both units are used simultaneously. Although mass concentration is analogous to amount of energy in heat transfer, it should be noted that in heat transfer only one variable, temperature, was sufficient while in mass transfer, we have to keep track of the concentration of each component. Generally, this results in increased complexity.

9.1.1 Concentrations in a Gaseous Mixture

In a mixture of perfect gases, the concentrations of individual gases can be calculated from their partial pressures. Using the ideal gas law for a mixture of gases of total volume V at temperature T, we can write for the component A

$$p_A V = n_A R_g T \tag{9.1}$$

where p_A is the partial pressure of gas A for which there is n_A moles are present and R_g is the gas constant. Using this equation, the molar concentration of gas A can be written as

$$c_A = \frac{n_A}{V} = \frac{p_A}{R_g T} \tag{9.2}$$

Thus, concentration of any gas species i can be written in terms of its partial pressure as

$$c_i = \frac{p_i}{R_g T} \tag{9.3}$$

The total concentration c of all the species is related to the total pressure, P. Using $PV = nRT$ where n is the total number of moles ($= \sum n_i$), the total concentration c is written as

$$c = \frac{P}{R_g T} \tag{9.4}$$

Example 9.1.1 Concentration of Individual Gases in Air

Consider an air-water vapor mixture at a total pressure of 1 atm and a temperature of 60°C having 20% water vapor, 17% oxygen, and 63% nitrogen. The percentages refer to the ratio of partial pressures to total pressure. For simplicity, we are ignoring the other constituent gases in air. Calculate the molar and mass concentrations of each of the three gases in air.

Known: Fractions of individual gases in a mixture

Find: The molar and mass concentrations

Schematic and given data:

1. $p_{vapor} = 0.2$ atm, $p_{oxygen} = 0.17$ atm, $p_{nitrogen} = 0.63$ atm
2. $P = 1$ atm
3. $T = 333$ K
4. $R_g = 8.315$ kJ/kmol·K

Assumptions: Air behaves as an ideal gas

Analysis: Using Equation 9.3, the concentrations are calculated as

$$c_{vapor} = \frac{p_{vapor}}{R_g T} = \frac{0.2 \times 1.013 \times 10^5 \text{ N/m}^2}{8.315 \text{ J/mol·K} \times 333 \text{ K}} = 7.32 \frac{\text{mol}}{\text{m}^3}$$

$$c_{oxygen} = \frac{p_{oxygen}}{R_g T} = \frac{0.17 \times 1.013 \times 10^5 \text{ N/m}^2}{8.315 \text{ J/mol·K} \times 333 \text{ K}} = 6.22 \frac{\text{mol}}{\text{m}^3}$$

$$c_{nitrogen} = \frac{p_{nitrogen}}{R_g T} = \frac{0.63 \times 1.013 \times 10^5 \text{ N/m}^2}{8.315 \text{ J/mol·K} \times 333 \text{ K}} = 23.05 \frac{\text{mol}}{\text{m}^3}$$

The mass concentrations are calculated by multiplying the molar concentrations by the respective molecular weights as

$$\rho_{vapor} = 7.32 \times 18 \times 10^{-3} \frac{\text{kg}}{\text{m}^3} = 0.1317 \frac{\text{kg}}{\text{m}^3}$$

$$\rho_{oxygen} = 6.22 \times 32 \times 10^{-3} \frac{\text{kg}}{\text{m}^3} = 0.1990 \frac{\text{kg}}{\text{m}^3}$$

$$\rho_{nitrogen} = 23.05 \times 28 \times 10^{-3} \frac{\text{kg}}{\text{m}^3} = 0.6454 \frac{\text{kg}}{\text{m}^3}$$

Comments: The total molar concentration of gases is given by

$$\begin{aligned} c &= c_{vapor} + c_{oxygen} + c_{nitrogen} \\ &= 7.32 + 6.22 + 23.05 \frac{\text{mol}}{\text{m}^3} = 36.59 \frac{\text{mol}}{\text{m}^3} \end{aligned}$$

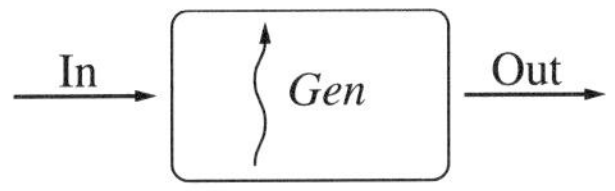

Figure 9.2: A control volume for mass conservation showing different components.

and the total mass concentration of the gases is given by

$$\begin{aligned}\rho &= \rho_{vapor} + \rho_{oxygen} + \rho_{nitrogen} \\ &= 0.1317 + 0.1990 + 0.6454 = 0.9761 \frac{\text{kg}}{\text{m}^3}\end{aligned}$$

Note that the total mass concentration, ρ, is another name for the density of the gas mixture.

9.2 Species Mass Balance (Mass Conservation)

Like total energy, total mass is conserved. In Section 3.3 (page 29) for energy transfer, we wrote a balance for thermal energy. Other forms of energy can convert into thermal energy and vice-versa. Likewise, one species of mass can convert to other species and vice-versa, for example, through chemical reactions. By analogy to the thermal energy conservation equation (Eqn. 3.3), consider mass conservation of a species for an arbitrary control volume shown in Figure 9.2. A word equation for mass conservation of a species can be written from this figure as

$$\underbrace{\text{Rate of Mass In}}_{\text{species } i} - \underbrace{\text{Rate of Mass Out}}_{\text{species } i} + \underbrace{\text{Rate of Mass Generation}}_{\text{species } i} = \underbrace{\text{Rate of Mass Storage}}_{\text{species } i} \quad (9.5)$$

Increase in storage is the increase in concentration of the species. Mass of a species can be generated or utilized in chemical reactions or biological processes. Note that this is a transformation of one mass species into another. For example, under suitable growth conditions the microbial mass in a fermentor increases. At the same time there is utilization of resources like O_2 in the fermentor in order to sustain growth.

9.3 Equilibrium

The transfer of mass within a phase or between two phases requires a departure from equilibrium. One needs to study interphase equilibrium in order to describe interphase mass transfer.

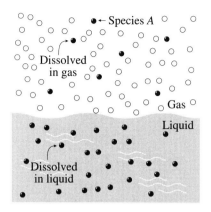

Figure 9.3: Schematic of equilibrium between a gas and a liquid phase.

9.3.1 Equilibrium Between a Gas and a Liquid

An example of equilibrium between gas and liquid is water saturated with dissolved oxygen in contact with air, as shown in Figure 9.3. Here the oxygen in water and the oxygen in air are considered to be in equilibrium with each other. Such equilibrium between a gas and a liquid phase can be described by the well known Henry's law:

$$p_A = H\, x_A \tag{9.6}$$

where p_A is the partial pressure of species A in gas phase at equilibrium, x_A is the concentration of species A in liquid phase at equilibrium, and H is the Henry's constant. Equation 9.6 is illustrated in Figure 9.5. One can think of Henry's law as the pressure dependence of solubility of a gas in a solution. It says that gas solubility should be directly proportional to pressure.

$$x_A = \frac{p_A}{H} \tag{9.7}$$

Henry's constants for several gases in water are given in Figure 9.4. In this figure, H increases with temperature, thus showing a decrease in solubility of gases in water at higher temperatures, as given by Eqn. 9.7.

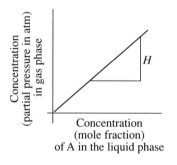

Figure 9.5: Equilibrium relationship between a gas phase and a liquid phase following Henry's law.

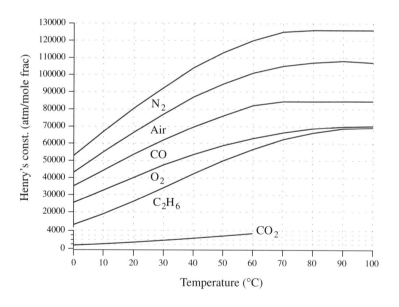

Figure 9.4: Henry's constant, H, for gases slightly soluble in water. Data abridged from Foust et al. (1960).

9.3. EQUILIBRIUM

Example 9.3.1 Bends from Diving in Deep Water

There are many biological applications of this pressure dependence of solubility. A condition known as caisson disease or bends or compressed air sickness results from a rapid change in atmospheric pressure from high to normal causing nitrogen bubbles to form in blood and body tissues. Divers can get the bends if they rise too rapidly.

Example 9.3.2 Oxygen Concentration in Water in Equilibrium with Air

Compared to air, there is little oxygen (dissolved) in water, even under the best of conditions. Thus, fish or other aquatic animals must either pump large quantities of water over their respiratory surface to obtain reasonable amounts of oxygen or else be limited to relatively low metabolic rates. This relatively low amount of dissolved oxygen is further reduced at higher temperature, as a direct consequence of the Henry's law just studied.

Water at 5°C is in contact with a large volume of ordinary air at a total pressure of 1 atmosphere. 1) How much oxygen is dissolved in the water in mg O_2/liter of water? Suppose this water got warmed up to 25°C. 2) What is the new amount of oxygen dissolved in the same units? 3) What happens if water at 25°C and having initially an amount of oxygen 10 mg O_2/liter of water is brought in contact with ordinary air at the same temperature?

Known: Oxygen in water is in equilibrium with oxygen in air

Find: The amount of oxygen dissolved in water at equilibrium

Schematic and given data:

1. Schematic is shown in Figure 9.3
2. H, Henry's law constant is found from Figure 9.4 to be 2.9×10^4 and 4.45×10^4 atm/mole fraction of oxygen in the water at temperatures 5°C and 25°C, respectively
3. Total pressure of air above the water is 1 atmosphere
4. Volume of air is very large, so the oxygen content in it does not change and the oxygen in water changes to come to equilibrium with this oxygen in the air

Assumptions:

1. Ordinary air has 21% oxygen
2. The air and water has been in contact with each other for a long enough time to have reached equilibrium

Analysis:

1. Since dry air contains 21% oxygen, partial pressure of oxygen is given by $p_{O_2} = 0.21$ atm. As the volume of air is very large, p_{O_2} is a constant. Using Henry's law to find the concentration of oxygen in the water, x_{O_2}, at equilibrium at 5°C:

$$\begin{aligned}
x_{O_2} &= \frac{p_{O_2}}{H} \\
&= \frac{0.21 \text{ atm}}{2.9 \times 10^4 \text{ atm/mole fraction}} \\
&= 7.241 \times 10^{-6} [\text{mol } O_2/\text{mol solution}] \\
&\approx 7.241 \times 10^{-6} [\text{mol } O_2/\text{mol water}] \\
&= 7.241 \times 10^{-6} \times \frac{32}{18} [\text{g } O_2/\text{g water}] \\
&= 12.87 [\text{mg } O_2/\text{liter of water}]
\end{aligned}$$

where mass of water with and without oxygen are considered approximately the same. Also $1 \text{ g} \approx 1 \text{ cm}^3 = 10^{-3}$ liter has been used for water.

2. Similarly, the amount of oxygen dissolved in water at 25°C is

$$\begin{aligned}
x_{O_2} &= \frac{p_{O_2}}{H} \\
&= \frac{0.21 \text{ atm}}{4.45 \times 10^4 \text{ atm/mole fraction}} \\
&= 4.719 \times 10^{-6} [\text{mol } O_2/\text{mol solution}] \\
&= 8.39 [\text{mg } O_2/\text{liter of water}]
\end{aligned}$$

3. Since the final equlibrium concentration of oxygen in water at 25°C is 8.39 mg O_2/liter of water, water having initially 10 mg O_2/liter of water will *lose* oxygen until it comes to the final equilibrium concentration of 8.39 mg O_2/liter of water.

Comments: Since Henry's constant increases with temperature, as shown in Figure 9.4, solubility of oxygen decreases with temperature and fish and aquatic life are expected to have greater difficulty in obtaining the needed oxygen at higher water temperature.

9.3.2 Equilibrium Between a Gas and a Solid (with Adsorbed Liquid)

The common example of a solid coming in equilibrium with a gas is the potato chip typically becoming soggy when left out of the package. The moisture in the potato chip (solid phase) comes to equilibrium with typically higher amount of water vapor

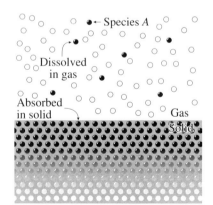

Figure 9.6: Schematic of equilibrium between a gas and a solid phase.

9.3. EQUILIBRIUM

in the outside air than inside the package, thus absorbing more moisture from the air and becoming soggy. Any wet solid, when brought into contact for sufficient time with a large amount of air having a certain concentration of moisture (water vapor), will reach a final moisture content that is the *equilibrium moisture content* of the solid for the particular concentration of moisture in the air. This equilibrium between moisture in the solid and moisture in the gas is illustrated in Figure 9.6. If the solid initially (before coming in contact with the air) had more moisture than the equilibrium moisture content corresponding to the concentration of moisture in air, it will reach equilibrium with the air by losing the excess moisture. Conversely, if the solid had less moisture than the corresponding equilibrium moisture content, it will gain moisture in coming to equilibrium with the air.

The relationship between the concentration[1] of moisture in air and the corresponding equilibrium moisture content of a solid can be expressed as an equilibrium moisture curve. A typical curve is shown in Figure 9.7. Equilibrium moisture content is also a function of temperature. Thus the curve in Figure 9.7 refers to a specific temperature, which is why the curve is often called an isotherm. Example of an empirical equation (Henderson, 1952) used to represent the equilibrium moisture content curve is:

$$1 - RH = b_1 \exp(-b_2 w) \tag{9.8}$$

where RH is the relative humidity (in fraction), b_1 and b_2 are constants, and w is the equilibrium moisture content (%). There are many such empirical equations. Equilibrium moisture content of a solid will also be typically different depending on whether moisture is being added or removed as the material comes to equilibrium with the surrounding air, i.e., whether it is a drying or a wetting process.

Examples of equilibrium moisture isotherms are shown in Figure 9.8 (see next page) for wood and soil. The moisture contents in this figure are on wet weight basis, i.e., kg of water per kg of total. To convert to dry basis, use the relationship $M_d = M_w/(1 - M_w)$ where M_w is the moisture content in fraction on wet basis and M_d is the moisture content in fraction on dry basis. Note that the moisture content in air is given in this figure in terms of relative humidity (see footnote).

Hygroscopic and non-hygroscopic solids

To understand the relative degree to which water is freely available in a solid, consider the same information as shown in Figure 9.7 except that now, instead of a large amount of air coming in contact with a solid, the solid is sealed inside a container together with a small amount of air. The moisture in air would come to equilibrium with moisture in the solid, as illustrated in Figure 9.9. This equilibrium relative humidity of the air is

---

[1]Concentration of moisture in air is typically measured in terms of humidity or relative humidity. Humidity denotes the absolute amount of moisture (water vapor) in air, in kg of water/kg of dry air. Relative humidity, on the other hand, refers to the amount of moisture in air relative to the maximum capacity of the air to carry water vapor. Symbolically, $RH = 100 p_v/p_{vs}$ where RH is the relative humidity, p_v is the partial pressure of water vapor and p_{vs} is the partial pressure of water vapor at saturation at the same temperature. Thus, when air is carrying the maximum possible amount of moisture (it is saturated), the relative humidity is 100%. Conversion between humidity and the relative humidity can be done using the psychrometric chart shown in Appendix C.9 on page 341.

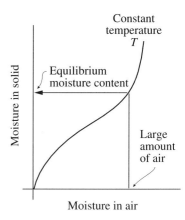

Figure 9.7: Typical equilibrium moisture isotherm for a solid. Superimposed on it is a process where the moisture in solid comes to equilibrium with the moisture in the gas, due to excess of the gas.

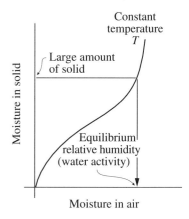

Figure 9.9: Typical equilibrium moisture isotherm for a solid. Superimposed on it is a process where the moisture in the gas comes to equilibrium with the moisture in the solid, due to excess of the solid.

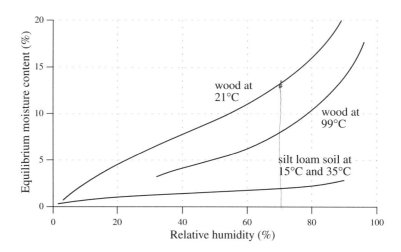

Figure 9.8: Equilibrium moisture isotherms for wood and soil. Wood data courtesy of USDA Forest Service and cited by Siau (1984). Soil data adapted from Berge (1990).

also called the *water activity* of the solid at its moisture content. This water activity of the solid is a measure of the relative ease at which water is available in the solid and is extremely important for biological activities such as growth of microorganisms.

The equilibrium moisture isotherms in Figure 9.8 show that the relative humidity and therefore the vapor pressure of air surrounding a solid in a closed system is a function of both moisture content and temperature of the solid. Solids showing such characteristics are called hygroscopic. When the vapor pressure surrounding a solid is a function of temperature only and is thus independent of the solid moisture level at all times, the solid is called non-hygroscopic. Examples of non-hygroscopic solids are sand, crushed minerals, polymer particles, and some ceramics. Water on the surface of these solids behave as if the solid was not there, and the vapor pressure above the surface as a function of temperature is given by the Clausius-Clapeyron equation:

$$\frac{d \ln p_v}{dT} = \frac{\Delta H_{vap}}{RT^2} \tag{9.9}$$

where p_v is the vapor pressure at absolute temperature T and ΔH_{vap} is the enthalpy of vaporization. The amount of *physically bound* water is negligble in a non-hygroscopic solid, while there can be a large amount of physically bound water in a hygroscopic solid. Once the hygroscopic solid is saturated with water, it can no longer bind more water and begins to behave like a non-hygroscopic solid. Thus, even in a hygroscopic solid, above a level of moisture content the vapor pressure is a function of temperature only. A non-hygroscopic solid does not shrink during drying. In a hygroscopic solid, shrinkage often occurs during drying due to the large amount of physically bound water. Many biological solids are hygroscopic.

9.3.3 Equilibrium Between Solid and Liquid in Adsorption

Consider a solid surface in contact with a liquid that has a dissolved species. As illustrated in Figure 9.10, some of the dissolved solute will bind to the solid surface, instead of remaining in solution. This phenomenon is called adsorption. An example is the adsorption of pollutant chemicals to a soil surface when water with dissolved pollutants is flowing through it. The dissolved species in the liquid reaches an equilibrium with the same species adsorbed to solid surfaces. Such equilibrium relationships are represented as adsorption isotherms and are often empirical. A common adsorption isotherm is the Freundlich isotherm shown in Figure 9.11. Other possible isotherms are Langmuir and BET, which are described well in numerous soil science and other textbooks.

The isotherm shown in Figure 9.11 can be represented mathematically as

$$c_{A,adsorbed} = K^* c_A^n \qquad (9.10)$$

where $c_{A,adsorbed}$ is the concentration of adsorbed solute A, c_A is the concentration of solute A in solution, and K^* and n are empirical constants. Note that even though the relationship expressed in Figure 9.11 is non-linear, it can be treated as linear in the ranges of small concentration. In the linear region of small concentrations (here $n = 1$), the slope of the curve in Figure 9.11 is sometimes referred to as the *partition coefficient* or *distribution coefficient*.

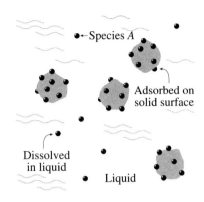

Figure 9.10: Schematic of equilibrium between dissolved and surface adsorbed species.

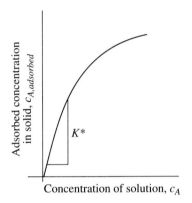

Figure 9.11: Equilibrium relationship between dissolved and surface adsorbed quantity, known as Freundlich isotherm.

9.4 Chemical Kinetics: Generation or Depletion of a Mass Species

So far in this chapter we have discussed the mechanisms of mass movement. In systems undergoing chemical changes, mass concentration at a location can also change due to chemical reaction, independent of any diffusive or convective transport. For example, a chemical transporting through soil can also degrade simultaneously. In heat transfer the term Q was used to denote the generation of heat within the system. Similarly with mass transfer, the term r_A is used to denote the rate of generation or depletion of a mass species A.

Chemical Kinetics is the study of the rate and mechanism by which one chemical species is converted to another species. The mechanism of a chemical reaction is the sequence of individual chemical events whose overall result produces the observed reaction. It is not always necessary to know the detailed mechanism of a reaction in order to quantify the observed rate of the overall reaction. A satisfactory equation for the intrinsic rate of the reaction is sufficient. Reaction mechanisms are reliably known for only a few systems. We will not deal with any mechanisms in this book.

Thermodynamics tells us the maximum extent to which a reaction can proceed, since the equations are correct only for equilibrium conditions. The maximum possible conversion, found at equilibrium, is important as a standard for evaluating the actual performance of the reaction. For practical use, we also need to know the time required for the reaction to proceed. The rate of a chemical reaction is defined by

$$\text{Rate} = \frac{\text{Mass of product produced or reactant consumed}}{(\text{Unit volume})(\text{Time})} \quad (9.11)$$

with the units being kg/m³·s. A few simple relationships describing rates of reactions are now discussed.

9.4.1 Rate Laws of Homogeneous Reactions

Rate laws describe the dependence of the reaction rates on concentration. Consider an irreversible reaction:

$$s_A A + s_B B \rightarrow s_C C + s_D D \quad (9.12)$$

where s_A, s_B, s_C and s_D are stoichiometric coefficients, respectively. The rate of disappearance of A may be written in terms of concentration c_A and c_B of A and B respectively, as:

$$r_A = -\frac{dc_A}{dt} = k'' c_A^\alpha c_B^\beta \quad (9.13)$$

where α is the order of reaction with respect to A, β is the order of reaction with respect to B, $\alpha + \beta$ is the overall order of reaction, and k'' is the *reaction rate constant*. Since r_A measures rate of disappearance, in the transport equation r_A will have (+) or (-) sign depending on whether A is generated or depleted, respectively.

The reaction rate constant k'' includes the effects of all variables other than concentration. While many variables may affect the reaction, temperature is the most important. However, if the reaction requires a catalyst, k'' might depend on the nature and concentration of the catalytic substance. We will now consider several special cases of Eqn. 9.13.

9.4.2 Zeroth Order Reaction

A reaction is of zeroth order when the rate of reaction is independent of the concentration of the species:

$$r_A = -\frac{dc}{dt} = k'' \quad (9.14)$$

where the concentration of species A, c_A has been replaced by c for simplicity. Integrating, and noting that c can never become negative:

$$c_0 - c = k'' t \quad for \ t \leq \frac{c_0}{k''} \quad (9.15)$$

Equation 9.15 is plotted in Figure 9.12. The dashed line in the figure signifies that the zeroth order reaction may not continue all the way. As a rule reactions are only zeroth order in certain high concentration ranges, because the small change in

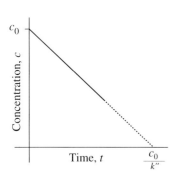

Figure 9.12: Linear change in concentration for a zeroth order reaction.

the amount of material caused by the reaction is negligible compared to the amount of material present. If the overall concentration of the material is decreased to a certain point, the reaction will usually become dependent on the concentration level, at which point the reaction order will increase from zero.

An example of a zeroth order reaction is the study of oxygen transport in human tissue. Augusu Krough won a Nobel prize by studying this mass transfer process. By considering a tissue cylinder surrounding each blood vessel, he proposed the diffusion of oxygen away from the blood vessel into the annular tissue was accompanied by a zeroth order reaction. This reaction was necessary to explain the metabolic consumption of oxygen to produce carbon dioxide.

9.4.3 First Order Reaction

A reaction is first order when the rate of reaction is linearly related to the concentration of the species; thus:

$$r_A = -\frac{dc}{dt} = k''c \qquad (9.16)$$

$$-\frac{dc}{c} = k'' \, dt$$

$$-\int_{c_0}^{c} \frac{dc}{c} = \int_{0}^{t} k'' \, dt$$

$$-\ln \frac{c}{c_0} = k''t$$

$$\boxed{c = c_0 \, e^{-k''t}} \qquad (9.17)$$

Equation 9.17 is plotted in Figure 9.13. The exponential decay of concentration is characteristic of the first order reaction. The concentration approaches zero asymptotically, or in other words, it never actually reaches zero even though it becomes very small.

Half life for a first order reaction is a convenient alternative description of reaction rate that is used widely in practice. It is defined as the time to change concentration by 50%. If $t_{1/2}$ denotes half life, substituting in Eqn. 9.17, we get

$$0.5c_0 = c_0 e^{-k''t_{1/2}}$$
$$-k''t_{1/2} = \ln(0.5)$$
$$= -0.693$$
$$k'' = \frac{0.693}{t_{1/2}} \qquad (9.18)$$

Equation 9.17 can be rewritten in terms of half-life as:

$$c = c_0 e^{-\frac{0.693}{t_{1/2}}t} \qquad (9.19)$$

An example of first order reaction is the degradation of certain pesticides in soil. Representative values of $t_{1/2}$ of some pesticides in soil are shown in Table 9.1.

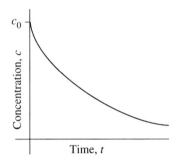

Figure 9.13: Exponential change in concentration for a first order reaction.

Table 9.1: Half life of some important pesticides. Here the health advisory level is the concentration level of a substance which is considered safe to be consumed daily throughout a person's lifetime.

Pesticide	$t_{1/2}$ in soil (days)	Health advisory level (μg/L or ppb)
Aldicarb	30	10
Atrazine	60	3
Carboxin	7	700

A larger half life means a lower reaction rate. For example, in the case of a pesticide, a longer half life means the pesticide remains in the soil longer. Using the half life data from Table 9.1 and with an assumed initial concentration of 1 unit, the concentrations are plotted in Figure 9.14 to show the relative rates of decay. It can be seen from the curves that a concentration of 0.5, half of the initial concentration, is reached for each chemical at times equal to their half life. During food processing, the thermal destruction of microorganisms, most nutrients and quality factors (texture, color and flavor) generally follow first order kinetics, although the exact mechnisms of these reactions are often complex and are still subjects of research. For example, bacterial death by heat is often attributed to the loss of reproductive ability due to the denaturation of one gene or a protein essential to reproduction. Data on kinetics of first order reactions in food systems is provided in Table D.7 (page 352).

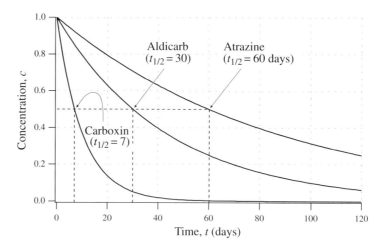

Figure 9.14: Concentration changes with time (due to decay) for the pesticides listed in Table 9.1.

9.4.4 *nth* Order Reaction

When the mechanism of a reaction is not known, we often attempt to fit the data with an *nth* order rate equation of the form:

$$r_A = -\frac{dc}{dt} = k'' c^n \tag{9.20}$$

Integrating:

$$\int_{c_0}^{c} \frac{dc}{c^n} = -\int_{0}^{t} k'' \, dt$$

$$\frac{c^{-n+1}\big|_{c_0}^{c}}{-n+1} = -k'' t$$

Putting the limits and rearranging:

$$c^{1-n} - c_0^{1-n} = (n-1) k'' t, \quad n \neq 1 \tag{9.21}$$

which provides concentration c as a function of time, t, for a *nth* order reaction.

9.4.5 Effects of Temperature

Higher temperature speeds up most reactions. Arrhenius in 1889 (Smith, 1970) and others later showed that for most chemical reactions, the temperature dependence of the reaction rate is of the form:

$$k'' = k_0'' e^{-E_a/(RT)} \tag{9.22}$$

where k_0'' is called the frequency factor, E_a is called the activation energy (J/mole), R_g is the gas constant (J/mole·K), T is the absolute temperature in K. The above equation is often referred to as the Arrhenius law of temperature dependency. It is an empirical equation that has been found to represent many chemical and biological processes. Attempts to arrive at this expression from a theoretical standpoint have had some success (e.g., the collision theory), but only for very idealized gaseous reactions. More details about the Arrhenius law can be found in most textbooks on physical chemistry or chemical kinetics.

9.5 Chapter Summary—Equilibrium, Mass Conservation, and Kinetics

- **Equilibrium**

  1. Equilibrium between a gas and a liquid is described by Henry's law (page 180).
  2. Equilibrium between gas and a solid is described by isotherms (page 182).

3. Equilibrium between a solid and liquid in absorption (page 185).

- **Kinetics of Chemical Reactions (page 185)**

    1. A reaction where the rate is constant is called a zeroth order reaction. Concentration change for this is given by Eqn. 9.15.

    2. A reaction is called first order whenever the reaction rate is proportional to the concentration of a single reactant. Concentration change for this is given by Eqn. 9.17. Many reactions are often approximated as first order reactions.

    3. Higher temperature speeds up most reactions. This temperature dependency of reaction rate for most reactions is given by Eqn. 9.22.

9.6 Concept and Review Questions

1. Why are zeroth order reactions not encountered often?

2. How is equilibrium different from steady state?

3. Why is equilibrium important in the context of transport?

4. At higher temperature, does the solubility of gases increase or decrease?

5. How long does it take for a process to come to complete equilibrium with the surroundings?

6. When considering the transport of pollutants through the soil to groundwater, is a higher partition coefficient more desirable?

9.7 Further Reading

Berge, H.F.M. ten. 1990. Heat and water transfer in bare topsoil and the lower atmosphere. Center for Agricultural Publishing and Documentation, Wageningen, the Netherlands.

Foust, A.S., L.A. Wenzel, C.W. Clump, L. Mans, and L.B. Anderson. 1960. *Principles of Unit Operations*. John Wiley & Sons, New York.

Henderson, S.M. 1952. A Basic Concept of Equilibrium Moisture. Agric. Eng. 33 (1) 29-31.

Masters, G.M. 1998. *Introduction to Environmental Engineering and Science*. Prentice Hall, New Jersey.

Siau, J.F. 1984. *Transport Processes in Wood*. Springer-Verlag, New York.

Smith, J.M. 1970. *Chemical Engineering Kinetics*. McGraw-Hill Inc., New York.

Smith, L.S. 1982. *Introduction to Fish Physiology*. T.F.H. Publications, Inc., Neptune City, New Jersey.

Welty, J.R., C.E. Wicks, R.E. Wilson, and G. Rorrer. 2001. *Fundamentals of Momentum, Heat, and Mass Transfer*. John Wiley & Sons, New York.

Yang, W.J. 1989. *Biothermal-Fluid Sciences: Principles and Applications*. Hemisphere Publishing Corporation, New York.

9.8 Problems

9.8.1 Concentration in Various Units

Calculate the molar concentration, mass concentration, mole fraction and mass fraction of nitrogen and oxygen in standard air (1 atmosphere, 20°C) assuming the air is made up of 79% nitrogen and 21% oxygen (percentages refer to partial pressure in relation to total pressure).

9.8.2 Mass Conservation: Salt Balance in an Agricultural Field

The salt content of the soil in a field is an important parameter in a crop's success. The minerals in these salts provide nutrients for the growth of the plants, but too much salt can be detrimental to the plants' health as well. 1) For the conditions given below, find the annual change in salt concentration per hectare (10,000 m^2) in the soil. 2) If the maximum salt concentration a particular field could tolerate is 5,000 kg/ha, has the field reached this limit?

The rainfall occurs entirely in winter and amounts to 30 cm, with a total salt concentration of 40 ppm. Capillary rise from shallow, saline ground-water occurs in the spring and fall, and amounts to 10 cm at a concentration of 1,000 ppm. 90 cm of irrigation water is applied during the summer growing season with a concentration of 400 ppm. Drainage occurs during the summer and amounts to 20 cm, with a soluble salt concentration of 800 ppm. Fertilizers and soil amendments add an additional 120 g/m^2 of soluble salts, while the crops use 100 g/m^2 of salts. Assume that the salt concentration at the beginning of the year was 1,000 kg/hectare and that there is no salt generation or decomposition within the soil.

9.8.3 Mass Conservation: Indoor Air Quality and Smoking

Tobacco smoke is often considered the most harmful and widespread contaminant of indoor air. Scientists have identified over 4,000 different chemical compounds in tobacco smoke of which at least 50 are known carcinogens. Long term exposure to tobacco smoke has been linked to heart diease, cancer, respiratory illness in young children and retardation of growth and development of fetuses during pregnancy.

Consider an enclosed space with a volume of 12 m × 12 m × 4 m that has 30 persons smoking two cigarettes every hour. One of the gases coming out of a cigarette is formaldehyde and each cigarette may be assumed to emit 1.35 mg of formaldehyde.

Conversion of formaldehyde to carbon dioxide can be assumed to be first order and the rate of reaction is 0.40 per hour. Outside (fresh) air enters the enclosed space at the rate of 800 m^3 per hour. Assume the smoke becomes completely mixed with air and this smoke-mixed air leaves the enclosed space at the same rate that the fresh air enters. 1) What is the concentration of formaldehyde in the room air at steady state? 2) If the threshold for eye irritation due to formaldehyde is 0.05 ppm (parts per million), what minimum flow rate of fresh air needs to be maintained for the enclosed space? At the assumed temperature of 25°C and pressure of 1 atm for the enclosed space, convert to ppm using the formula ppm = mg/m^3 × 24.45/M, where M is 30, the molecular weight of formaldehyde.

9.8.4 Effect of Altitude on Blood Oxygen Level

Consider a mountain climber at 10,000 m above sea level where the total air pressure is about 0.25 atm. What fraction of oxygen (compared to sea level) will be dissolved in the blood at this lower pressure? Assume the air surrounding blood vessels, etc., has the same composition as sea level, i.e., 21% oxygen and 79% nitrogen.

9.8.5 Equilibrium of O$_2$ in Blood

One of the primary functions of the blood is to carry oxygen from the lungs to cells in all parts of the body. One mechanism of transport is to dissolve the oxygen in the plasma as the blood passes through the lungs. The oxygen concentration equilibrates between the blood and the inhaled air.

1) Find the concentration of dissolved oxygen in the blood plasma. 2) Does this concentration seem reasonable? 3) What are the other factors involved in the transport of oxygen to the cells? Assume that blood plasma has the same material properties as water, and that the concentration of oxygen in the inhaled air is not affected by the transport of oxygen into the blood plasma. In addition the blood plasma is well-mixed so the concentration of dissolved oxygen is constant throughout the vessel. Mole fraction of O$_2$ at 1 atm and 38°C air is 0.145, Henry's constant is 5.21×10^4 atm/mole fraction of O$_2$ in water, molecular weight of O$_2$ is 32, and molecular weight of H$_2$O is 18.

9.8.6 The Deep Diver and the Bends

Consider a diver in water down to a level where the total pressure is 3 atm. Consider air in tissue and everywhere surrounding the blood vessels of the diver to be still at the same composition as the air before entering the water, i.e., 21% oxygen and 79% nitrogen. 1) How much more nitrogen will be dissolved in the blood at this higher pressure of 3 atm? Leave your answer in terms of mole fractions. 2) What happens to this dissolved nitrogen as the diver starts to go up rapidly?

9.8. PROBLEMS

9.8.7 Oxygen Availability for Fish in Aquaculture

Consider an aquaculture system where water containing fish is surrounded by air. Calculate the fraction of original oxygen that will be retained in the water as temperature of the water and the air increases from 5°C to 15°C.

9.8.8 Dissolved Oxygen in Water

The amount of oxygen dissolved in water is an important parameter for industrial waste treatment. Often it is desired to know the dissolved oxygen level in the efflux. Determine the saturation concentration of oxygen (in mg/liter) in pure water which is exposed to dry air (0.21 mole fraction O_2) at 1 atm and 20°C.

9.8.9 The "zing" in the Soft Drinks

In the making of carbonated soft drinks, dissolved CO_2 is an important ingredient which provides the necessary "zing" to the beverage. We need to maintain a desirable level of CO_2 concentration in the drink. At a temperature of 10°C, what partial pressure of CO_2 must be kept in the gas to keep the CO_2 concentration in the water at 0.9×10^{-4} kg CO_2/kg of water? Henry's constant for CO_2 in water at 10°C is 1040 atm/mole fraction.

9.8.10 Moisture Equilibration During Storage

Potato that has reached equilibrium with drying air at 60°C and a humidity of 0.04 kg of moisture per kg of dry air is packaged in a bag that is impermeable to moisture. The bag initially had air at 60°C and 0.008 kg of moisture per kg of dry air. We would like to know what will be the final moisture content of the potato in the sealed bag. Consider the package to have 95% by weight of potato and 5% by weight of air. 1) Formulate the problem, i.e., write the equation from which the final moisture content of the potato in the package can be calculated. 2) Solve for the final moisture content in step 1. The equilibrium moisture content of potato at 60°C is shown in Figure 9.15.

9.8.11 Moisture Loss During Storage

Consider Figure 9.16 on equilibrium moisture content of soap at approximately 25°C. Note that the moisture content is given on dry basis. Consider a soap of weight 100 g that has 40 g water in it initially (in the package) and was left unpackaged in summer in your bathroom. The air temperature in the bathroom stayed pretty much the same at approximately 25°C. However, the relative humidity in the summer was about 60% in the bathroom while in the winter it is down to 10%. 1) Determine the weight loss in soap after unpacking it in summer. 2) Determine the weight loss between summer and winter.

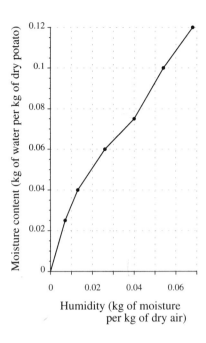

Figure 9.15: Equilibrium moisture content of potato.

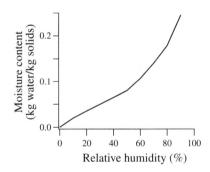

Figure 9.16: Equilibrium moisture content of soap.

9.8.12 Change in Moisture Content in a Hygroscopic Material During Storage

Calculate the moisture that will be gained per unit of dry weight of a certain wood when stored at 21°C if the relative humidity changes from 20% to 80%. Note that the equilibrium moisture content of the wood, given in Figure 9.8, is water content per unit of total (water + drywood) weight.

9.8.13 Half-life in a First Order Reaction

If a first-order reaction has an activation energy of 25,000 cal/mol and, in the equation $k'' = k_0'' e^{-E_a/RT}$, k_0'' has a value of 5×10^{13}/s, at what temperature will the reaction have a half-life of (a) 1 min; (b) 1 month (30 days)?

9.8.14 Decay of Pesticides in Soil

Consider two pesticides aldicarb and atrazine in the soil degrading independently as a first order reaction given by $c = c_0 e^{-k''t}$. They were applied in equal amounts initially. The soil was tested 30 days after application and the relative concentration $c_{\text{aldicarb}}/c_{\text{atrazine}}$ was found to be 0.70. Calculate the half life of aldicarb in the soil if the half life of atrazine in the soil is 60 days.

9.8.15 Heating Shorter at Higher Temperatures Can Keep More Nutrients in Food

Destruction of bacteria and nutrient occuring during a heat sterilization process are often assumed to follow first order reactions. Consider a bacterial death reaction given by $k_0'' = 4.669 \times 10^{38}$/s and a $E/R = 35,751$ K and a vitamin C destruction given by $k_0'' = 8.885 \times 10^{11}$/s and a $E/R = 14,300$ K. Consider the entire material to be at the same temperature during heating, i.e., it can be considered lumped. 1) Calculate the fraction of initial concentrations of bacteria and nutrients that would be retained by a heating process that maintains the food temperature at 120°C for 10 minutes. 2) If the food is maintained at 140°C instead, how long should it be heated to destroy the same fraction of the initial bacterial population as in step 1? 3) For the duration of heating calculated in step 2 at the temperature 140°C, what would be the fraction of the nutrients retained? 4) Would you prefer to heat the food at 120°C or 140°C? Why?

9.8.16 Adjustment of Sterilization Time

Sterilization of foods and biomaterials involves killing of bacteria that is often assumed to follow first order reaction kinetics. A particular food material was to be heated at 121°C for 6 minutes to reduce the bacterial concentration to a very low value. Due to the failure of the heating system, the food could only be heated at 111°C. To reduce the bacterial concentration to the same low final value, how many minutes should the food be heated at this lower temperature? The E/R value for the bacterial death is 35,751 K.

Chapter 10

MODES OF MASS TRANSFER

CHAPTER OBJECTIVES

After you have studied this chapter, you should be able to:

1. Explain the process of molecular diffusion and its dependence on molecular mobility.

2. Explain the process of capillary diffusion

3. Explain the process of dispersion in a fluid or in a porous solid.

4. Understand the process of convective mass transfer as due to bulk flow added to diffusion or dispersion.

5. Explain saturated flow and unsaturated capillary flow in a porous solid

6. Have an idea of the relative rates of the different modes of mass transfer.

7. Explain osmotic flow.

KEY TERMS

- **diffusion, diffusivity, and diffusion coefficient**
- **dispersion and dispersion coefficient**
- **hydraulic conductivity**
- **capillarity**
- **osmotic flow**
- **mass and molar flux**
- **Fick's law**
- **Darcy's law**

In this chapter, we will study the fundamental ways mass can be transferred. Figure 10.1 shows how the contents of this chapter relate to other chapters on the subject of mass transfer. We study the processes of molecular and capillary diffusion, dispersion, and bulk flow or convection. Of these processes, diffusion and dispersion can be formulated the same way and are treated together. These two processes together are also the subject of later chapters. Hydraulic or Darcy flow in a porous media is introduced in this chapter as a description for bulk flow through such media.

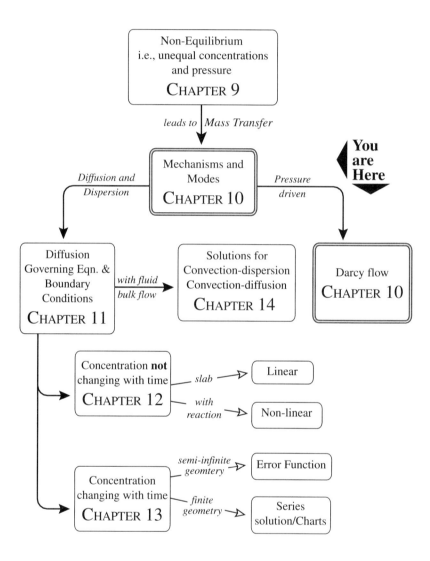

Figure 10.1: Concept map showing how the contents of this chapter relate to other chapters in mass transfer.

10.1 A Primer on Porous Media Flow

Before we introduce the three modes of mass transfer— diffusion, dispersion and convection in the following sections, it is important to note that convection mode of mass transfer requires information about the velocities, that we can obtain from studying pressure driven flow in a course on fluid mechanics. The important case of pressure driven flow through porous media is not typically studied in an undergraduate course. Flow through a porous media is introduced here due to its importance in applications as varied as pollutant transport through soil in a field and water transport in a food material during drying. The introduction to porous media covered here will provide the foundations for studying transport due to capillary diffusion and convection-dispersion through such a porous media.

Movement of liquid in a porous material can be described by Darcy's law, written as:

$$n^v = -K \frac{\partial \mathcal{H}}{\partial s} \qquad (10.1)$$

where n^v is the volumetric flux (also called specific discharge) in m^3/m$^2 \cdot$s, $\mathcal{H}$ is the hydraulic or water potential in m that is causing the flow and s is distance along flow in m. The proportionality constant K is termed hydraulic conductivity. The units of K can be calculated as:

$$[K] = \frac{[n^v]}{[\partial \mathcal{H}/\partial s]} = \frac{\text{m/s}}{\text{m/m}} = \text{m/s} \qquad (10.2)$$

Hydraulic potential is the sum of pressure and matric potential h, and gravitational potential z:

$$\mathcal{H} = h + z \qquad (10.3)$$

The matric potential measures the physical forces, such as capillarity, which bind the water to the porous matrix. The retention of water is a result of attractive forces between the solid and liquid phases. In a soil, for example, these matric forces enable it to hold water against such forces as gravity, evaporation, and uptake by plant roots. In soil, there are three mechanisms for binding of water to the solid matrix: direct adhesion of water molecules to solid surfaces by London-van der Waals forces, capillary binding, and osmotic binding in double layers. This matric potential is important as a driving force for flow in unsaturated soil and other systems such as in the cell walls of root cortex and leaf mesophyll tissue. Matric potential is always negative or zero. The pressure potential is due to hydrostatic or pneumatic pressure applied to water. The gravitational component of the water potential is simply due to a difference in depth z in the vertical (parallel to g) direction from a reference point, usually taken as soil surface or surface of a water table.

Like Fourier's and Fick's laws, Darcy's law (Equation 10.1) is an empirical relationship. The hydraulic conductivity, K, represents the ease with which fluid can be transported through a porous matrix, and is discussed more fully in section 10.1.1. The units for the volumetric flux n^v in Equation 10.1 are m^3/m$^2 \cdot$s or m/s. Since the units for n^v are the same as those for velocity, it is also called Darcy velocity. Note that, even though n^v has the units of velocity, it is not the true average velocity of fluid through the pores as flow takes place through only the porous part of the cross-sectional area

A. Let ϕ be the volumetric porosity, defined as the ratio of volume of void space (pore volume) to the bulk volume of a porous medium. Average areal porosity–that is the ratio of void area to the total area at a cross section–can be considered the same as average volume porosity ϕ. Thus, the true average velocity of fluid through the pores is

$$v_{average} = \frac{n^v}{\phi} \qquad (10.4)$$

Example 10.1.1 Flow of Water Through a Column of Packed Sand

Consider the flow of water through a column of packed sand. The column has a cross-sectional area of 250 cm^2, and a length of 150 cm. The packed sand has a porosity of 0.30 and a hydraulic conductivity of 15 m/day. For a pressure drop of 200 cm through the column, calculate 1) volumetric flux (also called *specific discharge*) n^v, 2) volumetric flow rate, 3) average velocity of fluid in the pores.

Solution

1. Volumetric flux or specific discharge is calculated using Darcy's law:

$$\begin{aligned} n^v &= K \frac{\Delta \mathcal{H}}{\Delta x} \\ &= 15 \left[\frac{m}{day}\right] \frac{200}{150} \left[\frac{m}{m}\right] \\ &= 0.023 \text{ cm/s} \end{aligned}$$

2. The volumetric flow rate can be calculated from the volumetric flux as:

$$\begin{aligned} n^v A &= 0.023 [\text{cm/s}] \times 250 [\text{cm}^2] \\ &= 5.75 \text{ cm}^3/\text{s} \end{aligned}$$

3. The *average velocity of fluid in the pores* is given by

$$v(\text{average velocity}) = \frac{n^v}{\text{porosity}} = \frac{0.023}{0.3} = 0.077 \text{ cm/s}$$

10.1.1 Physical Interpretation of Hydraulic Conductivity K and Permeability k

Since hydraulic conductivity K relates the flux to the potential causing the flow, it must depend on both matrix and fluid properties. As a first approximation, a porous medium can be considered as a bundle of tubes of varying diameter embedded in the solid matrix, as shown by the schematic in Figure 10.2. Using this simple approximation, we explicate the physical meaning of K. From fluid mechanics, the volumetric flow rate Q_i in a tube of uniform radius r_i (Poiseuille's flow) is

10.1. A PRIMER ON POROUS MEDIA FLOW

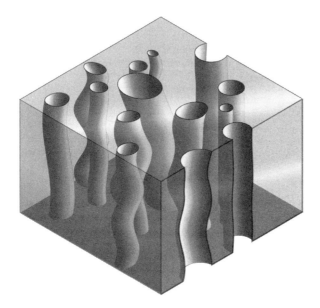

Figure 10.2: Idealization of a porous media as bundle of tubes of varying diameter and tortuosity.

$$Q_i = -\frac{\pi r_i^4 \rho g}{8\mu}\frac{\partial h}{\partial s}$$

If ω_i represents the number of pores (tubes) in the ith pore size class with radius r_i, and Q_i is the discharge rate per pore in that group, then the total volumetric flux, n^v, through the collection of pores is given by

$$\begin{aligned}n^v &= \frac{\sum_i \omega_i Q_i}{A}\\ &= -\frac{\rho g}{8\mu}\frac{\sum_i \pi \omega_i r_i^4}{A}\frac{\partial h}{\partial s}\end{aligned}$$

where A is the total cross-sectional area. All pores do not run parallel to the flux direction. To account for the lengthened distance, a tortuosity factor τ is introduced that is the ratio of the actual roundabout path along the pore to the apparent, or straight, flow path. In soils, for example, the value of τ is typically between 1 and 2. Introducing tortuosity, the above equation is written as

$$n^v = -\frac{\rho g}{8\mu\tau}\frac{\sum_i \pi \omega_i r_i^4}{A}\frac{\partial h}{\partial s} \qquad (10.5)$$

If $\Delta\beta_i$ is the volume fraction of pores with radius r_i, $\Delta\beta_i$ can be approximated as an area fraction

$$\Delta\beta_i = \frac{\omega_i \pi r_i^2}{A} \tag{10.6}$$

Using Equation 10.6, the volumetric flux n^v is written as

$$n^v = -\underbrace{\frac{\rho g}{8\mu\tau} \sum_i \Delta\beta_i r_i^2}_{K} \frac{\partial h}{\partial s} \tag{10.7}$$

Comparing Equation 10.7 with Darcy's law (Equation 10.1), we can write

$$K = \frac{\rho g}{8\mu\tau} \sum_i \Delta\beta_i r_i^2 \tag{10.8}$$

which can be rearranged as

$$K = \underbrace{\frac{\rho g}{\mu}}_{\text{fluid property}} \underbrace{\frac{1}{8\tau} \sum_i \Delta\beta_i r_i^2}_{\text{matrix property}} \tag{10.9}$$

Equation 10.8 shows that, as expected, K depends on both fluid and matrix properties. The relevant fluid properties are density ρ and viscosity μ. The relevant solid matrix properties, as given by Equation 10.8 are pore size distribution, shape of pores, porosity, and tortuosity. The sole effect of matrix property can be included in *permeability* or *intrinsic permeability* k such that

$$k = \frac{1}{8\tau} \sum_i \Delta\beta_i r_i^2 \tag{10.10}$$

so that

$$K = \frac{k\rho g}{\mu} \tag{10.11}$$

The ratio ρ/μ represents the effect of fluid properties. To show the units of the various quantities in the above equation, we can write

$$K \left[\frac{\text{m}}{\text{s}}\right] = \frac{k\,[\text{m}^2]\,\rho\left[\frac{\text{kg}}{\text{m}^3}\right]\,g\left[\frac{\text{m}}{\text{s}^2}\right]}{\mu\left[\frac{\text{kg}}{\text{ms}}\right]}$$

Using this definition of permeability, k, Darcy's law can be rewritten as

$$\begin{aligned} n^v &= -K\frac{\partial h}{\partial s} \\ &= -\frac{k}{\mu}\frac{\partial(\rho g h)}{\partial s} \\ &= -\frac{k}{\mu}\frac{\partial P}{\partial s} \end{aligned} \tag{10.12}$$

since pressure is related to head as $P = \rho g h$. Equation 10.12 is the alternate form of Darcy's law written in terms of gradient in pressure, P, instead of head, h. The units of k can be seen from Equation 10.10 to be m^2.

When the solid is saturated with liquid, as in saturated soil where pores are completely filled with water, both the matrix contribution and the fluid contributions are constant. This leads to a constant hydraulic conductivity of a given solid saturated by a particular fluid. For an unsaturated solid, the hydraulic conductivity can vary dramatically, as described in section 10.2.2.

10.1.2 Capillarity and Unsaturated Flow in a Porous Media

Capillary flow is due to the difference between the relative attraction of the molecules of the liquid for each other and for those of the solid. As an example, such difference causes the rise of water in an open tube of small cross-section (Figure 10.3). For the column of water of height h at equilibrium shown in this figure, the hydrostatic pressure, $P(= \rho g h)$, matches the capillary attraction given by $2\gamma/r$, where γ is the surface tension. Equating the two pressures results in the following relationship for the height of capillary rise

$$h = \frac{2\gamma}{\rho g r} \qquad (10.13)$$

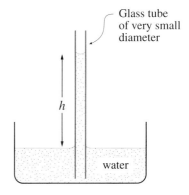

Figure 10.3: Capillary attraction between the tube walls and the fluid causes the fluid to rise.

Data on surface tension of water is given on page 353 in Appendix D.8. As the radius becomes very small, capillary rise increases significantly. Capillarity is the reason, for example, that the soil does not get completely drained by gravity. Capillarity is also the mechanism by which water can rise to the top of a tree as tall as 30 m, and be available for transpiration. The capillarities in a tree which are formed by numerous interstices of the cell wall of the xylem vessels aid the tree in transporting water from the roots and base to the upper branches and leaves. A representative radius of these channels in the cell was estimated as 5×10^{-9} m (Nobel, 1974). Using Eqn. 10.13, it can be seen that a capillary of this size could support a water column 3 km in height.

In a porous solid, capillarity will cause the liquid to be attracted more strongly or held more tightly when there is less of it, i.e., at lower concentrations of the liquid. Conversely, the liquid will be held less tightly when there is more of it. This sets up a situation where differences in capillary action result in a flow of liquid from higher concentration (relatively loosely held) to lower concentration (more tightly held). This is referred to as unsaturated flow. Unsaturated flow is extremely important, for example, in drainage of soils or drying of food materials. In an unsaturated solid, capillarity can be the primary mode of transport for the liquid. In contrast to the positive pressures (applied or gravitational) of liquid in a saturated solid, water pressure is negative or the water is attracted to the solid. As an example, Figure 10.4 shows this negative matric potential due to capillarity and other effects in a soil as a function of moisture content. This difference in the negative pressures drives the flow when the soil is unsaturated. As will be shown in Section 10.2.2, this capillary flow can be treated mathematically as analogous to molecular diffusion.

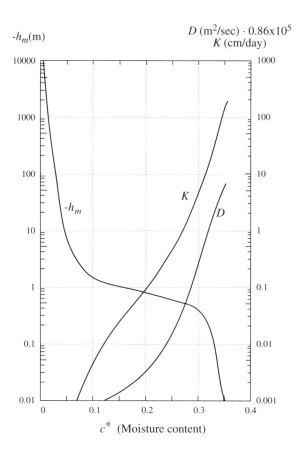

Figure 10.4: Typical relationship of capillary diffusivity to the moisture content in a porous media. Shown is data for soil where moisture diffusivity D is calculated from soil water characteristics (matric potential) h and hydraulic conductivity K.

10.1.3 Osmotic Flow in a Porous Media

Osmotic flow, or osmosis, is the transport of a solvent from a region of low solute concentration to a region of high solute concentration through a semi-permeable membrane. A semi-permeable membrane is a porous structure that allows some of the species (here the solvent) to go through but stops the passage of other species (here the solute). Consider two solutions separated by a semi-permeable membrane, as shown in Figure 10.5. The solvent moves from low solute concentration to high solute concentration until a hydraulic pressure of magnitude Π is developed that exactly opposes the osmotic flow, stopping further flow. The pressure magnitude (Π) is called the osmotic pressure and it depends on the concentration c of the solute. This relationship is given by the Van't Hoff law

$$\Pi = cRT \tag{10.14}$$

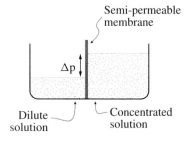

Figure 10.5: Osmotic flow from a dilute to a concentrated solution through a semi-permeable membrane.

10.2. DIFFUSIVE MASS TRANSFER

where c is the total concentration of the solutes in a solution, T is the absolute temperature, and R is the gas constant. Note the similarity of this equation with the ideal gas law (Section 9.1.1). Thus, osmosis is in the direction to reduce the solute concentration gradient, i.e., toward equalizing the concentration. Since cell membranes are permeable to water, water can flow through them. An example of osmotic flow is how a cell can shrink during freezing of a tissue, as explained in Figure 7.10 on page 130.

Since osmotic pressure is equivalent to hydraulic pressure, as shown in Figure 10.5, Darcy's law for flux can be generalized to include both hydraulic and osmotic pressures in a porous medium as

$$n^v = -\frac{k}{\mu}\frac{\partial}{\partial s}(P - \Pi) \qquad (10.15)$$

In Eqn. 10.15 for osmotic flux, if the applied or hydraulic pressure P does not change,

$$n^v = \frac{k}{\mu}\frac{\partial \Pi}{\partial s} \qquad (10.16)$$

In biological literature, this equation is often written in terms of an osmotic pressure difference $\Delta \Pi$ as

$$n^v = L_p \Delta \Pi \qquad (10.17)$$

where L_p is the membrane permeability that lumps together the effects of the porous structure of the membrane, the thickness of the membrane, and the fluid properties. Note that membrane permeability, L_p, (unit m/Pa·s) is different from hydraulic permeability k, (unit m^2) although the term permeability is used in either case. The direction for the flux in Eqn. 10.16 is from low solute concentration (low osmotic pressure) to high solute concentration (high osmotic pressure).

10.2 Diffusive Mass Transfer

In the previous section, we developed an understanding of bulk flow through a porous media. As mentioned earlier, bulk flow in a fluid general is not covered here and is the subject of fluid mechanics. In this section and the following one, we will study how transport of a mass species can occur due to differences in concentration, in contrast with transport due to bulk flow.

10.2.1 Molecular Diffusion

In a material with two or more mass species whose concentrations vary within the material, there is tendency for mass to move. Diffusive mass transfer is the transport of one mass component from a region of higher concentration to a region of lower concentration. Examples of diffusive mass transfer are plentiful. For instance, perfume from one corner of the room can eventually be smelled from everywhere in the room, even if there is not much air flow in the room.

Diffusive mass transfer is analogous to diffusive heat transfer described in section 2.1. However, it is inherently more complicated since it deals with a mixture with at least two species. Diffusion is a natural, dynamic molecular process which tends to equilibrate the differences in concentration. There is net movement of a species from higher to lower concentration, simply due to the random molecular movement. Note the similarity between the mechanisms of molecular and thermal diffusion.

For molecular diffusion whose driving force is a concentration gradient, the rate at which mass is transported per unit area (diffusive flux) is related to the concentration gradient as:

$$\underbrace{j_{A,x}}_{\text{Diffusive flux}} = -\underbrace{D_{AB}}_{\text{Diffusivity}} \underbrace{\left(\frac{dc_A}{dx}\right)}_{\substack{\text{Concentration} \\ \text{gradient}}} \quad (10.18)$$

where $j_{A,x}$ is the diffusive flux in kg/m^2·s of species A at point x, c_A is the concentration of A in kg/m^3, x is distance in m, and D_{AB} is the mass diffusivity of A in B in m^2/s which is also called the *diffusion coefficient*. This is known as the *Fick's law of diffusion*. In molar units, $j_{A,x}$ is in kmol/m^2·s and c_A is in kmol/m^3. The units for mass diffusivity D_{AB} are m^2/s, as can be seen from:

$$[D_{AB}] = \frac{[j_{A,x}]}{\left[\frac{dc_A}{dx}\right]} = \frac{\left[\frac{\text{kg}}{\text{m}^2\text{s}}\right]}{\left[\frac{\text{kg}}{\text{m}^3\text{m}}\right]} = \left[\frac{\text{m}^2}{\text{s}}\right]$$

Physical interpretation of diffusivity

A solute introduced into a fluid at a certain point diffuses out. If we consider a 1-D diffusion process along x, then the resulting concentration profile is a Gaussian distribution (see Figure 10.6) in x at any time, with some molecules having moved further out from the origin than others. Since the molecules would move in the +ve as well as -ve direction, an average of the square of the displacements, $<x^2>$, can be defined as a measure of the movement of the molecules. This average is termed the mean-square displacement.

From the Gaussian distribution, the mean-square displacement $<x^2>$ can be evaluated with respect to the diffusivity by the relation (see, e.g., Barrow, 1981):

$$D = \frac{<x^2>}{2t} \quad (10.19)$$

The diffusivity is thus one half of the mean-square displacement per unit time. Using the square root of the mean square displacement, a diffusion velocity can be defined as:

$$u_{diff} = \frac{\left(<x^2>\right)^{1/2}}{t} = \sqrt{\frac{2D}{t}} \quad (10.20)$$

Note that the velocity u_{diff} is many orders of magnitude smaller than the molecular velocity. The molecules' movement in random directions is known as Brownian motion. Although molecular velocities are very large, any molecule encounters a large number of collisions per unit time. As a result, the net distance a molecule moves is

10.2. DIFFUSIVE MASS TRANSFER

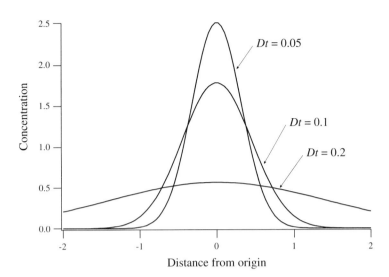

Figure 10.6: Concentration profiles at different times from an instantaneous source placed at zero distance. Here units of Dt and x are arbitrary, but related so that Dt/x^2 is dimensionless.

much smaller, as shown by the diffusion velocity. For example, the root mean square displacement of a N_2 molecule at atmospheric pressure and 25°C is about 0.56 cm in one second. However, the total distance traveled along a zigzag path during this one second is about 475 m!

The mass diffusivity defined above is completely analogous to the thermal diffusivity defined earlier in the case of heat transfer (see Chapter 2). As thermal diffusivity was the proportionality constant between heat flux and energy gradient, mass diffusivity is the proportionality constant between diffusive mass flux and concentration gradient. Diffusional mass transfer is generally a slower process than diffusional heat transfer. To see this, compare mass diffusivities in Figure 10.7 with thermal diffusivities in Figure 2.6.

The mass diffusivity depends on the pressure, temperature, and composition of the system. Since the mass diffusivity is a measure of molecular mobility, it is expected to be higher for gases than for liquids or solids. The typical ranges of diffusivities are given in Figure 10.7. Most available diffusivity values are experimental. Theoretical expressions for very idealized systems are discussed in the next section.

Example 10.2.1 Diffusivity of the Protein Myoglobin

At 20°C the diffusivity of the protein myoglobin (size 10 μm) in water is 11.3×10^{-11} m^2/s. What is the mean time required for a myoglobin molecule to diffuse a distance of 10 μm?

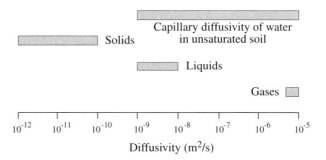

Figure 10.7: Typical ranges of diffusivity values.

Analysis: The analysis here simply involves the correspondence between diffusivity and the physical distances moved, i.e., use of Eqn. 10.19. Thus, we get

$$\begin{aligned} t &= \frac{<x^2>}{2D} \\ &= \frac{(10 \times 10^{-6})^2 \text{m}^2}{2(11.3 \times 10^{-11})\text{m}^2/\text{s}} \\ &= 0.44 \text{s}. \end{aligned}$$

In spite of their large size, macromolecules like myoglobin (mol wt 17000) diffuse rapidly in water. Note, however, to move a distance of 1 cm, it will take

$$\begin{aligned} t &= \frac{<x^2>}{2D} \\ &= \frac{(1 \times 10^{-2})^2 \text{m}^2}{2(11.3 \times 10^{-11})\text{m}^2/\text{s}} \\ &= 5.12 \text{ days}. \end{aligned}$$

Thus, for macroscopic distances, molecular diffusion is a slow process.

Diffusivity for gases

Here, we discuss the diffusive movement of one gas inside another gas. For binary mixtures of low density gases that are non-polar and non-reacting, diffusivity can be estimated using the kinetic theory of gases. One such estimation of diffusivity is:

$$D_{AB} = \frac{0.001858 T^{3/2} \left(1/M_A + 1/M_B\right)^{1/2}}{p \sigma_{AB}^2 \Omega_{D,AB}} \qquad (10.21)$$

where D_{AB} is the diffusivity of A through B, T is the absolute temperature, M_A and M_B are molecular weights, p is absolute pressure in atm, σ_{AB} is the collision diameter in Å, and $\Omega_{D,AB}$ is a dimensionless function of the temperature and the intermolecular potential. As expected, Equation 10.21 predicts an increase in diffusivity with temperature due to increased energy and a decrease in diffusivity for larger molecules.

Diffusivity for liquids

Here, we discuss the diffusivity of liquids or of dissolved solutes in other liquids. When studying diffusivity of liquids, it is useful to make a distinction between electrolytes and non-electrolytes. The diffusional coefficient of non-electrolytes can be predicted using hydrodynamical theory. A rigid sphere translating through a viscous medium experiences a frictional force with the frictional coefficient f given by the Stokes equation as:

$$f = 6\pi \mu r \qquad (10.22)$$

where μ is the medium viscosity and r is the radius of the sphere. Einstein derived a relation between the macroscopic diffusion coefficient, D, and the frictional coefficient f. This relation is known as the Stokes-Einstein relation, and is expressed as:

$$D = \frac{\kappa T}{f} \qquad (10.23)$$

where κ is the Boltzmann constant, and T is the absolute temperature. This relation has been used to estimate the radius of macromolecules, such as proteins. However, a certain amount of solvent is usually associated with a macromolecule in solution (termed solvation), which increases the effective radius of the molecule and therefore increases the frictional coefficient. Even so, the Stokes-Einstein relation is the most common basis for estimating diffusion coefficients in liquids. Substituting Equation 10.22 in Equation 10.23, we get:

$$D = \frac{\kappa T}{6\pi \mu r} \qquad (10.24)$$

Thus, the diffusivity is related to the solute molecule's mobility, i.e., the net velocity of the molecule. The reduction of diffusivity with size can be seen in the data for proteins in Figure 10.8.

Diffusivity for solids

Diffusion in solids can be quite complex. Also, diffusion of solids in solids is very slow (Figure 10.7) and does not seem to play a major role in biological or environmental systems. Thus, we discuss here only the diffusion of gases and liquids in solids. For simplicity, we can divide the solids into two groups: porous and non-porous. In the non-porous solids, the liquid or the gas is considered dissolved in the solid and diffuses through the solid. This type of diffusion is described by the Fick's law (Equation 10.18). In a porous solid, liquid movement is primarily due to capillarity and other forces. This is not molecular diffusion, although there can be some analogy in special cases, as discussed in Section 10.2.2. The rest of this section considers only gaseous molecular diffusion in a porous solid.

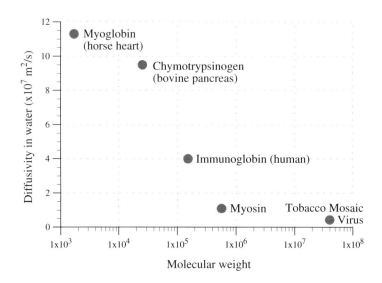

Figure 10.8: Diffusivity of some proteins in water at 20°C, as related to their molecular size.

Diffusion of gases in a porous solid

The diffusion of gases in a porous solid plays an important role in biological systems. Most biological materials, such as a cell membrane or a tissue, and agricultural material, such as soil, can be described as capillary porous materials. Capillarity refers to small pore sizes. For example, Figure 10.9 shows such pores in wood. Liquid water transport in these material is by capillarity and is described in Section 10.1.2. The vapor transport is typically by diffusion through the air in the pores. In soil, the metabolic activity of plant roots and microorganisms consumes oxygen and generates carbon dioxide. Oxygen, carbon dioxide, and water vapor all diffuse into and out of soil. A sufficient concentration of oxygen (adequate aeration) needs to be maintained in the soil for the absorption of nutrients by roots and for the beneficial activity of microorganisms.

Diffusion of gases in a porous solid is more complex than diffusion in a homogeneous non-porous solid as the structure of the material plays a role. For example, as the pores become more tortuous, it takes longer for molecules to move between two given locations. This longer path decreases diffusivity. In addition, when the pores in the solid have diameters comparable to, or smaller than, the molecular mean free path, molecules collide with the walls of the pores (Figure 10.10). Such molecule-wall collisions change the nature of diffusion, and this is known as Knudsen diffusion. The effective diffusivity (D_{eff}) values for this type of diffusion can be significantly different, although Equation 10.18 is still used to describe the process.

10.2. DIFFUSIVE MASS TRANSFER

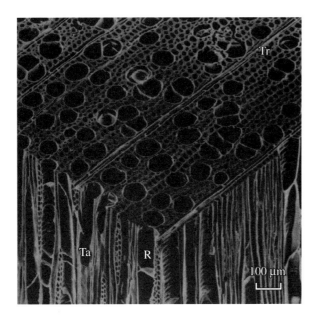

Figure 10.9: A scanning electron micrograph of wood showing pores with diameters ranging from 50-100 μm. Shown is a diffuse-porous hardwood Liriodendron tulipifera. Photograph courtesy of N. C. Brown Center for Ultrastructural Studies, College of Environmental Science and Forestry, State University of New York at Syracuse, New York.

Transport in a cell membrane

A cell membrane is another example of a porous material. Cellular membranes are of fundamental importance to the cell because membranes act as selective barriers, controlling the intracellular contents and providing the cell with the capacity to produce internal compartments with specialized functions. Transport across cell membranes is quite a complex process, as illustrated in Figure 10.11 and passive molecular diffusion (subject of this text) is only one of several different transport mechanisms. Other mechanisms such as convective transport through water channels, transport through gated ion channels, carrier mediated transport and ion pumps are very important for transport and the reader is referred to the excellent treatise by Weiss (1996) for details of these transport mechanisms. Using the mechanism of diffusion, some solutes dissolve in and diffuse through the lipid bilayer portion of a membrane (Figure 10.11). One of the important molecules that diffuses through the lipid bilayer is water.

Anisotropy in a biomaterial

Complexities in structures of biomaterials and soils can lead to *anisotropy* of their transport properties. Anisotropy refers to when the material's physical properties vary along different directions. As an example, Figure 10.9 shows the pore structure in a wood that has the pores aligned in one direction. Therefore, in wood, resistance to

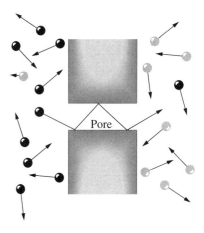

Figure 10.10: Schematic of Knudsen diffusion of a gas in a solid.

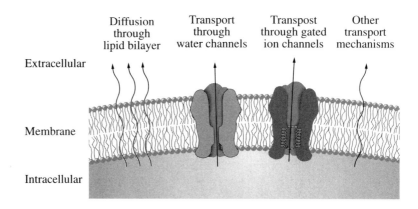

Figure 10.11: Schematic diagram of the cross-section of a cell membrane showing some of the transport mechanisms.

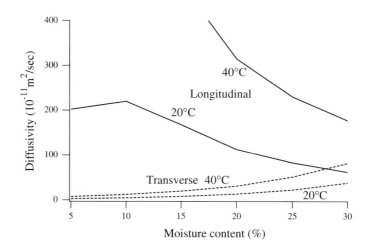

Figure 10.12: Directional and temperature variation of diffusivity in a biomaterial. Shown is data for wood at a oven dry specific gravity of 0.5, calculated using experimental data adapted from Siau (1984).

diffusion in the longitudinal direction (along the pith) is expected to be different from resistance in the transverse direction (perpendicular to the pith). This is shown in Figure 10.12 where the diffusivity of moisture in wood is much higher in the longitudinal direction than in the transverse direction. Note that for the moisture content range in this figure, the pith has mostly vapor, as opposed to pure liquid which would move by capillarity. So this data refers to diffusion of vapor through the air in the pith and

10.2. DIFFUSIVE MASS TRANSFER

through the cell walls. The difference in diffusivities is explained by the much larger values in air than in solid cell wall material and by the presence of more of cell wall in the transverse direction.

Effect of temperature and moisture content

Several other characteristics of diffusion can be illustrated from Figure 10.12. Diffusivities generally increase with moisture content as more moisture is available for diffusion. This is seen for the transverse component. However, for the longitudinal component, the effect of the moisture isotherm between air and wood makes it decrease. All diffusivities increase with temperature as molecules become more mobile.

Example 10.2.2 Transport Through Skin: Medication Without Injection or Ingestion

The pharmaceutical industries would like to send medication through skin, instead of via the traditional route of an injection or ingestion. Not having to use the needles would simplify delivery, and it would avoid worries of intestinal degradation. Such topical application would offer a direct route for medication. A knowledge of transport properties through skin is required for such applications.

Substances can penetrate skin by two major routes (see Figure 5.14 on page 85 for a cross section of skin). The main path being direct diffusion through the stratum corneum (dried epidermal cells). Substances can also bypass the stratum corneum and penetrate via sweat ducts, hair follicles, or perhaps through minor breaches in the stratum corneum. However, for the most part, the effect of this second mode of transport is minimal owing to their relatively small fractional area (10^{-2} to 10^{-3} per unit skin area) and to the fact that they have significant diffusional resistance themselves. Nonetheless, for very hairy skin and for substances that penetrate the stratum corneum slowly, diffusion through these pathways can be significant. Diffusion through these pathways appears particularly important for highly polar, pharmacologically potent molecules that can act quickly on local structures of the skin. The permeability of intact stratum corneum to most electrolytes is extremely low and, for all practical purposes, these shunts provide the only means of access for such substances.

10.2.2 Capillary Diffusion

Capillarity in a porous solid was studied in Section 10.1.2 using Darcy's law (Eqn. 10.1). Recall that the hydraulic potential, $\mathcal{H}$ is composed of matric and gravitational potentials. When a porous solid is unsaturated, i.e., pores are not completely filled with liquid, the matric potential, h, due to the binding of water from capillary and attractive forces, is negative and much stronger than the gravitational potential. Under such conditions, gravitational potential can be ignored and the capillary flux (volumetric) of liquid, n^v, can be written as

$$n^v = -K \frac{\partial h}{\partial s} \quad (10.25)$$

In terms of mass flux, this equation can be written as

$$n = -\rho_{liquid} K \frac{\partial h}{\partial s} \qquad (10.26)$$

where n has the units of kg/m$^2\cdot$s. Let c^* be the volume fraction of liquid in m^3 of liquid per m^3 of solid and c be the concentration of liquid in kg of liquid per m^3 of solid, such that,

$$c = c^* \rho_{liquid} \qquad (10.27)$$

Using this relationship, equation 10.26 can be transformed as:

$$\begin{aligned} n &= -\rho_{liquid} K \frac{\partial h}{\partial c} \frac{\partial c}{\partial s} \\ &= -\frac{K}{\frac{\partial (c/\rho_{liquid})}{\partial h}} \frac{\partial c}{\partial s} \\ &= -\frac{K}{\left(\frac{\partial c^*}{\partial h}\right)} \frac{\partial c}{\partial s} \\ &= -\underbrace{D_{cap}}_{\substack{\text{capillary}\\\text{diffusivity}}} \underbrace{\frac{\partial c}{\partial s}}_{\substack{\text{concentration}\\\text{gradient}}} \qquad (10.28) \end{aligned}$$

where D_{cap} is given by

$$D_{cap} = \frac{K}{\left(\frac{\partial c^*}{\partial h}\right)} \qquad (10.29)$$

Since the second term in Equation 10.28 is a concentration gradient, D_{cap} can be interpreted as "diffusivity". It is important to note that, although the term "diffusivity" is being used here for movement of a liquid (water) through a solid, the mechanism of flow is primarily capillarity, not molecular diffusion. It so happens that the Darcy's law for capillary flow can be cast in a form similar to the molecular diffusion, as shown in Eqn. 10.28. Thus, diffusivity D_{cap} here is really capillary diffusivity. However, in practice, often it is referred to simply as diffusivity and the subscript cap is dropped.

The term $\partial c^*/\partial h$ is termed differential water capacity. The relationship between the matric potential $h(c^*)$ and the volumetric water content c^* is a moisture characteristic curve that is used in soil science literature. Both $h(c^*)$ and permeability $K(c^*)$ are obtained from experiments. Figure 10.4 shows an example of such relationships for a particular type of soil. In this figure, $\partial c^*/\partial h$ is the slope of the soil moisture characteristic curve at any particular value of moisture level, c^*.

Thus, capillary diffusivity, D, can be shown to be a ratio of a transport coefficient K and differential capacity $\partial c^*/\partial h$. Note the similarity of this relationship (Eqn. 10.29) with other diffusivities such as thermal diffusivity:

$$\begin{aligned} \alpha \begin{pmatrix} \text{thermal} \\ \text{diffusivity} \end{pmatrix} &= \frac{k_{thermal}}{\rho c_p} \\ &= \frac{\text{thermal transport parameter}}{\text{thermal capacity}} \qquad (10.30) \end{aligned}$$

The capillary diffusivity $D(c^*)$ is very much material specific and depends on $h(c^*)$ and $K(c^*)$. It can be calculated from experimentally measured $h(c^*)$ and $K(c^*)$ and its value drops significantly (as illustrated in Figure 10.4) as the concentration of water decreases or the material dries. Such capillary diffusivity of water as a function of water content is also needed for engineering study of other important biological processes such as drying of foods, although detailed moisture variation of diffusivity is typically hard to find (Kiranoudis et al., 1994).

10.3 Dispersive Mass Transfer

Dispersion is the spreading of a mass component from higher concentrations to lower concentrations, for example, the spreading of smoke in air from a chimney shown in Figure 10.13. Dispersion has qualitative effects similar to diffusion, but it is a different effect. Dispersion depends on *flow*, being generally an effect of turbulent flow. Molecular diffusion, due to the thermal-kinetic energy, is always present. In a turbulent flow, the random mixing effect, in addition to molecular diffusion, is defined to be dispersion (Cussler, 1984).

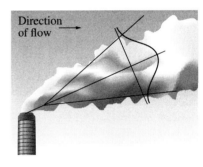

Figure 10.13: Convection-dispersion of smoke from a chimney.

Hydrodynamic dispersion or dispersion in fluid systems is the spreading of a liquid or gas from the path it would follow due to the bulk flow or the convective hydraulics of the system. At the microscopic level, dispersion is not understood in detail. The lateral spreading of a plume from a smokestack (Figure 10.13) relative to its bulk flow in the wind direction is due to turbulence in air.[1] In surface-water regimes, dispersion is also primarily due to turbulence. The net result is that some of the water molecules travel more rapidly than the average linear velocity, and some travel more slowly.

Mechanical dispersion in porous media, on a microscopic scale, is caused by three mechanisms (see Figure 10.14). The first occurs in individual pore channels because molecules travel at different velocities at different points in the channel due to the drag exerted on the fluid by the roughness of the pore surfaces. The second mechanism is due to the difference in pore sizes along the flow paths followed by the water molecules. Because of the differences in surface area and roughness relative to the volume of water in individual pore channels, different pore channels have different bulk velocities. The third mechanism leading to a dispersive process is related to the tortuosity, branching, and inter-fingering of pore channels.

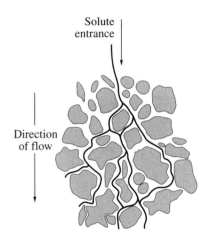

Figure 10.14: Convection-dispersion of fluid in a porous media.

Dispersion is discussed in the context of diffusion since the two processes can usually be described with very similar mathematics. Dispersive mass flux is written analogously to diffusion, defined in Equation 10.18:

$$\underbrace{j_{A,x}}_{\text{Dispersive flux}} = -\underbrace{E_x}_{\substack{\text{Dispersion}\\\text{coefficient}}} \underbrace{\left(\frac{dc_A}{dx}\right)}_{\substack{\text{Concentration}\\\text{gradient}}} \quad (10.31)$$

---

[1] Spreading of a plume in air pollution is a convection-dispersion process (dispersion, combined with bulk flow in the wind direction), analogous to processes in liquid systems discussed in Section 14.1. However, spreading of a plume is inherently three-dimensional and its analysis is more complex. For further details, the reader is referred to books on air pollution, such as Stern (1976).

where c_A is the mass concentration of species A being dispersed and the diffusivity is now replaced by a dispersion coefficient E_x. The dispersion coefficient has the units of m²/s, which is the same as that of diffusivity, as can be seen by substituting the units in the above equation:

$$[E_x] = \frac{[j_{A,x}]}{\left[\frac{dc_A}{dx}\right]} = \frac{\left[\frac{kg}{m^2 s}\right]}{\left[\frac{kg}{m^3 m}\right]}$$
$$= \left[\frac{m^2}{s}\right] \qquad (10.32)$$

Unlike the diffusivity, the dispersion coefficient is not a strong function of the chemistry, e.g., the molecular weight or structure. Instead, as the subscript x in E_x indicates, the dispersion coefficient is a strong function of position. This is unlike the diffusivity which typically does not strongly depend on position. The dispersion coefficient is almost always measured experimentally.

Although dispersion is defined to be the random mixing component in addition to the molecular diffusion, unless the velocities are quite low, the dispersion effect is much stronger than diffusion. For example,[2] a crop canopy loses water vapor or absorbs carbon dioxide due to turbulent airflow (leading to dispersion) over the canopy. The dispersion coefficient for the rate of dispersive transport of water vapor or carbon dioxide between the plant and the bulk air can be 10,000 - 100,000 times larger than molecular diffusivities. Thus, in practice, if a dispersion coefficient is available, it generally includes the small or insignificant contribution from molecular diffusion. Moreover, it is nearly impossible to separately quantify the two processes.

10.4 Convective Mass Transfer

Convective mass transfer is studied the same way as convective heat transfer. It is the added effect of bulk flow on diffusion or dispersion. Convective mass transfer is the movement of mass through a medium as a result of the net motion of a material in the medium (e.g. air flow over water surface which carries the water vapor away). This text will cover two scenarios of convection— 1) Convection-diffusion over a surface and 2) Convection-dispersion in a fluid or porous media.

10.4.1 Convection-Diffusion Mass Transfer Over a Surface

Convection-diffusion mass transfer over a surface is completely analogous to Section 2.2 for convective heat transfer, and is therefore characterized by a *convective mass transfer coefficient*. This is illustrated in Figure 10.15. As in heat transfer, we can have forced or free convection. Forced convection is due to an external force such as a fan, while free convection is driven by a density difference in the fluid (created by

---

[2]See Nobel (1974).

10.4. CONVECTIVE MASS TRANSFER

concentration or temperature differences, analogous to Figure 2.7 on page 22 in heat transfer). Convective mass transfer over a surface is described by:

$$N_{A_{1-2}} = \underbrace{h_m}_{\text{Convective coefficient}} \underbrace{A}_{\text{Area}} \underbrace{(c_1 - c_2)}_{\text{Concentration difference}} \quad (10.33)$$

where the underbraced $N_{A_{1-2}}$ is Mass flow rate.

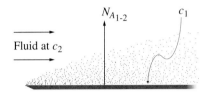

Figure 10.15: Schematic of convection-diffusion over a surface.

where $N_{A_{1-2}}$ is the mass flow rate *from* 1-2, A is the area normal to the direction of mass flow, $c_1 - c_2$ is the concentration difference between surface and fluid, h_m is the convective mass transfer coefficient, also called the film coefficient. Equation 10.33 is not a law but a defining equation for h_m, like the equation for convective heat transfer coefficient h. The units of h_m can be shown to be in m/s:

$$[h_m] = \frac{[N_{A_{1-2}}]}{[A][(c_1 - c_2)]} = \frac{\left[\frac{\text{kg}}{\text{s}}\right]}{[\text{m}^2]\left[\frac{\text{kg}}{\text{m}^3}\right]} = \left[\frac{\text{m}}{\text{s}}\right]$$

Convective mass transfer coefficient h_m includes the effects of diffusion and bulk flow. Like h, the heat transfer coefficient, h_m depends on system geometry, fluid properties, flow situation, and the magnitude of the concentrations. Details of calculation of h_m are provided in Chapter 14.

Example 10.4.1 Water Loss from a Reservoir

Estimate the evaporative loss of water from a reservoir where surface temperature, as well as the ambient air temperature, is 40°C. The air blowing over the reservoir leads to a surface mass transfer coefficient of 0.25 m/s. Assume the water surface is saturated with water vapor, and that the air is dry before contact with water.

Known: Temperature of a water surface and the surface mass transfer coefficient for the transport of water vapor from the water surface into air blowing over it

Find: Rate of loss of water due to evaporation

Schematic and Given Data:

1. Water surface temperature is 40°C
2. Surface mass transfer coefficient for blowing air is 0.25 m/s

Assumptions:

1. Air is at water surface temperature, hence no heat transfer
2. Water surface is still and stays flat

3. The mass transfer coefficient value provided is an average over the entire surface

4. Concentration of water vapor in the bulk air is zero

Analysis: The concentration of water vapor at the surface can be found from data on partial pressure of water vapor at saturation at 40°C. Using the steam table (see Table C.12 on page 344), this value is 0.07318×10^5 Pa. The concentration corresponding to this partial pressure is calculated using the ideal gas law (see Section 9.1.1) as:

$$\begin{aligned} c_{vapor,surface} &= \frac{p_{vapor,surface}}{RT} \\ &= \frac{0.07318 \times 10^5 \, [\text{Pa}]}{8.314 \left[\frac{\text{J}}{\text{mol K}}\right] 313 \, [\text{K}]} \\ &= 2.812 \, \frac{\text{mol}}{\text{m}^3} \end{aligned}$$

The flux of the evaporation water loss is given by

$$\begin{aligned} n_{water} &= h_m(c_{vapor,surface} - c_{vapor,bulk\ air}) \qquad (10.34) \\ &= \left(0.25 \, \frac{\text{m}}{\text{s}}\right)\left(2.812 \, \frac{\text{mol}}{\text{m}^3} - 0\right) \\ &= 0.703 \, \frac{\text{mol}}{\text{m}^2 \, \text{s}} \\ &= 1.265 \times 10^{-2} \, \frac{\text{kg}}{\text{m}^2 \cdot \text{s}} \end{aligned}$$

Comments: Depending on the application, this rate of water loss may or may not be significant.

10.4.2 Convection-Dispersion Mass Transfer

Like convection-diffusion mass transfer was the addition of bulk flow to the process of diffusion, convection-dispersion mass transfer is the added effect of bulk flow on the process of dispersion. Unlike the convection-diffusion mass transfer that we studied over a surface and focused on a thin layer called the boundary layer, we will study convection-dispersion mass transfer in the bulk fluid. Specifically, we will study in Chapter 13, convection-dispersion in a stream and in flow through a porous media such as soil.

10.5 Comparison of the Modes of Mass Transfer

Molecular diffusion is the spontaneous movement of atoms and molecules of liquids, gases, and solids. Movement of mass occurs because of a concentration gradient, in contrast with bulk flow due to pressure gradient.

10.5. COMPARISON OF THE MODES OF MASS TRANSFER

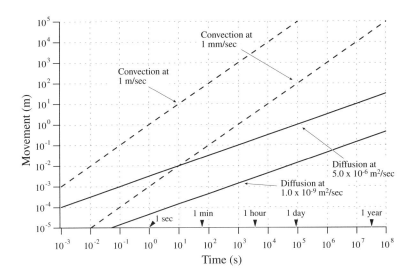

Figure 10.16: Relative movements of mass due to diffusion/dispersion and convection.

Capillary "diffusion" is movement of mass due to capillary action in a porous media. Such capillary transport can be treated mathematically similar to molecular diffusion. In practice, capillary diffusion is often included under the general term diffusion. Expected range of capillary diffusivities in soils is illustrated in Figure 10.7.

Dispersion is the spreading of the fluid from its bulk flow path. Although its effects are quite similar to molecular diffusion, the mechanisms are entirely different. In fluids, for example, dispersion is primarily due to turbulence. In porous media, it is due to the random orientation of the shape, size, direction, and tortuosity of the pores. Generally, dispersion coefficients, as reported in the literature, include any molecular diffusion.

Convection refers to the addition of bulk flow to diffusion/dispersion. In practice, a convection-diffusion process is often referred to as simply convection. Surface convective mass transfer coefficient provides a measure of this combined convection-diffusion process for flow over a surface.

Advection is the transport of mass due to the bulk flow itself, without the diffusion or dispersion transport mechanism. It is due to pressure gradient. Note, however, the terms convection and advection are sometimes used interchangably in practice and there is some ambiguity in their use. The term advection will not be used any further in this text.

It is useful to have a qualitative idea of the relative speeds of diffusion, dispersion, and advection. As shown in Figure 10.16, generally speaking, advective processes are relatively much faster and dominate if they are present simultaneously with diffusion.

10.6 Chapter Summary—Modes of Mass Transfer

• (Bulk) Flow Through Porous Media (page 197)

1. It is the bulk movement of a fluid in a porous media due to hydraulic forces (pressure, matric attractive forces, and gravity).

2. It is described by Darcy's law (Eqn. 10.1) which states that the volumetric flux of a fluid is proportional to the gradient of the hydraulic potential. The proportionality constant is termed the hydraulic conductivity.

3. Hydraulic conductivity depends on the fluid property called permeability (that depends on density and viscosity) and matrix property such as the pore size distribution, shape of pores, porosity, and tortuosity.

4. When the porous media is unsaturated, the flow is driven primarily by the matric attractive (capillary) forces. Flow in such a media can be written in terms of Fick's law and an apparent (capillary) diffusivity.

• Molecular Diffusion (page 203)

1. It is the movement of mass from a higher to a lower concentration due to random molecular motion.

2. It is described by Fick's law (Eqn. 10.18) which states that the flux of a mass species is proportional to the concentration gradient of that species. The proportionality constant is called the diffusivity.

3. Diffusivity measures speed of movement— it is one half of the mean square displacement of a number of molecules per unit time through a given medium.

4. Diffusivity for gases are the highest due to their highest mobility. They are lower for liquids and the lowest for solids.

5. Diffusivity of a gas in porous solid depends on the porous structure.

6. Diffusivity of a liquid in a porous media is really a measure of capillary flow.

• Dispersion (page 213)

1. Dispersion is analogous to diffusion and is described by Eqn. 10.31. However, dispersion is due to completely separate mechanisms— turbulent flow in a fluid or flow through uneven and tortuous paths in a porous media.

2. Dispersion coefficients are generally much higher than diffusion coefficients.

• Convection (page 214)

1. It is the effect of adding bulk flow to diffusion or dispersion. Convection-diffusion mass transfer over a surface is described by Eqn. 10.33 in complete analogy to heat transfer.

10.7 Concept and Review Questions

1. What are the sources of matric forces in a porous media?

2. Is the average fluid velocity inside the pores the same as the average velocity obtained by dividing volumetric flow rate by the cross-sectional area of flow?

3. Does intrinsic permeability, k, of a porous media depend on fluid properties?

4. In an unsaturated capillary porous material, why would water move or "diffuse" from higher to lower concentration?

5. What is molecular diffusion? How is it different from capillary flow? Dispersion? Convection?

6. What factors may cause molecular diffusion to occur?

7. In what ways are heat and mass transfer analogous and in what ways do they differ?

8. What are the units of mass diffusivity and thermal diffusivity? How do you physically interpret these diffusivity values?

9. Why are the diffusivities of gases larger than liquids and liquids larger than solids?

10. Most convective processes also simultaneously involve molecular diffusion. Why does it make sense typically to ignore the molecular diffusion?

11. Why does the diffusivity typically increase with temperature?

12. Why does a larger molecule have a smaller diffusivity?

13. What bulk physical properties are related by the Stokes-Einstein equation? Verify that the equation is dimensionally consistent.

14. Do you expect gas diffusivity to increase as a material becomes more porous? What will be the limiting value of this diffusivity when the material is very very porous?

10.8 Further Reading

Barrow, G.M. 1981. *Physical Chemistry for the Life Sciences*. McGraw-Hill, New York.

Campbell, G.S. 1985. *Soil Physics with Basic: Transport Models for Soil-Plant Systems*. Elsevier Science Publishing Company, Amsterdam.

Canny, M. J. 1990. Rates of apoplastic diffusion in wheat leaves. New Phytologist 116(2):263-268.

Cussler, E.L. 1997. *Diffusion Mass Transfer in Fluid Systems*. Cambridge University Press, New York.

Dainty, J. 1985. Water transport through the root. Acta Horticulturae 171:21-31.

Goldstick, T.K. and I. Fatt. 1970. Diffusion of oxygen in solutions of blood proteins. Chem. Eng. Progress 66(99):101-113.

Ingham, D.B. and I. Pop. (Editors) 1998. *Transport Phenomena in Porous Media*. Oxford Publishing Company, Danvers, Massachusetts.

Ishiguro M. 1991. Solute transport through hard pans of paddy fields. 1. Effect of vertical tubular pores made by rice roots on solute transport. Soil Science 152(6):432-439.

Kiranoudis, C.T., Z.B. Maroulis, D. Marinos-Kouris, and G.D. Saravacos. 1994. Estimation of the effective moisture diffusivity from drying data. application to some vegetables, in *Developments in Food Engineering: Proceedings of the 6th International Congress on Engineering and Food* (T. Yano, R. Matsuno, and K. Nakamura, eds.), vol. 1, (New York), pp. 340–342, Blackie Academic & Professional.

McGrath, J.J. 1997. Quantitative measurement of cell membrane transport: Technology and applications. Cryobiology 34(4):315-334.

Monteith, J.L. and M.H. Unsworth. 1990. *Principles of Environmental Physics*. Edward Arnold, London.

Nobel, P.S. 1974. *Biophysical Plant Physiology*. W.H. Freeman and Company, San Francisco.

Siau, J.F. 1984. *Transport Processes in Wood*. Springer-Verlag, New York.

Stern, A.C. 1976. *Air Pollution*, 3rd Ed., Vol. I, Academic Press, New York.

Taura, T., Y. Iwaikawa, M. Furumoto, and K. Katou. 1988. A model for radial water transport across plant roots. Protoplasma 144:170-179.

Weiss, T.F. 1996. *Cellular Biophysics, Volume 1: Transport*. The MIT Press, Cambridge, Massachusetts.

10.9 Problems

10.9.1 Saturated Flow of Water in a Porous Media

A layered soil column is positioned horizontally as shown in Figure 10.17. It consists of 35 cm of a loam soil of saturated hydraulic conductivity of 5 cm/day followed by 100 cm of a sandy soil of saturated hydraulic conductivity 25 cm/day. The right end of the column is open to the atmosphere, as shown. What is the volumetric flux through the soil column? Consider saturated flow at steady state.

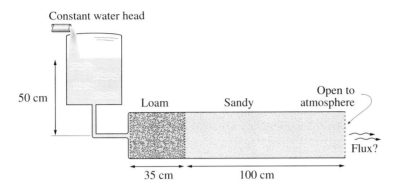

Figure 10.17: Schematic for Problem 10.9.1.

10.9.2 How Water Rises to the Top of a Tree

The numerous interstices in the cell wall of the xylem vessels of a tree form the capillaries by which water can rise to the top of tall trees. If a representative radius of these capillary channels is 5×10^{-9} m, show that the capillary suction in theses channels can, in fact, raise the water to the top of any tree, at ambient temperature ($25°C$).

10.9.3 Molecular and Average Distances Moved During Diffusion

Consider gaseous diffusion with a diffusion coefficient of $10^{-6} m^2/s$. 1) What is the average distance travelled by a group of molecules after 1 hour? 2) Is this the actual distance travelled by a single molecule during the same time?

10.9.4 Distance Moved by Diffusion and Dispersion

Calculate the average distance moved in one day by smoke particles in air for the following cases: 1) Smoke diffuses in air with a diffusion coefficient of $10^{-7} m^2/s$. 2) Smoke disperses in air with a dispersion coefficient of $10^{-5} m^2/s$.

10.9.5 Diffusivity Increase with Temperature

Approximately what temperature change is needed for diffusivity of gases (dilute, ideal) to go up by a factor of 10?

10.9.6 Estimating Diffusivity of Solutes in a Liquid

Calculate the diffusion coefficients of 1) a protein macromolecule bovine serum albumin (BSA) and 2) urea in water at $23°C$. Consider the effective radius of BSA is 50 nm and that of urea is 0.16 nm. The viscosity of water at this temperature is 10^{-3}

Pas. Compare these two diffusivities using the fact that BSA has a molecular weight of 67,500 while urea has a molecular weight of 60.1.

10.9.7 Estimating Diameter from Diffusivity

The diffusivity of human immunoglobin (a protein) is $4 \times 10^{-7} \mathrm{m}^2/\mathrm{s}$ in water at $20°\mathrm{C}$. The viscosity of water at this temperature is 1×10^{-3} Pas. Assuming the protein to be spherical, what is its effective radius?

Chapter 11

GOVERNING EQUATIONS AND BOUNDARY CONDITIONS OF MASS TRANSFER

CHAPTER OBJECTIVES

After you have studied this chapter, you should be able to:

1. Identify the terms describing storage, convection, diffusion (or dispersion) and generation of mass species in the general governing equation for mass transfer.

2. Specify the three common types of mass transfer boundary condition.

KEY TERMS

- Fick's law, modified for convection
- diffusive mass flux
- convective mass flux
- total mass flux
- storage term
- convection term
- diffusion term
- generation term
- boundary concentration specified
- boundary flux specified
- boundary convective mass transfer

In this chapter we will develop a general equation that describes concentration of a mass species as a function of time and position in a material. Later, in Chapters 12 through 14, we will solve this general equation for specific situations. Relationship of this chapter to other chapters in mass transfer is shown in Figure 11.1.

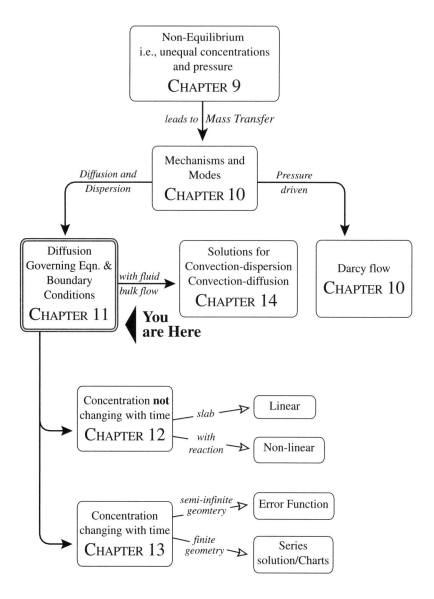

Figure 11.1: Concept map of mass transfer showing how the contents of this chapter relates to other chapters in the part on mass transfer.

11.1 Modified Fick's Law for Bulk Flow or Convection

The Fick's law (Eqn. 10.18) introduced in the previous chapter provides the expression for diffusive flux. When bulk flow or convection is present, the expression for total flux needs the additional term due to convection, which is developed in this section.

11.1. MODIFIED FICK'S LAW FOR BULK FLOW OR CONVECTION

Since diffusional mass transfer involves at least two mass species unlike diffusional heat transfer, certain additional quantities need to be defined for mass transfer.

11.1.1 Velocity and Mass Average Velocity

We know that the individual molecules move randomly, producing an average velocity given by (Eqn. 10.20)

$$u_{diff} = \sqrt{\frac{2D}{t}}$$

Note that u_{diff} is the diffusion velocity of a species, for an aggregate of many molecules. It is not the velocity of one molecule, it is the average velocity of a large number of molecules contained in a small enough volume. The total velocity u_i with which a species i moves is this diffusion velocity superimposed on the bulk or mass average velocity, u. In other words, the velocity difference, $u_i - u$, of a species i is due to the diffusion velocity. The mass average velocity u of the fluid is found using the mass concentration of each species i

$$u = \frac{\sum c_i u_i}{\sum c_i} \qquad (11.1)$$

where

$$c_i = \text{mass concentration of species } i$$
$$= \text{mass of species } i \text{ per unit volume}$$

The total mass concentration of all species, c, is given by

$$c = \sum_i c_i \qquad (11.2)$$

11.1.2 Flux Equation for a Convective Situation

If u_A is the velocity of mass species A and u is the mass average velocity of all the species, $u_A - u$ is the diffusion velocity of species A relative to mass average velocity and is due to diffusive movement relative to the bulk flow.

$$j_{A,x} = \text{mass flux of } A \text{ in } x \text{ direction due to diffusion only}$$
$$= c_A \left(u_{A,x} - u \right)$$
$$= -D_{AB} \frac{\partial c_A}{\partial x} \text{ by Fick's law} \qquad (11.3)$$

Alternatively:

| Total velocity of species A in x direction | = | Velocity due to bulk flow in x direction | + | Movement relative to bulk flow in x direction |

$$u_{A,x} = u + u_{diffusion} \tag{11.4}$$

Multiplying both sides by c_A:

$$c_A u_{A,x} = c_A u + \underbrace{c_A u_{diffusion}}_{\text{Diffusive flux}}$$

$$= c_A u - D_{AB}\frac{\partial c_A}{\partial x} \tag{11.5}$$

Equation for *total mass flux*

$$n_{A,x} = \text{total mass flux of } A \text{ in } x \text{ direction}$$

$$= c_A u_{A,x}$$

$$= \underbrace{-D_{AB}\frac{\partial c_A}{\partial x}}_{\text{Diffusive flux}} + \underbrace{c_A u}_{\text{convective flux}} \tag{11.6}$$

Equation 11.6 also shows how the Fick's law is modified to include the effect of bulk flow or convection. Compare this with the *total energy flux* in heat transfer when convection is present:

$$q''_x = \underbrace{-k\frac{\partial T}{\partial x}}_{\text{Conductive flux}} + \underbrace{\rho c_p (T - T_R) u}_{\text{Convective flux}} \tag{11.7}$$

Note that the quantity $\rho c_p (T - T_R)$ in the second term of Eqn. 11.7, having the units of kJ/m^3, is the energy per unit volume and compares with c_A in Eqn. 11.6 which is mass per unit volume. Note that in terms of flux symbols, Eqn. 11.6 can be written as

$$\underbrace{n_{A,x}}_{\text{Total}} = \underbrace{j_{A,x}}_{\text{Diffusive}} + \underbrace{c_A u}_{\text{Convective}} \tag{11.8}$$

When convective flux is zero or the flux is due to diffusion only, $n_{A,x} = j_{A,x}$. In such cases, either symbol $n_{A,x}$ or $j_{A,x}$ can be used to describe flux. In this text, $n_{A,x}$ will be used for flux in general and $j_{A,x}$ will be used only when there is a need to distinguish between diffusive and total flux.

Using the definition for the mass average velocity u for a binary system

$$u = \frac{c_A u_{A,x} + c_B u_{B,x}}{c_A + c_B} = \frac{c_A u_{A,x} + c_B u_{B,x}}{c} \tag{11.9}$$

an alternative expression for total mass flux can be developed as

$$n_{A,x} = -D_{AB}\frac{\partial c_A}{\partial x} + \frac{c_A}{c}(c_A u_{A,x} + c_B u_{B,x}) \tag{11.10}$$

$$\underbrace{n_{A,x}}_{\substack{\text{Total flux} \\ \text{of } A \text{ in } x \\ \text{direction}}} = \underbrace{-D_{AB}\frac{\partial c_A}{\partial x}}_{\substack{\text{Diffusive} \\ \text{flux}}} + \underbrace{\underbrace{\omega_A}_{\substack{\text{mass fraction} \\ \text{of } A}} \underbrace{(n_{A,x} + n_{B,x})}_{\text{total flux}}}_{\substack{\text{convective flux due} \\ \text{to bulk motion}}} \tag{11.11}$$

11.2. GOVERNING EQUATION FOR MASS TRANSFER

where ω_A is the mass fraction of A in the system having the units of kg of species A per kg of total mass. Equation 11.11 is an alternative expression for mass flux, showing how the convective mass flux of species A is simply the mass fraction multiplied by the total flux. The units of fluxes $j_{A,x}$ and $n_{A,x}$ are in kg of $A/m^2 \cdot s$. For the most part, we will use Eqn. 11.6 as the expression for total mass flux.

11.2 Governing Equation for Mass Transfer Derived from Conservation of Mass Species and Fick's Law

In studying heat transfer, we said energy moves in the direction of decreasing temperature. The rate of energy movement was given by Fourier's law. We used energy conservation combined with Fourier's law to arrive at the differential equation of energy transfer that describes temperature as a function of position and time.

We will follow a completely analogous procedure using mass conservation and the modified Fick's law (Equation 11.6) for mass flux to describe movement of mass in the direction of decreasing concentration. Consider an elemental control volume as shown in Figure 11.2 fixed in space with mass flux into and out of the control volume. The total mass flux includes diffusive mass flux arising from the diffusion of the mass species A and convective mass flux due to the bulk movement of the fluid as a whole. The total mass flux $n_{A,x}$ is written from Eqn. 11.6 as

$$n_{A,x} = \underbrace{-D_{AB}\frac{\partial c_A}{\partial x}}_{\text{Diffusive or Dispersive flux}} + \underbrace{c_A u}_{\text{Convective flux}} \quad (11.12)$$

where c_A is the concentration of mass species A in mass of A per unit volume. As the stored mass of species A increases in the control volume, it is reflected in increased value of c_A. Conversely, if the stored amount decreases, the concentration decreases. Within the control volume, mass species A is generated at the rate of r_A in mass per unit volume. Like in heat transfer, generation refers to transformation of other forms of mass into mass species A, as in chemical reactions. For example, O_2 in the fluid (mass species) can be used up due to biochemical reactions in the fluid, in which case r_{O_2} has a negative value. Using conservation of mass species A (Eqn. 9.5)

$$\begin{array}{c} \text{Mass in} \\ \text{(of species } A\text{)} \end{array} - \begin{array}{c} \text{Mass out} \\ \text{(of species } A\text{)} \end{array} + \begin{array}{c} \text{Mass generated} \\ \text{(of species } A\text{)} \end{array} = \begin{array}{c} \text{Mass stored} \\ \text{(of species } A\text{)} \end{array} \quad (11.13)$$

and referring to Figure 11.2 we can write the various quantities in Eqn. 11.13 as

$$\begin{array}{c} \text{Amount of } A \text{ in} \\ \text{during time } \Delta t \end{array} = n_{A,x} \Delta y \Delta z \Delta t$$

$$\begin{array}{c} \text{Amount of } A \text{ out} \\ \text{during time } \Delta t \end{array} = n_{A,x+\Delta x} \Delta y \Delta z \Delta t$$

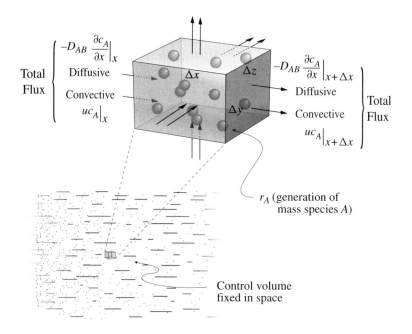

Figure 11.2: Control volume showing inflow and outflow of a mass species by diffusion and convection.

$$\begin{aligned}\text{Amount of } A \text{ generated during time } \Delta t &= r_A \Delta x \Delta y \Delta z \Delta t \\ \text{Amount of } A \text{ stored during time } \Delta t &= \Delta c_A \Delta x \Delta y \Delta z\end{aligned}$$

Substituting in Eqn. 11.13 for mass balance on the control volume,

$$\left(\left(n_{A,x} - n_{A,x+\Delta x} \right) \Delta y \Delta z + r_A \Delta x \Delta y \Delta z \right) \Delta t = \Delta c_A \Delta x \Delta y \Delta z$$

$$\frac{n_{A,x} - n_{A,x+\Delta x}}{\Delta x} + r_A = \frac{\Delta c_A}{\Delta t}$$

$$-\frac{\partial}{\partial x}\left(n_{A,x} \right) + r_A = \frac{\partial c_A}{\partial t} \quad (11.14)$$

which is the differential equation for mass balance for species A. Substituting Equation 11.12 for $n_{A,x}$ in Equation 11.14:

$$-\frac{\partial}{\partial x}\left(-D_{AB}\frac{\partial c_A}{\partial x} + c_A u \right) + r_A = \frac{\partial c_A}{\partial t}$$

11.2. GOVERNING EQUATION FOR MASS TRANSFER

$$-\frac{\partial}{\partial x}\left(-D_{AB}\frac{\partial c_A}{\partial x}\right) - \frac{\partial}{\partial x}(c_A u) + r_A = \frac{\partial c_A}{\partial t}$$

If the diffusivity D_{AB} can be considered a constant,

$$D_{AB}\frac{\partial^2 c_A}{\partial x^2} - \frac{\partial}{\partial x}(c_A u) + r_A = \frac{\partial c_A}{\partial t}$$

Rearranging,

$$\frac{\partial c_A}{\partial t} + \frac{\partial}{\partial x}(c_A u) = D_{AB}\frac{\partial^2 c_A}{\partial x^2} + r_A \qquad (11.15)$$

Although Eqn. 11.15 is for conservation of mass species A, the total mass of all species together is also conserved. The conservation of total mass provides an additional relationship that can be used to rewrite Eqn. 11.15. Writing conservation of total mass between the locations x and $x + \Delta x$,

$$(cuA)_x = (cuA)_{x+\Delta x}$$

where c is the total mass concentration. If the total mass concentration c does not change with x, we can rewrite this conservation equation for constant areas A along the flow in the control volume of Figure 11.2 as

$$u_x = u_{x+\Delta x}$$

or

$$\frac{\partial u}{\partial x} = 0 \qquad (11.16)$$

Substituting Eqn. 11.16 in Eqn. 11.15, we get

$$\underbrace{\frac{\partial c_A}{\partial t}}_{\text{storage}} + \underbrace{u\frac{\partial c_A}{\partial x}}_{\substack{\text{flow or}\\\text{convection}}} = \underbrace{D_{AB}\frac{\partial^2 c_A}{\partial x^2}}_{\text{diffusion}} + \underbrace{r_A}_{\text{generation}} \qquad (11.17)$$

Note that each term in Eqn. 11.17 has the dimensions of mass per unit volume per unit time or kg/m³·s. Compare this with Eqn. 3.5, where, upon multiplying throughout by ρc_p, the terms have dimensions of energy per unit volume per unit time or J/m³·s. Equation 11.17 is the general governing equation for mass transfer in one dimension cartesian coordinate system with constant diffusivity. Compare this equation with Eqn. 3.5 for heat transfer. Like in Section 3.5, we can develop more general forms of Eqn. 11.17 in other coordinate systems and in multiple dimensions as presented later in this chapter. An excellent reference book for more general forms of the mass and heat transfer equations is Bird et al. (1960).

11.2.1 Meanings of Each Term in the Governing Equation

Note the similarity of Eqn. 11.17 with Eqn. 3.5, which is repeated here for convenience:

$$\underbrace{\frac{\partial T}{\partial t}}_{\text{storage}} + \underbrace{u\frac{\partial T}{\partial x}}_{\substack{\text{flow or}\\\text{convection}}} = \underbrace{\frac{k}{\rho c_p}\frac{\partial^2 T}{\partial x^2}}_{\text{conduction}} + \underbrace{\frac{Q}{\rho c_p}}_{\text{generation}} \qquad (11.18)$$

Both equations represent the four basic terms of diffusion, convection, generation and storage. Like the energy equation, the mass transfer equation is quite general and is useful for a variety of situations. Following the discussion in Section 3.1.1 on heat transfer, we only need to keep a few terms for the physical situations discussed in this text. The meanings of each term in Eqn. 11.17 are summarized in Table 11.1.

Table 11.1: Various terms in the governing equation for mass transport (Eqn. 11.17) and their interpretations.

Term	What does it represent	When can you ignore it
Storage	Rate of change of stored mass	Steady state (no variation of concentration with time)
Convection	Rate of net mass transport due to convection or bulk flow	Typically in a solid, with no bulk flow through it
Diffusion	Rate of net mass transport due to diffusion	Slow diffusion in relation to generation or convection. For example, bacterial species in sterilization of a solid food
Generation	Rate of generation of mass	No chemical reaction leading to conversion from or to species A

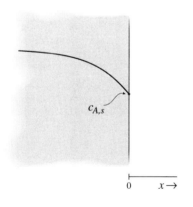

Figure 11.3: A surface concentration specified boundary condition.

11.3 General Boundary Conditions

This section parallels Section 3.2 on heat transfer. Like the description of a heat transfer problem is not complete without specifying the thermal conditions at the boundary, concentration conditions at the boundary are necessary for the full description of a mass transfer problem. The following are the three most common types of boundary conditions present in a mass transfer situation:

1. Surface concentration is specified A simple concentration condition that can occur on a surface is a specified concentration. For a one-dimensional mass transfer, as shown in Figure 11.3, this boundary condition can be expressed as

$$c_A \big|_{x=0} = c_{A,s} \qquad (11.19)$$

where $c_{A,s}$, the concentration at the surface, can be specified as a constant or a function of time. At a solid-fluid interface, if we are setting up the mass transfer problem for the solid, we need $c_{A,s}$ for the solid. Sometimes, at such an interface, the boundary condition on the fluid side is provided instead. In such a case, $c_{A,s}$ for the solid is obtained from the boundary condition on the fluid side and equilibrium relations such as those in Figure 9.8. For example, suppose air at temperature 21°C and relative

11.3. GENERAL BOUNDARY CONDITIONS

humidity 20% is blowing at a high velocity over the surface of a wet wood. From Figure 9.8, this maintains the surface of the wet wood at equilibrium with the air, at a moisture content value of 4.5%. Thus, the boundary condition at the wood surface will be given by

$$c_{water}\Big|_{x=0} = 0.045 \text{ kg/kg of total}$$

On the other hand, if we are setting up a mass transfer problem for a fluid (i.e., we need to calculate concentrations in the fluid), we need the boundary condition for the fluid. An example (see also Section 14.4) is the evaporation from the surface of free water or a very wet solid into air. Here the concentration of vapor in air at the boundary will be maintained at the vapor pressure of water. Thus, at a temperature of 30°C, this boundary condition for the fluid is written from Table C.10 on page 342 as

$$c_{vapor}\Big|_{x=0} = \frac{0.0424 \times 10^5 \text{ [N/m}^2\text{]}}{RT}$$
$$= 0.03 \frac{\text{kg}}{\text{m}^3}$$

Another example of a concentration specified boundary condition is the disappearance of a species at a surface, e.g., diffusing species O_2 is completely consumed at a surface, leading to $c_{O_2} = 0$ at the surface.

2. Surface mass flux is specified

Sometimes it is possible to know and specify the rate of mass transfer or mass flux on a surface. For a one-dimensional mass transfer, as shown in Figure 11.4, this boundary condition is expressed as

$$-D_{AB} \frac{\partial c_A}{\partial x}\Big|_{x=0} = n_{A,s} \quad (11.20)$$

Here $n_{A,s}$, the surface mass flux, can be specified as a constant or a function of time. Two important special cases of surface mass flux occur frequently in practice. These are now discussed.

2a) Special case: Impermeable condition

Surfaces are sometimes made impermeable in an attempt to stop a particular type of mass movement. When the surface is highly impermeable to a particular species, the mass flux of that species is very small and can be approximated as zero, as illustrated in Figure 11.5. This boundary condition is given by

$$-D_{AB} \frac{\partial c_A}{\partial x}\Big|_{x=0} = 0 \quad (11.21)$$

For example, when a biomaterial containing water is wrapped with aluminum foil to reduce moisture loss, the flux of water at the foil surface can be considered zero, leading to the above boundary condition.

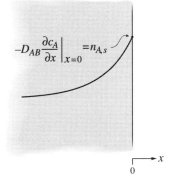

Figure 11.4: A mass flux specified boundary condition.

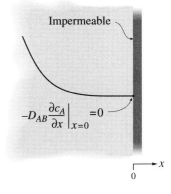

Figure 11.5: An impermeable (zero mass flux specified) boundary condition.

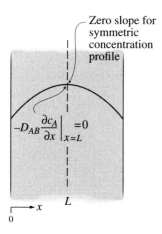

Figure 11.6: A symmetry (zero mass flux) boundary condition at the centerline.

2b) Special case: Symmetry condition

Another common situation arises in a mass transfer process when the geometry as well as the boundary conditions are symmetric, as shown in Figure 11.6 for the drying of a slab. Here the slab is of uniform thickness (making it symmetric about the centerline) as well as it is being dried symmetrically (same boundary conditions on both faces). The resulting concentration profile in the slab will also be symmetric about the centerline, having a zero slope at the centerline. Thus, the symmetry boundary condition at the centerline can be written as

$$\left.\frac{\partial c_A}{\partial x}\right|_{x=L} = 0 \qquad (11.22)$$

Note that this boundary condition resembles the impermeable condition mentioned above, since to maintain symmetry, mass flux has to be zero at the line of symmetry.

3. Convection at the surface

Another common mass transfer boundary condition occurs when a fluid is flowing over a surface, as shown schematically in Figure 11.7. At the surface ($x = 0$), amount of mass diffused through solid or liquid is equal to the amount of mass convected away. This boundary condition is written simply as a mass balance at the surface, which is:

$$\underbrace{-D_{AB}\left.\frac{\partial c_A}{\partial x}\right|_{x=0}}_{\text{mass diffusion to the left}} = \underbrace{h_m(c_{A,x=0}^{\text{fluid}} - c_{A,\infty}^{\text{fluid}})}_{\text{mass convection to the right}} \qquad (11.23)$$

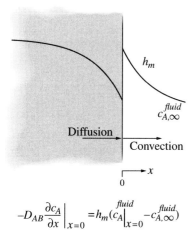

Figure 11.7: A convection mass transfer boundary condition.

Here h_m is mass transfer coefficient, defined analogous to heat transfer coefficient and is explained in later chapters. In Eqn. 11.23, it is important to note that the left hand side uses concentration of species A in the solid, c_A, while the right hand side uses concentration of species A in the fluid, c_A^{fluid}. A common example of this type of boundary condition is that at the surface of a wet solid with air flowing over it, after the solid has experienced some drying, i.e., when the surface is not very wet. In this case, the left hand side of Eqn. 11.23 uses concentration of water in the solid, c_{water}, while the right hand side uses concentration of water vapor in air, c_{vapor}. Suppose the corresponding mass transfer coefficient for a situation is $h_m = 0.01$m/s. Equation 11.23 is written for this specific case as

$$\underbrace{-D_{water,solid}\left.\frac{\partial c_{water}}{\partial n}\right|_{\text{surface}}}_{\text{diffusion in wet solid}} = \underbrace{0.01(c_{vapor,\text{surface}} - c_{vapor,\infty})}_{\text{convection in air}}$$

At the surface, concentration of water in the solid, $c_{water,surface}$ and concentration of water vapor in air, $c_{vapor,surface}$ are not the same, but they are related by isotherms, as discussed in Section 9.3.2 on page 182.

Also, in analogy to the boundary conditions in Section 3.2 on page 34 of heat transfer, when $h_m \to \infty$ (such as in high fluid velocity), Eqn. 11.23 leads to:

$$c_{A,\text{surface}}^{\text{fluid}} = c_{A,\infty}^{\text{fluid}}$$

or the concentration in fluid at the surface becomes equal to the concentration in the bulk fluid. Thus, the boundary condition given by Eqn. 11.23 reverts to the first kind of boundary condition given by Eqn. 11.19 with the surface concentration in the solid being specified as that in equilibrium with surface concentration in the fluid, $c_{A,\text{surface}}^{\text{fluid}}$.

11.4 Governing Equations for Mass Diffusion in Various Coordinate Systems

The governing equation for diffusion mass transfer is shown here in some familiar coordinate systems. Control volumes for the various coordinate systems can be seen in Section 3.5. For a more general version of these governing equations, see Bird et al., 1960.

Cartesian

$$D_{AB}\left[\frac{\partial^2 c_A}{\partial x^2} + \frac{\partial^2 c_A}{\partial y^2} + \frac{\partial^2 c_A}{\partial z^2}\right] + r_A = \frac{\partial c_A}{\partial t} \qquad (11.24)$$

Cylindrical

$$D_{AB}\left[\frac{1}{r}\frac{\partial}{\partial r}\left(r\frac{\partial c_A}{\partial r}\right) + \frac{1}{r^2}\left(\frac{\partial^2 c_A}{\partial \phi^2}\right) + \frac{\partial^2 c_A}{\partial z^2}\right] + r_A = \frac{\partial c_A}{\partial t} \qquad (11.25)$$

Spherical

$$D_{AB}\left[\frac{1}{r^2}\frac{\partial}{\partial r}\left(r^2\frac{\partial c_A}{\partial r}\right) + \frac{1}{r^2 \sin^2\theta}\frac{\partial^2 c_A}{\partial \phi^2} + \frac{1}{r^2 \sin\theta}\frac{\partial}{\partial \theta}\left(\sin\theta \frac{\partial c_A}{\partial \theta}\right)\right]$$
$$+ r_A = \frac{\partial c_A}{\partial t} \quad (11.26)$$

Symbolically (Any coordinate system)

$$D_{AB}\nabla^2 c_A + r_A = \frac{\partial c_A}{\partial t} \qquad (11.27)$$

11.5 Chapter Summary—G.E. & B.C. of Mass Transfer

- **Governing Equation of Mass Transfer (page 227)**

    1. Governing equation of mass transfer given by Eqn. 11.17 comprises of four terms representing storage, convection, diffusion (or dispersion), and generation.
    2. Depending on the physical situation, some terms may be dropped.
    3. Depending on the appropriate geometry of the physical problem, choose a governing equation in a particular coordinate system from the equations in Section 11.4.

- **Boundary Conditions of Mass Transfer (page 230)**

    One of the following three types of boundary condition typically holds at a surface

    1. Concentration is specified (Eqn. 11.19)—for example, when the convective mass transfer coefficient is high or when there is evaporation at the surface.
    2. Mass flux is specified (Eqn. 11.20)—for example, it is zero at an impermeable surface.
    3. Convective mass transfer (Eqn. 11.23)—neither concentration, nor mass flux is specified. This is the most general situation.

11.6 Concept and Review Questions

1. How is the mass generation different from storage of mass? Explain using an example.
2. If we keep only the diffusion term in the governing equation, it becomes the Laplace equation. What other physical phenomena (besides mass or heat transfer) can be described by this equation?
3. Show that when the mass transfer coefficient is high ($h_m \to \infty$), boundary condition given by Eqn. 11.23 approaches the boundary condition given by Eqn. 11.19.
4. In what ways are heat and mass transfer analogous and in what ways do they differ?

11.7 Further Reading

Bird, R.B., W.E. Stewart, and E.N. Lightfoot. 1960. *Transport Phenomena*. John Wiley & Sons, New York. (Comprehensive source for governing equations).

Crank, J. 1975. *The Mathematics of Diffusion*. Oxford University Press, Oxford, UK.

Shitzer, A. and R.C. Eberhart. 1985. *Heat Transfer in Medicine and Biology* Vol. 1, 2. Plenum Press, New York.

Welty, J.R., C.E. Wicks, R.E. Wilson, and G. Rorrer. 2001. *Fundamentals of Momentum, Heat, and Mass Transfer*. John Wiley & Sons, New York.

Zaritzky, N.E. and A.E. Bevilacqua. 1988. Oxygen diffusion in meat tissues. International Journal of Heat and Mass Transfer, 31(5):923-930.

11.8 Problems

11.8.1 Governing Equation in Cylindrical Coordinate System

Derive the governing equation for radial mass transfer in cylindrical coordinate (Eqn. 11.25 with only the r dependence).

11.8.2 Problem Formulation: Cell Division with Diffusion of Microorganism in a Gel

Starting from the general governing equations in Section 11.4 on page 233, formulate the governing differential equation for mass transfer to describe the diffusion of a microorganism placed in a stagnant fluid or gel, where the microorganism reproduces following a first-order reaction. Give two boundary conditions that could be used to solve the differential equation. (Adapted from Welty, et. al., 1984).

11.8.3 Problem Formulation: Oxygen Diffusion in Meat Tissue

The color of meat is determined by the relative proportions of myoglobin derivatives. For example, oxygen take up by myoglobin converts the purple reduced pigment to the bright red oxygenated form, oxymyoglobin. Zaritzky et al. (1988) considered oxygen penetration in meat tissue by diffusion with simultaneous consumption of oxygen by the tissue. Consumption follows zeroth order rate constant that decays exponentially with (post-mortem) time. (Contrast this with zeroth order consumption of oxygen in living tissue of Problem 12.6.5.) Formulate a one-dimensional problem with a constant dissolved oxygen concentration at the meat surface exposed to air. At the maximum distance the gas penetrates, a zero flux boundary condition can be used. Do not need to solve the problem (solution provided in Zaritzky et al., 1988).

Chapter 12

DIFFUSION MASS TRANSFER: STEADY STATE

CHAPTER OBJECTIVES

After you have studied this chapter, you should be able to:

1. Formulate and solve for a diffusive mass transfer process in a slab geometry where concentrations do not change with time (steady-state).

2. Extend the steady-state mass transfer concept for a slab to a composite slab using the overall mass transfer coefficient.

3. Extend the steady-state diffusive mass transfer in a slab to include a first order decay of the mass species.

KEY TERMS

- **steady-state diffusion**
- **mass transfer resistances in series**
- **overall mass transfer coefficient**
- **diffusion mass transfer with chemical reaction**

This chapter will deal with mass diffusion under steady-state conditions, closely following Chapter 4 for steady-state heat conduction. We will solve the general governing equation derived in Chapter 11 for the particular situation of steady-state. The relationship of this chapter to the other chapters in mass transfer is shown in Figure 12.1. Like in Chapter 4 (page 45) for heat transfer, steady-state is defined as the situation when mass concentrations are not changing with time. Note that steady-state is not the same as equilibrium, when the system is at same concentration everywhere. In steady-state, mass can continue to move if there are gradients in concentration or other driving forces. However, the concentration does not change with time.

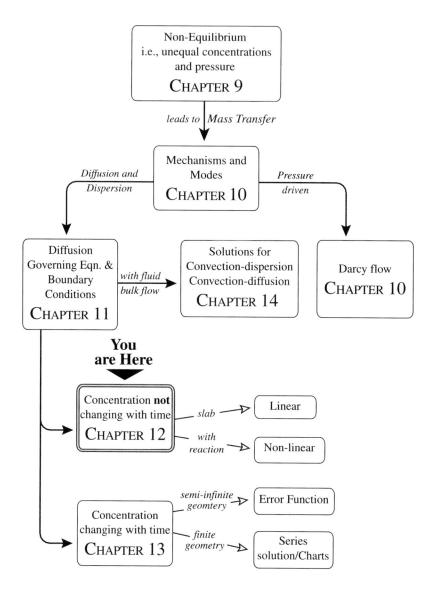

Figure 12.1: Concept map of energy transfer showing how the contents of this chapter relates to other chapters in the part on mass transfer.

12.1 Steady-State Mass Diffusion in a Slab

As discussed previously, there are many parallels between the systems of heat and mass transfer. This chapter deals with the solution to a one-dimensional, steady-state diffusion/dispersion problem. The topics closely parallel those of Chapter 4 for heat

12.1. STEADY-STATE MASS DIFFUSION IN A SLAB

transfer, as the governing equations are almost identical. Thus, in a slab geometry as shown in Figure 12.2, if the concentrations are uniform over the surfaces at $x = 0$ and $x = L$, at locations not too close to the edges the concentration would vary only along the thickness of the slab. In other words, the mass transfer is one-dimensional. At steady state with no mass generation and no bulk flow, the general governing equation is:

$$\underbrace{\frac{\partial c_A}{\partial t}}_{\text{steady state}}\!\!{}^{0} + \underbrace{u\frac{\partial c_A}{\partial x}}_{\text{no convection}}\!\!{}^{0} = D_{AB}\frac{\partial^2 c_A}{\partial x^2} + \underbrace{r_A}_{\text{no reaction}}\!\!{}^{0} \qquad (12.1)$$

which becomes

$$\frac{d^2 c_A}{dx^2} = 0 \qquad (12.2)$$

after eliminating the terms. For the simplest kind of boundary condition where constant concentration c_{A_1} and c_{A_2} can be assumed at the two surfaces, we can write the boundary conditions as

$$c_A(x = 0) = c_{A_1}$$
$$c_A(x = L) = c_{A_2} \qquad (12.3)$$

The solution to Eqn. 12.2 is given by

$$c_A = \frac{c_{A_2} - c_{A_1}}{L}x + c_{A_1} \qquad (12.4)$$

which shows a linear change in concentration from c_{A_1} to c_{A_2} at steady state. Note that this equation is identical with Eqn. 4.5 on page 47, both equations showing a linear change (in concentration or temperature) in a slab under steady state. The diffusive mass flow rate at a position x is given by:

$$\begin{aligned} N_{A,x} &= -D_{AB}A\frac{dc_A}{dx} \\ &= -D_{AB}A(c_{A_2} - c_{A_1})/L \quad \text{using Eqn. 12.4} \\ &= D_{AB}A(c_{A_1} - c_{A_2})/L \end{aligned} \qquad (12.5)$$

which is the same at all locations x along the thickness, signifying steady-state (why?).

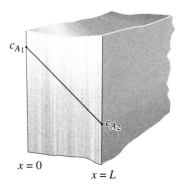

Figure 12.2: A linear concentration profile at steady state in a slab geometry.

Example 12.1.1 Mass Transfer Through a Biofilm

A biofilm is a layer-like aggregation of microscopic animals, plant or bacteria attached to a solid surface. In many natural aquatic systems, especially those having a high specific surface area and low nutrient concentrations, biofilms constitute 90% to 99.9% of bacteria. Biofilms are found in or on stream-beds, ground-water aquifers, water pipes, ship hulls, piers, and aquatic plants and animals. Engineered processes that utilize films of bacteria include trickling filters, biologically activated carbon beds and land

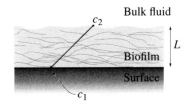

Figure 12.3: Schematic for Example 12.1.1.

treatment systems. Sometimes the biofilm can shield a surface undergoing a reaction, i.e., protection against corrosion. For simplicity, here we will consider such a biofilm where the the reactant is consumed only at the solid surface, maintaining a particular value of surface concentration. We want to calculate the reactant flux.

Calculate the steady-state reactant flux into a biofilm of thickness 0.01 cm, if the diffusivity of the reactant in the biofilm is 0.8 cm^2/day. Assume the reactant concentration in the bulk fluid to be uniform at 3.2 mg/liter. At the surface, reactant concentration is 0.25 mg/liter.

Known: The concentration difference of reactant over the biofilm thickness, i.e., the concentration gradient

Find: The flux of reactant through the biofilm

Schematic and Given Data: Schematic for the problem is shown in Figure 12.3. The given data are:

1. Reactant concentration at the two surfaces of the biofilm are 3.2 mg/liter and 0.25 mg/liter, respectively
2. Thickness of biofilm, L, is 0.01 cm
3. Diffusivity of the reactant in the biofilm is 0.8 cm^2/day

Assumptions: 1. Diffusion is one-dimensional

2. Thickness of the biofilm is uniform

Analysis: The flux at steady state is given by the Fick's law:

$$n_{reactant} = D_{reactant, biofilm} \frac{c_{reactant_2} - c_{reactant_1}}{L}$$

$$= 0.8 \left[\frac{cm^2}{day}\right] \frac{3.2 - 0.25 \, [mg/liter]}{0.01 \, [cm]} \times 10^{-3} \left[\frac{liter}{cm^3}\right]$$

$$= 0.236 \, mg/cm^2 \cdot day$$

Comments: If the reactant is a substrate for the biofilm, such that it is consumed throughout the biofilm, a different formulation of the problem results, as discussed in Section 12.2.

12.1.1 One-Dimensional Mass Diffusion Through a Composite Slab—Overall Mass Transfer Coefficient

In practice, we often work with layers of two or more different materials, as was discussed under heat transfer (page 48). In fact, mass transfer through layers of skin, fat, and muscle, each having its own diffusivity, is an example of mass transfer through a composite material. For a slab material, with uniform surface concentrations, the concentration varies essentially in only one dimension. If there are no reactions (loss or

12.1. STEADY-STATE MASS DIFFUSION IN A SLAB

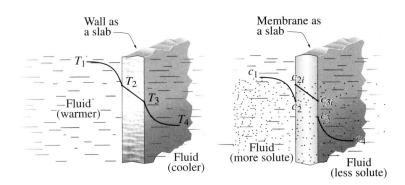

Figure 12.4: Distribution of concentrations around a slab with convection on both sides with analogy to distribution of temperatures around a slab with convection on both sides.

Table 12.1: Analogy of heat and mass fluxes for a slab with convection on both sides

Flux	Mass Transfer		Heat Transfer Analog
Convection in fluid on left face	$n_{A,x} = \dfrac{c_1 - c_2}{\dfrac{1}{h_{m_1}}}$		$q''_x = \dfrac{T_1 - T_2}{\dfrac{1}{h_1}}$
Diffusion inside the slab	$n_{A,x} = \dfrac{c_{2i} - c_{3i}}{\dfrac{\Delta L}{D_{AB}}}$	$= \dfrac{c_2 - c_3}{\dfrac{\Delta L}{K^* D_{AB}}}$	$q''_x = \dfrac{T_2 - T_3}{\dfrac{\Delta L}{k_{solid}}}$
Convection in fluid on right face	$n_{A,x} = \dfrac{c_3 - c_4}{\dfrac{1}{h_{m_2}}}$		$q''_x = \dfrac{T_3 - T_4}{\dfrac{1}{h_2}}$

generation of mass), for steady state mass transfer, the same constant mass flux $n_{A,x}$ appears at any position x.

We can write this mass flux in terms of concentrations with analogy to heat transfer (see Figure 12.4) as shown in Table 12.1. In the flux expressions for mass transfer, note that the concentrations of the transporting species c_{2i} in the solid and c_2 in the liquid are not the same. These two concentrations at the interface are, however, related by the distribution coefficient discussed under solid-liquid equilibrium in page 185. For convenience, Figure 9.11 from page 185 is repeated here as Figure 12.5. Similarly for c_{3i} and c_3.

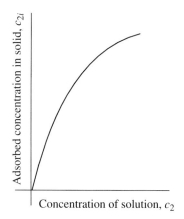

Figure 12.5: Equilibrium relationship between dissolved and surface adsorbed quantity.

In the lower concentration region, the relationship is linear, as seen in Figure 12.5. In this region, the concentrations in the liquid and in the solid are related by

$$\frac{c_{2i}}{c_2} = K^* \quad \frac{c_{3i}}{c_3} = K^* \tag{12.6}$$

The concentration equations are rewritten as:

$$c_1 - c_2 = \frac{n_{A,x}}{h_{m_1}}$$

$$c_2 - c_3 = \frac{n_{A,x}}{K^* D_{AB}} \Delta L$$

$$c_3 - c_4 = \frac{n_{A,x}}{h_{m_2}}$$

and their left and the right hand sides are added to obtain:

$$c_1 - c_4 = n_{A,x} \left(\frac{1}{h_{m_1}} + \frac{\Delta L}{K D_{AB}} + \frac{1}{h_{m_2}} \right)$$

Rewriting in terms of flux, we get

$$n_{A,x} = \frac{(c_1 - c_4)}{\underbrace{1/h_{m_1}}_{\text{convective}} + \underbrace{\Delta L/(K^* D_{AB})}_{\text{diffusive}} + \underbrace{1/h_{m_2}}_{\text{convective}}} \tag{12.7}$$

$$= \frac{\text{concentration difference}}{\Sigma \text{ mass transfer resistances}}$$

As was done in section 4.1.1, the terms in the denominator of Eqn. 12.7 are identified as the mass transfer resistances. Thus, $1/h_m$ is the convective mass transfer resistance and $\Delta L/K^* D_{AB}$ is the diffusive mass transfer resistance. Since the inverse of resistance is termed conductance (by electrical analogy), h_m and $K^* D_{AB}/\Delta L$ are sometimes called conductances. An example of conductance values can be seen in Appendix D.4 for transport of water vapor during transpiration in a leaf.

Overall mass transfer coefficient The conductive and convective resistances just discussed are often combined and reported in terms of an overall mass transfer coefficient, U_m, such that

$$1/U_m = 1/h_{m_1} + \Delta L/(K^* D_{AB}) + 1/h_{m_2} \tag{12.8}$$

In terms of this overall mass transfer coefficient, flux can be written as:

$$n_{A,x} = U_m(c_1 - c_4) \tag{12.9}$$

Thus, if U_m is known, flux can be calculated directly, without knowing the individual mass transfer resistances. An example of mass transfer through combined resistances is now shown.

12.1. STEADY-STATE MASS DIFFUSION IN A SLAB

Example 12.1.2 Dialysis to Remove Urea from Blood

The main function of the kidneys is to regulate the volume, composition, and the pH of body fluids. When kidneys fail to function, a process known as *hemodialysis* can be used. In this process, the person's blood is rerouted across an artificial membrane that "cleanses" it, removing substances that would normally be excreted in the urine. A patient must typically use this artificial kidney three times a week, for several hours each time. Figure 12.6 shows the cross-section of one tube in a capillary dialyzer. Such a dialyzer has a large number of tubes in a parallel arrangement. Blood flows through the inside of the tube and the dialyzing fluid flows on the outside. The tube wall acts as a semi-permeable membrane, letting some of the components (mostly the waste product) of blood to pass. The various mass fluxes are shown in Figure 12.6.

Calculate the flux and the rate of removal (for the total area, in g/hour) of urea from blood at steady state using a cellophane membrane dialyzer at 37°C. The membrane is 0.025 mm thick and has an area of 2.0 m². The mass transfer coefficient on the blood side is estimated as $h_{m_1} = 1.25 \times 10^{-5}$ m/s and that on the dialyzing fluid side is $h_{m_2} = 3.33 \times 10^{-5}$ m/s. The diffusivity of urea through the membrane is 1×10^{-10} m²/s. The distribution coefficient is 2, i.e.,

$$2 = \frac{\text{concentration of urea in membrane at surface}}{\text{concentration of urea in solution at surface}}$$

The concentration of urea in the blood is 0.02g/100 cc and that in the dialyzing fluid is assumed to be zero.

Solution

Known: Steady state mass transfer through a layer of solid with convection on both sides.

Find: The flux through the walls of the tube

Schematic and given data: Schematic of the problem is shown in Figure 12.7. The given data are

1. concentration on blood side = 0.02 g/100 cc
2. concentration on the dialyzing fluid side = 0
3. mass transfer coefficient on blood side, $h_{m_1} = 1.25 \times 10^{-5}$ m/s
4. mass transfer coefficient on the dialyzing fluid side, $h_{m_2} = 3.33 \times 10^{-5}$ m/s
5. $D_{urea,membrane} = 1 \times 10^{-10}$ m²/s
6. thickness of membrane = 0.025 mm
7. Distribution coefficient = 2

Assumptions: For a thin wall, we can approximately use the equations for a slab.

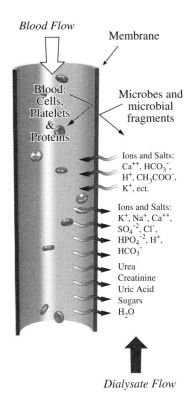

Figure 12.6: Magnified representation of a single hollow fiber in a dialysis system. Picture courtesy of Prof. J. Hunter, Dept. of Biological and Environmental Engineering, Cornell University.

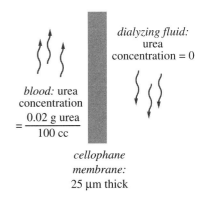

Figure 12.7: Schematic for Problem 12.1.2.

Analysis: $n_{urea,x}$

$$= \frac{(0.02 \times 10^4 - 0)\left[\frac{g}{m^3}\right]}{\underbrace{1/(1.25 \times 10^{-5})\left[\frac{s}{m}\right]}_{\text{convective}} + \underbrace{0.025 \times 10^{-3}/(2 \cdot 1 \times 10^{-10})\left[\frac{s}{m}\right]}_{\text{diffusive}} + \underbrace{1/(3.33 \times 10^{-5})\left[\frac{s}{m}\right]}_{\text{convective}}}$$

$$= \frac{(0.02 \times 10^4 - 0)\left[\frac{g}{m^3}\right]}{\underbrace{80000\left[\frac{s}{m}\right]}_{\text{convective}} + \underbrace{125000\left[\frac{s}{m}\right]}_{\text{diffusive}} + \underbrace{30030\left[\frac{s}{m}\right]}_{\text{convective}}}$$

$$= 0.00085\left[\frac{g}{m^2 s}\right]$$

Therefore, the rate of removal is

$$= 0.00085\left[\frac{g}{m^2 s}\right] \times 2\left[m^2\right] \times 3600\left[\frac{s}{\text{hour}}\right]$$

$$= 6.127\left[\frac{g}{\text{hour}}\right]$$

Comments: Note the higher contribution to resistance from diffusion through the membrane.

12.2 Steady-State Diffusion in a Slab with Chemical Reaction

Chemical reactions often occur simultaneously with diffusion. For example, as oxygen is diffusing in a tissue, it is also consumed by the tissue for metabolism. We will consider a slightly more complicated situation of steady-state mass transfer in a slab when the diffusing species decays with time, as illustrated in Figure 12.8. A simple but quite common situation of first-order decay, as explained under kinetics in Section 9.4.3 on page 187, will be considered. For steady-state diffusion in one-dimension with a first order decay, the governing equation is:

$$\underbrace{\cancel{\frac{\partial c_A}{\partial t}}^{0}}_{\text{at steady state}} + \underbrace{\cancel{u\frac{\partial c_A}{\partial x}}^{0}}_{\text{for no convection}} = D\frac{\partial^2 c_A}{\partial x^2} - k'' c_A$$

which simplifies to the governing equation:

$$\frac{d^2 c_A}{dx^2} - \frac{k''}{D}c_A = 0 \qquad (12.10)$$

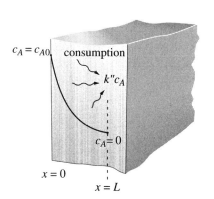

Figure 12.8: Schematic of one dimensional diffusion of a mass species in a slab with simultaneous consumption of the species.

12.2. STEADY-STATE DIFFUSION IN A SLAB WITH CHEMICAL REACTION

We consider the boundary conditions:

$$c_A(x = 0) = c_{A0} \qquad (12.11)$$
$$c_A(x = L) = 0 \qquad (12.12)$$

Here L is the distance from the surface at which the concentration becomes zero. Note that the governing equation (Eqn. 12.10) and the boundary conditions are almost identical to those for heat transfer in a fin, as described in Section 4.4 on page 58, with only a difference in the boundary condition given by Eqn. 12.12. The analogy between the heat and mass transfer situations is quite straightforward. In heat transfer, energy is diffusing as well as being lost from the surface (decaying). In mass transfer, the species is diffusing as well as it is decaying as a first order reaction. The solution to Eqn. 12.10 is obtained following a process similar to the solution process for heat transfer (see Section 4.4 on page 58). Defining

$$m = \sqrt{\frac{k''}{D}} \qquad (12.13)$$

Eqn. 12.10 is written as:

$$\frac{d^2 c_A}{dx^2} - m^2 c_A = 0$$

We assume a solution of the form

$$c_A = c_1 e^{-mx} + c_2 e^{mx} \qquad (12.14)$$

where c_1 and c_2 are constants to be determined using the boundary conditions. Substituting the boundary conditions, we get

$$c_{A0} = c_1 + c_2$$
$$0 = c_1 e^{-mL} + c_2 e^{mL}$$

Solving for c_1 and c_2 (see Appendix G.5 on page 373 for details) and substituting into Eqn. 12.14, we get the final solution for concentration c_A as a function of position as:

$$\frac{c_A}{c_{A,0}} = \frac{-e^{-mL}}{e^{mL} - e^{-mL}} \left(e^{mx} - e^{-mx} \right) + e^{-mx} \qquad (12.15)$$

The solution when a reaction is present (Eqn. 12.15) is plotted in Figure 12.9 and compared with the situation when no reaction is present.

For a material that is quite thick (semi-infinite), the solution given by Eqn. 12.15 can be simplified by taking its limits as $L \to \infty$. Note that as $L \to \infty$,

$$\frac{e^{-mL}}{e^{mL} - e^{-mL}} = \frac{1}{e^{2mL} - 1} \to 0$$

So that Eqn. 12.15 can be simplified to

$$\frac{c_A}{c_{A,0}} = e^{-mx} \qquad (12.16)$$

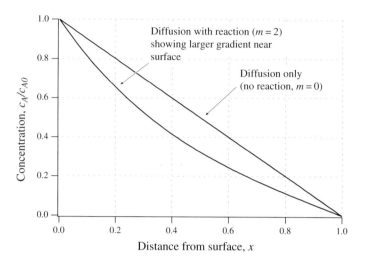

Figure 12.9: Presence of chemical reaction (decay) produces a larger gradient of concentration near the surface and a higher flux. The figure is a plot of Eqn. 12.15.

which shows an exponential decay of concentration from the surface for a thick material.

The rate of mass transfer (the total quantity taken up) at the surface is often an important parameter. The flux at the surface can be compared to the situation with no reaction to find the effect of reaction. The flux at the surface is calculated from Eqn. 12.15 for a finite thickness L as:

$$\begin{aligned} n_{A,x} &= -D\frac{dc_A}{dx}\bigg|_{x=0} \\ &= -Dc_{A,0}\left(\frac{-e^{-mL}}{e^{mL}-e^{-mL}}\left(me^{mx}+me^{-mx}\right)-me^{-mx}\right)_{x=0} \\ &= -Dc_{A,0}\left(\frac{-e^{-mL}}{e^{mL}-e^{-mL}}(2m)-m\right) \\ &= -Dc_{A,0}\cdot\frac{m}{e^{mL}-e^{-mL}}\left(-2e^{-mL}-e^{mL}+e^{-mL}\right) \\ &= mDc_{A,0}\frac{e^{mL}+e^{-mL}}{e^{mL}-e^{-mL}} \\ &= \frac{Dc_{A,0}}{L}mL\frac{e^{mL}+e^{-mL}}{e^{mL}-e^{-mL}} \end{aligned}$$

when no reaction is present (diffusion only), the solution is a straight line given by

Eqn. 12.4. From this equation, the flux can be calculated as

$$n_{A,x} = -D\frac{(0 - c_{A,0})}{L} = \frac{Dc_{A,0}}{L} \qquad (12.17)$$

which is lower than when there is reaction, since

$$mL\frac{e^{mL} + e^{-mL}}{e^{mL} - e^{-mL}} > 1 \qquad (12.18)$$

The two concentration profiles for reaction and no reaction are shown in Figure 12.9. In the presence of reaction, the gradient at the surface is higher, leading to higher flux.

12.2.1 Photosynthesis and the Transport of Water Vapor and CO_2 in a Leaf

The important process of *photosynthesis* can be modeled as diffusion with chemical reaction. Photosynthesis is the manufacture of carbohydrates from carbon dioxide and water by chlorophyll containing plant cells. It involves uptake of carbon dioxide as well as efflux of water vapor (*transpiration*) through the stomatal opening (Figure 12.10). Carbon dioxide from the atmosphere enters the leaf through diffusion in the air just outside the leaf surface and through the water vapor and air inside the leaf (in its substomatal cavity) and is ultimately absorbed by the mesophyll cell walls of the cavity. Water vapor evaporated from the mesophyll cell walls exits the leaf, by diffusion through the air and CO_2 in the substomatal cavity and by convection outside the leaf surface.

The concentrations of CO_2 and water vapor along the substomatal cavity are shown in Figure 12.11. For more details, see Rand (1977) and Rand and Cooke (1997). The figure shows that CO_2 is absorbed into mesophyll cell walls generally throughout the deep interior of the leaf, while water vapor essentially evaporates from those cell walls near the substomatal cavity.

The flux of water vapor out of the leaf and the flux of CO_2 into the leaf has several resistances in series (see Section 12.1.1 on page 240). Intercellular air space is relatively unstirred and it offers primarily diffusive resistance. The stomatal opening through which gases must diffuse offers the second resistance. Finally, on the outside surface, there is the boundary layer resistance. When water vapor passes through the cuticle which is a waxy layer, it has two resistances—the diffusive resistance of the waxy layer and the boundary layer resistance. The waxy cuticle layer greatly reduces the water loss (water in both liquid and vapor phases has difficulty diffusing through the wax). Also, the resistance of the stomata is less on sunny days when the stomata are fully opened.

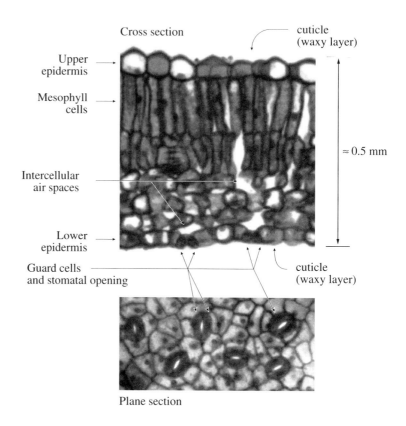

Figure 12.10: Cross section of a leaf showing opening and pathways for transport of CO_2 and water vapor.

12.3 Chapter Summary—Steady-State Mass Diffusion with and Without Chemical Reaction

- Steady-state diffusion in a slab produces a linear concentration profile (Eqn. 12.4)

- Steady-state diffusion in a composite slab is described by overall mass transfer coefficient (Eqn. 12.8).

- Presence of chemical reaction (decay) in the slab increases surface flux since a higher concentration gradient is maintained. The concentration in the slab is given by Eqn. 12.16.

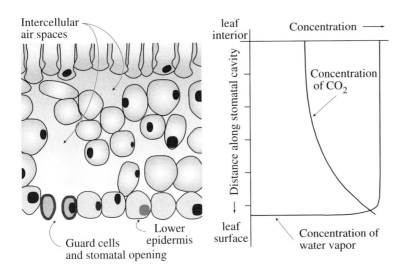

Figure 12.11: Schematic of a leaf cross-section and the concentrations of CO_2 and water vapor in the air space along the substomatal cavity.

12.4 Concept and Review Questions

1. By analogy to heat transfer, explain the concentration profile in a slab at steady-state when no reaction is present.

2. Compare thermal resistance with mass transfer resistance. What parameter in mass transfer resistance replaces the thermal conductivity?

3. Why is the overall mass transfer coefficient a useful parameter?

4. Provide a physical example of diffusion with reaction, outside of what has already been discussed.

5. What role does distribution coefficient play in mass transfer involving a solid and a liquid phase (see Example 12.1.2).

12.5 Further Reading

Armstrong, W. 1989. Aeration in roots. NATO ASI Series, Series G, Ecological Sciences 19:197-206.

Armstrong, W., M.T. Healy, and T. Webb. 1982. Oxygen diffusion in pea I: Pore space resistance in the primary root. New Phytologist 91(4):647-659.

Cooke, J.R. and R.H. Rand. 1980. Diffusion resistance models. In: *Predicting Photosynthesis for Ecosystem Models*, Vol. 1. J.D. Hesketh and J.W. Jones, Ed. CRC Press.

Dainty, J. 1985. Water transport through the root. Acta Horticulturae 171: 21-31.

DeWilligen, P. and M. Van Noordwijk. 1984. Mathematical models on diffusion of oxygen to and within plant roots, with special emphasis on effects of soil-root contact. Plant and Soil 77(2/3):215-231.

Heath, M.S., S.A. Wirtel, and B.E. Rittmann. Simplified design of biofilm processes using normalized loading curves. Research Journal WPCF 62(2):185-192.

Kim, B.R. and M.T. Suidan. 1989. Approximate algebraic solution for a biofilm model with the Monod kinetic expression. Water Research 23(12):1491-1498.

Nobel, P.S. 1996. Leaves and Fluxes. In: *Physiochemical and Environmental Plant Physiology*. Academic Press, San Diego.

Rand, R.H. 1977. Gaseous diffusion in the leaf interior, Transactions of the ASAE 20(4):701-704.

Rand, R.H. and J.R. Cooke. 1996. Fluid mechanics in plant biology. In *Handbook of Fluid Dynamics and Fluid Machinery*, Edited by J.A. Schetz and A.E. Fuhs, John Wiley & Sons.

Welty, J.R., C.E. Wicks, and R.E. Wilson. 1984. *Fundamentals of Momentum, Heat, and Mass Transfer*. John Wiley & Sons, New York.

12.6 Problems

12.6.1 Diffusion of Oxygen Through Soil

Calculate the rate of diffusion of O_2 gas in air at steady state through the pores of a column of dry sandy soil, in $g/m^2 \cdot$ hour. The temperature is 25°C and the total pressure of the system is 1 atmosphere. The bed depth is 1 m and the void fraction is 0.20. The partial pressure of O_2 at the top of the column is 0.23 atmosphere while at the bottom, the partial pressure is 0 atm. A tortuosity of 2 can be assumed. (Hint: Due to the tortuous nature of the paths, the diffusivity in air will be reduced by a factor equal to the tortuosity.)

12.6.2 Drug Delivery Through Skin

For the simplified case of transport through skin, assume total diffusional resistance of skin is entirely due to the stratum corneum (see Figure 5.14) and none through the sweat ducts, etc. Consider application of some medication to the surface of skin, maintaining a concentration of 10 μg/cc of the medication at the skin surface. The inner surface of the stratum corneum is assumed to be maintained at essentially zero concentration since the molecules are removed as soon as they reach the microcirculation by a sufficiently high peripheral blood flow through skin. 1) Calculate the flux of medication through the skin at steady state. 2) Calculate how much of the medication resides in the stratum corneum per unit skin area at steady state. The thickness of the stratum corneum is 1 micron (10^{-6}m). The diffusivity through the stratum corneum for the medication (a low-molecular weight non-electrolyte) is 10^{-10} cm^2/s.

12.6.3 Moisture Loss of Food Through Packages

Consider food that is wrapped to keep it from drying out. The moisture in the small amount of air inside the package will be at equilibrium with the moisture in the food. The system is at 60°C and follows the equilibrium moisture curve of Figure 9.15 on page 193. The moisture in the food is 0.04 kg of water per kg of dry solids. The outside air is quite dry, i.e., has zero moisture and is blowing such that the package outside surface is also maintained at zero moisture. Calculate the loss of water vapor in g/day at steady state for a wrapping of 0.08 mm thickness and a total surface area of 10 cm $\times$ 10 cm. Assume the diffusivity of water vapor in packaging material is 0.25×10^{-11} m^2/s. Density of dry air is 1.07 kg/m^3.

12.6.4 Diffusion of Gas Through the Walls of a Tube

A gas diffuses through the walls of a cylindrical tube whose cross-section is shown in Figure 12.12. 1) Derive a relationship for the rate of diffusion of the gas through the tube in terms of the diffusivity and tube dimensions. 2) Setting up a resistance formulation, show the mass transfer resistance of the tube. Assume the inside and outside walls are at gas concentration c_i and c_0, respectively. Hint: Follow the derivations in section 4.2 for heat transfer.

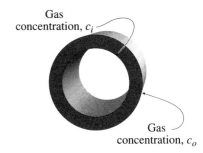

Figure 12.12: Tube cross section.

12.6.5 Oxygen Transport in Human Tissue

Oxygen transport in the human tissue has been considered as diffusion from the blood vessels into tissue cylinders accompanied by a zero-order reaction (Welty et al., 1984). The reaction represents the metabolic consumption of the oxygen to produce carbon dioxide. The tissue cylinders can be considered to be arranged in a bundle such that the boundary condition at the inner surface of the tissue cylinder is a constant c_i corresponding to the oxygen concentration in the arterial blood. Due to symmetry at the outer surface of these tissue cylinders, a zero flux condition can be used. 1) Formulate the steady state governing equation and the boundary conditions for oxygen concentration c in the tissue where the reaction rate for zero order reaction is $-k''$. 2) Determine the oxygen concentration profile as a function of radius, r. 3) Determine the oxygen consumption in the tissue at steady state.

12.6.6 Oxygen Diffusion and Consumption in the Cornea

The cornea of the mammalian eye can be considered, as a first approximation, to be a membrane 0.05 cm thick. The inside of the cornea is bathed by a dilute salt solution, the aqueous humor. The aqueous humor maintains a small oxygen concentration at the corneal inner surface which we assume for this problem to be zero. The cornea uses up oxygen at the rate of 8.045×10^{-10} mol O$_2$/cm^3 cornea per second. The oxygen consumption follows a zeroth order reaction. The diffusivity of oxygen in the cornea is 1.08×10^{-9} m^2/s. Assume the thickness of the cornea to be much smaller than its surface area so it can be treated as a slab. 1) Calculate the concentration of oxygen in the corneal outer surface in mol/cm^3. Assume the eye is open and the outer surface of the cornea is in equilibrium with the oxygen concentration (21%) in the air. The equilibrium Henry's law constant between oxygen at the corneal surface and that in the air is 47,000 atm/mole fraction. For conversion, use 1 mole fraction of oxygen in cornea as approximately equal to 55.56×10^{-3} moles of O$_2$/cm^3 of cornea. 2) Calculate the oxygen concentration profile in the cornea *under steady state*. 3) Roughly sketch the oxygen concentration profile calculated in step 1. 4) Calculate the oxygen flux at the corneal outer surface.

12.6.7 Drug Delivery to the Brain

Since traditional methods often cannot be used for delivering drugs to the brain, researchers are working on implanting drug containing polymers directly in the brain tissue. Consider steady one-dimensional diffusion in the x direction and first order elimination of drug dopamine near an implant, as shown in Figure 12.13. The boundary conditions of drug concentration are c_0 at the implant surface and zero far from the surface into the brain tissue. The reaction rate, k'', for elimination is 0.5/hour and diffusivity of dopamine in brain tissue is 30×10^{-3} m^2/s. 1) Write the solution for drug concentration as a function of position x from the implant surface. 2) Find the distance from the implant surface in cm at which the drug concentration becomes 10% of its value at the implant surface. The volume corresponding to this distance is termed the tissue treatment volume. 3) If the *average* concentration over the distance calculated in item 2 above needs to be 5 μg/g of tissue for dopamine for a certain treatment, what should be the concentration c_0 at the implant surface for this treatment?

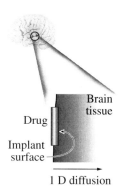

Figure 12.13: Schematic of drug delivery in a tissue.

12.6.8 Diffusion-Driven Aeration in Plants

Consider a plant growing in a water-logged plot, as shown in Figure 12.14. In such water-logged soils, plants can obtain little if any oxygen from the soil. It is desired to find the oxygen concentration in the submerged root that receives atmospheric oxygen by diffusion mechanism, so that we can know the portion that is in need of oxygen. For simplicity, assume steady one-dimensional diffusion. The volumetric respiratory activity of the plants is constant at k''. Neglect any generation of CO$_2$. 1) Write the governing equation for the problem. 2) Write the three boundary conditions for this problem (the additional condition is needed since the submerged length, p, at which the concentration becomes zero is unknown). At this submerged length there are two boundary conditions— both concentration and fluxes are zero. The third boundary condition is constant concentration c_0 at the soil surface. 3) Find the submerged length (where concentration becomes zero) as a function of diffusivity, c_0 and k''.

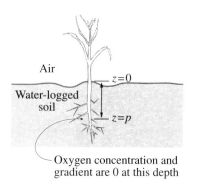

Figure 12.14: Schematic for Problem 12.6.8.

Chapter 13

DIFFUSION MASS TRANSFER: UNSTEADY-STATE

CHAPTER OBJECTIVES

After you have studied this chapter, you should be able to formulate and solve for a time-varying (unsteady) diffusion mass transfer process for the following situations:

1. Where concentrations do not change with position.

2. In a simple slab geometry where concentrations vary also with position.

3. Near the surface of a large body (semi-infinite region).

KEY TERMS

- internal resistance
- external resistance
- mass transfer Biot number
- lumped parameter analysis
- 1D and multi-dimensional diffusion
- Heisler charts
- semi-infinite region

In this chapter, like Chapter 5 on heat transfer, we will consider mass transfer situations where concentration is a function of both position and time. Such situations are typically more complex as opposed to steady state studied in Chapter 12 where concentration did not vary with time. As mentioned in heat transfer, time-varying processes are called unsteady or transient. Relationship of this chapter to the other chapters in mass transfer is shown in Figure 13.1.

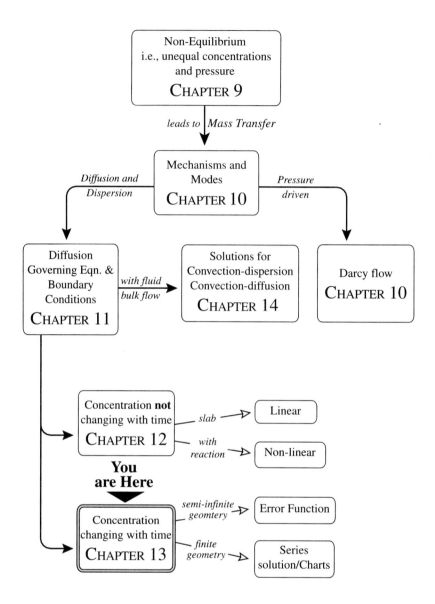

Figure 13.1: Concept map of mass transfer showing how the contents of this chapter relates to other chapters in mass transfer.

13.1 Transient Mass Transfer when Internal Diffusive Resistance is Negligible: Lumped Parameter Analysis

In Chapter 11, we derived the general governing equation for mass transfer that was valid irrespective of a small or large diffusivity. If the diffusivity in a particular solid is

13.1. TRANSIENT MASS TRANSFER WITH NO INTERNAL DIFFUSIVE RESISTANCE

large, the governing equation for mass transfer in the solid can be described in a much simpler way. This approach is analogous to the lumped parameter analysis covered in heat transfer in Section 5.1 on page 71. As an example, consider the drying of a very wet solid such as a moist food due to dry air flowing over it, as shown in the schematic in Figure 13.2. Moisture evaporates from the surface of the solid and is removed by the flowing air. The convective resistance to mass (water vapor) transfer is the *external resistance*. Moisture removed from the surface of the solid creates a moisture gradient inside the solid that will cause capillary diffusion of water from the inside of the solid to the surface. The (capillary) diffusional resistance of the water inside the solid is the *internal resistance*. In this example, when the solid is very wet (there is abundance of water), water can diffuse very rapidly and equilibrate the concentration everywhere inside. Under such condition, moisture concentration is *spatially uniform* at any instant. In other words, internal resistance to diffusive movement is negligible. The assumption of spatial uniformity or negligible concentration gradients is the essence of a *lumped parameter* approach. This is only an approximation, though, as according to Fick's law there cannot be moisture transport if the gradient is zero. But in the above example of a moist solid, we know intuitively that the solid will keep losing moisture especially when it is very wet. Thus, there must be moisture gradients inside the solid for moisture transport to occur, it is just that the gradients are small.

We will now formulate a simple governing equation for mass transfer situations that can be described by the simple lumped parameter approach. If a solid of dry mass m_s and moisture content w (see Figure 13.2) is placed in air with moisture concentration c_∞, moisture loss over the surface will be described by convection. A mass balance on water content in the solid over time Δt during which its moisture content drops by Δw leads to:

$$\underbrace{-m_s \Delta w}_{\text{moisture lost}} = \underbrace{h_m A (c_s - c_\infty) \Delta t}_{\text{moisture convected away at surface}} \quad (13.1)$$

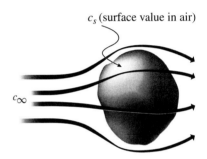

Figure 13.2: A moist solid with air flow over it.

where h_m is the convective mass transfer coefficient over the surface, A is the surface area of the solid, and c_s is the moisture (vapor) concentration in the air next to the solid surface. Note that c here denotes concentration of water vapor in air, $c_{vapor,air}$, where the subscripts have been dropped for simplicity. Dividing both sides by $-m_s \Delta t$,

$$\frac{\Delta w}{\Delta t} = -\frac{h_m A}{m_s}(c_s - c_\infty)$$

Taking the limit as $\Delta t \to 0$,

$$\frac{dw}{dt} = -\frac{h_m A}{m_s}(c_s - c_\infty) \quad (13.2)$$

Eqn. 13.2 is the governing equation describing concentration as a function of time for a lumped parameter mass transfer situation. Since the concentration w is only a function of time, we only need the initial condition

$$w(t = 0) = w_i \quad (13.3)$$

to solve the equation. To solve Eqn. 13.2, we note two different cases for which the surface concentration c_s will be obtained differently.

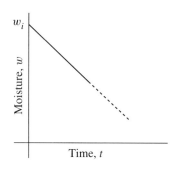

Figure 13.3: Linear decrease in concentration with time during drying of a very high moisture solid, illustration of Eqn. 13.5.

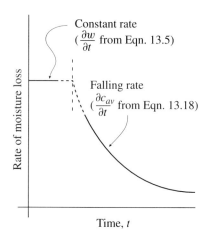

Figure 13.4: Initial rate of drying in a very high moisture solid is a constant corresponding to Figure 13.3, followed by a falling rate of drying when the moisture levels have reduced.

Case I: When solid surface has free water This assumption remains valid for small amounts of drying within a very high moisture range. Since the surface has free water, concentration of water vapor at the surface will be given by the vapor pressure of water at the surface temperature, making the surface concentration a constant. Under such condition, Eqn. 13.2 can be integrated as

$$\int_{w_i}^{w} dw = \int_0^t -\underbrace{\frac{h_m A}{m_s}(c_s - c_\infty)}_{\text{constant}} dt \qquad (13.4)$$

with the final solution:

$$w - w_i = -\frac{h_m A}{m_s}(c_s - c_\infty)t \qquad (13.5)$$

Equation 13.5 shows that concentration w would change linearly with time, as illustrated in Figure 13.3. Thus, the rate of moisture change, dw/dt, is a constant as illustrated in the first segment of the rate curve in Figure 13.4. The second segment of the rate curve will be discussed in the next section. When drying a solid, this situation is referred to as the constant rate of drying. Note that the analysis resulting in Eqn. 13.5 corresponds to high rate of evaporation, same as evaporation from a free surface of water. This constant rate of moisture loss will not continue for too long. In the wet cloth example, as the moisture levels are reduced in the cloth, water will be held more tightly by the cloth and the diffusivity will decrease substantially. This decrease in diffusivity will increase the internal resistance to movement of water, and eventually the internal diffusive resistance will dominate over the external convective resistance, thus decreasing the rate of drying.

Case II: When solid surface has no free water In this case, the vapor at the surface is in equilibrium with the moisture at the solid surface. Thus, c_s will be obtained from the equilibrium moisture relationships described on page 182 for air-wet solid equilibrium. If a linear relationship can be assumed between c_s and w for a small range of moisture, Eqn. 13.2 becomes completely equivalent to Eqn. 5.2 for heat transfer and the solution will be an exponential decay of moisture with time. However, this situation may not exist in reality for reasons mentioned in the previous paragraph, i.e., when moisture level becomes lower, moisture diffusivity decreases (and therefore internal resistance increases) and lumped parameter assumption is no longer valid.

Deciding When to Use Lumped Parameter Analysis The lumped parameter analysis being so simple, it is preferred over other more complex analysis such as those described later. Obviously, lumped parameter analysis would not be appropriate in every situation. Thus, it is important to know the conditions under which we can apply the lumped parameter approach. In an analogous way as described on page 72 for heat transfer, the parameter that compares the internal and the external resistance is defined as the mass transfer Biot number, Bi_m given by

$$Bi_m = \frac{h_m K^* L}{D_{AB}} = \frac{L/D_{AB}}{1/(h_m K^*)} = \frac{\text{diffusive resistance}}{\text{convective resistance}} \qquad (13.6)$$

13.2. TRANSIENT MASS TRANSFER WITH INTERNAL DIFFUSIVE RESISTANCE

Here D_{AB} is the mass diffusivity and L is the characteristic length (same discussion on characteristic length as on page 73 holds). Note the additional variable K^*, the distribution coefficient, introduced here when the mass transfer is between two phases, one internal (i.e., solid) and one external (i.e., fluid flowing over the solid). Need for K^* in the Biot number equation for mass transfer can be seen by comparing Eqns. 13.22 and 13.25 described later. For a mass transfer Biot number $Bi_m < 0.1$, i.e.,

$$\frac{h_m K^* L}{D_{AB}} < 0.1 \tag{13.7}$$

the error in concentration calculation using lumped parameter analysis is less than 5%, and such analysis is often used. For diffusive mass transfer in wet solids, Biot numbers are typically high. For example, for a diffusivity value given by $D_{AB} = 10^{-8} \text{m}^2/\text{s}$, a mass transfer coefficient given by $h_m = 0.01 \text{m/s}$, a K^* value of 0.05[1] and a slab half thickness of $L = 10$ cm, the mass transfer Biot number is

$$Bi_m = \frac{0.01[\text{m/s}] \times 0.05 \times 0.1[\text{m}]}{10^{-8}[\text{m}^2/\text{s}]}$$
$$\approx 5 \times 10^3$$

which is quite large and the primary resistance to mass transfer is the internal diffusive resistance (see Eqn. 13.6). As mentioned in heat transfer, the lumped parameter approach is typically suitable (Eqn. 13.7 is satisfied) when surface to volume ratio is large or the diffusivity is large with respect to the convective mass transfer coefficient.

13.2 Transient Diffusion when Internal Resistance is Not Negligible: Example of a Slab Geometry

As stated in the previous section, when the mass transfer Biot number is large, i.e., internal diffusional resistances are significant, a lumped parameter analysis is no longer applicable. This section parallels the analogous section on heat transfer (Section 5.3 on page 75). In this section, we will learn how to formulate and solve for the spatial distribution of concentration in the solid. Since concentration variation in the solid is significant, we will need to solve this problem by starting with the general governing equation on page 229. For simplicity, we will consider a one-dimensional (1D) slab where the concentration varies along the thickness, as shown in Figure 13.5. To keep things simple, we will consider the boundary condition at the surface to be a constant concentration, i.e., external fluid resistance is negligible. This corresponds to a very high convective mass transfer coefficient (see discussion on boundary conditions at the end of Section 11.3). The governing equation and boundary condition for symmetric drying or wetting of an infinite slab that has no chemical reaction are:

---

[1] value corresponds approximately to equilibrium between wet wood and moist air at 20°C, data from ICT, 1929.

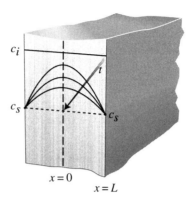

Figure 13.5: Schematic of a slab showing the line of symmetry at $x = 0$ and the two surfaces at $x = L$ and at $x = -L$ maintained at concentration c_s.

$$\frac{\partial c_A}{\partial t} + \underbrace{u\frac{\partial c_A}{\partial x}}_{\text{no bulk flow}}{}^{0} = D\frac{\partial^2 c_A}{\partial x^2} + \underbrace{r_A}_{\text{no reaction}}{}^{0} \qquad (13.8)$$

Thus, the simplified governing equation is

$$\frac{\partial c}{\partial t} = D\frac{\partial^2 c}{\partial x^2} \qquad (13.9)$$

where we have dropped the subscript A in c_A for simplicity. Due to the second order in spatial (x) variation, two boundary conditions in space will be needed. For symmetric drying or wetting, the conditions are:

$$\left.\frac{\partial c}{\partial x}\right|_{x=0,t} = 0 \quad \text{(due to symmetry)} \qquad (13.10)$$

$$c(x = L, t) = c_s \quad \text{(surface concentration is specified)} \qquad (13.11)$$

and the initial condition is given by

$$c(x, t = 0) = c_i \qquad (13.12)$$

where c_i is the constant initial concentration and c_s is the constant concentration at the two slab surfaces at time $t > 0$. Note that although c_s is at the surface, it is concentration in the solid, not in the surrounding fluid. Typically, concentration in the fluid is more readily available and is used to obtain c_s, concentration in the solid, using equilibrium relationships such as in Figure 9.7. Because the slab started at uniform initial concentration and was subjected to the same concentration on both sides, the concentration profile will remain symmetric during the transient process. Following the analogy from heat transfer on page 75 and the solution given in Appendix G.1 (page 365), the

13.2. TRANSIENT MASS TRANSFER WITH INTERNAL DIFFUSIVE RESISTANCE

solution to the governing equation and the boundary and initial conditions mentioned above is

$$\frac{c - c_s}{c_i - c_s} = \sum_{n=0}^{\infty} \frac{4(-1)^n}{(2n+1)\pi} \cos \frac{(2n+1)\pi x}{2L} e^{-D\left(\frac{(2n+1)\pi}{2L}\right)^2 t} \qquad (13.13)$$

The non-dimensional quantity Dt/L^2 in the exponential term of Eqn. 13.13 is the Fourier number

$$Fo = \frac{Dt}{L^2}$$

where the diffusivity D replaces the thermal diffusivity, α, from heat transfer.

13.2.1 How Concentration Changes with Time

As discussed on page 76 for heat transfer, for large times the exponential terms in Eqn. 13.13 drop off (decay) more rapidly and the equation can be simplified to:

$$\frac{c - c_s}{c_i - c_s} = \frac{4}{\pi} \underbrace{\cos \frac{\pi x}{2L}}_{\text{spatial}} \underbrace{e^{-D\left(\frac{\pi}{2L}\right)^2 t}}_{\text{time}} \qquad (13.14)$$

This simplified solution at large times shows that for a given position, the concentration-time relationship is exponential. It can never reach the steady-state concentration. As the concentration difference decreases, the rate of mass flow decreases. If we take the natural logarithm of both sides of Eqn. 13.14, we get:

$$\ln \frac{c - c_s}{c_i - c_s} = \ln \left(\frac{4}{\pi} \cos \frac{\pi x}{2L} \right) - D\left(\frac{\pi}{2L}\right)^2 t \qquad (13.15)$$

This linear relationship between $\ln(c-c_s)/(c_i-c_s)$ and time t can be seen in Figure B.1 on page 327 for various values of position x/L. Note that at times close to $t = 0$, they are not linear. This is expected since by dropping terms in the series we fail to satisfy the initial condition.

13.2.2 Concentration Change with Position and Spatial Average

It can be easily seen from Equation 13.14 that eventually, at a given time, the concentration varies as a cosine function. Note that it stays as a cosine function for all large values of time, except the amplitude of the cosine wave drops exponentially with time. Often, in practice, a spatial average concentration is of interest. A spatial average concentration would be defined by

$$c_{av} = \frac{1}{L} \int_0^L c \, dx \qquad (13.16)$$

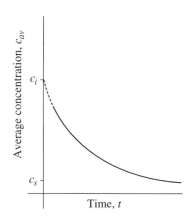

Figure 13.6: Plot of Eqn. 13.17 showing average concentration decays exponentially with time. Note that the equation is not valid for times close to initial time.

Applying this definition of average to Eqn. 13.15, we obtain an equation for the average concentration as

$$\ln \frac{c_{av} - c_s}{c_i - c_s} = \ln \frac{8}{\pi^2} - D\left(\frac{\pi}{2L}\right)^2 t \quad (13.17)$$

which shows that the average concentration changes exponentially with time, as illustrated in Figure 13.6. The time rate of change of the average concentration is obtained by differentiating Eqn. 13.17 as:

$$\frac{dc_{av}}{dt} = -\frac{2D}{L^2}(c_i - c_s)e^{-D(\pi/2L)^2 t} \quad (13.18)$$

which shows that the rate of change of average concentration drops exponentially with time. This is illustrated in Figure 13.4 for a typical drying process, along with the initial constant rate of drying discussed under lumped parameter analysis (Eqn. 13.5). As mentioned earlier, for a typical biomaterial, constant rate of drying occurs for only a small time and the drying rate goes down for the rest of the time.

13.2.3 Concentration Change with Size

Consider Eqn. 13.17 for average concentration as a function of time for large values of time. It can be rewritten as:

$$\frac{Dt}{L^2} = -\frac{4}{\pi^2} \ln\left[\frac{\pi^2}{8}\left(\frac{c_{av} - c_s}{c_i - c_s}\right)\right] \quad (13.19)$$

This implies that for a given change of concentration (measured in terms of fractional change of the total possible change $c_i - c_s$), the time required increases with the square of the thickness. This observation can be generalized for other systems by saying the time required is proportional to the square of the characteristic dimension, i.e.,

$$t \propto L^2 \quad (13.20)$$

where L is the characteristic dimension— it is half thickness for a slab and radius for a long cylinder or a sphere.

13.2.4 Charts Developed from the Solutions: Their Uses and Limitations

The solution given by Eqn. 13.13 is identical to the solution given by Eqn. 5.13 and therefore the same Heisler charts as shown in pages 327-329 can be used to calculate the concentration instead of temperatures. The vertical axis would now stand for $(c - c_\infty)/(c_i - c_\infty)$. Note that like in heat transfer, these charts only provide the approximate solution at longer times given by Eqn. 13.14.

Since the charts were developed from the analytical solution, the assumptions that went into the analytical solution are implicit in the charts. It is important to remind ourselves of these conditions that have to be satisfied in order to be able use the charts. These are:

1. Uniform initial concentration
2. Constant boundary fluid concentration
3. Perfect slab, cylinder or sphere
4. Far from edges
5. No chemical reaction ($r'_A = 0$)
6. Constant diffusivity D

The discussion in Section 5.3.6 on using numerical methods to overcome the limitations of charts is also applicable for mass transfer.

13.2.5 When Both Internal and External Resistances are Present: Convective Boundary Condition

So far in Section 13.2 we have considered a negligible external fluid resistance to mass transfer (resistance $1/h_m \to 0$ for $h_m \to \infty$) represented by the boundary condition of specified surface concentration. For a convective boundary condition on the slab surface given by:

$$-D_{AB} \left.\frac{\partial c_A}{\partial n}\right|_{\text{surface}} = h_m(c_{A,\text{surface}}^{\text{fluid}} - c_{A,\infty}^{\text{fluid}}) \tag{13.21}$$

where h_m would be finite (instead of $h_m \to \infty$), external resistance of the fluid needs to be considered. This boundary condition appears similar to the convective heat transfer boundary condition

$$-k \left.\frac{dT}{dx}\right|_{x=0} = h(T|_{x=0} - T_\infty) \tag{13.22}$$

However, there is one important difference. Unlike temperature T, the variables c_A and c_A^{fluid} on either side of the equation are not the same, as illustrated in Figure 13.7. This complexity arises in mass transfer calculations since concentration on the surface in the solid, c_A, is different from concentration on the surface in the fluid, c_A^{fluid}, as discussed under boundary conditions in Section 11.3. This is unlike the case of heat transfer, where there is only one value of temperature at the surface. When c_A and c_A^{fluid} can be related in a simple manner as

$$\frac{c_A^{\text{fluid}}}{c_A} = K^* \tag{13.23}$$

Eqn. 13.21 can be rewritten as

$$-D_{AB} \left.\frac{\partial c_A}{\partial n}\right|_{\text{surface}} = h_m(K^* c_{A,\text{surface}} - c_{A,\infty}^{\text{fluid}}) \tag{13.24}$$

$$-D_{AB} \left.\frac{\partial c_A}{\partial n}\right|_{\text{surface}} = h_m K^*(c_{A,\text{surface}} - c_{A,\infty}^{\text{fluid}}/K^*) \tag{13.25}$$

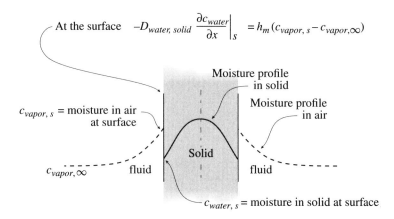

Figure 13.7: In a convective boundary condition, surface concentration is not the same as bulk fluid concentration, signifying additional resistance of the fluid to mass transfer. Also, surface concentrations in the fluid, c_{vapor}, and surface concentration in the solid, c_{water}, are not the same, but related through equilibrium relationship.

Equation 13.25 is now completely analogous to the convective heat transfer boundary condition (Eqn. 13.22). Thus, the solution to the diffusion mass transfer problem with convective boundary condition will follow that in heat transfer (see page 80), and the same charts on pages 327-329 can be used with appropriate replacement of parameters. Like in the case of convective heat transfer, the concentration-time lines for a finite h_m value will become a function of h_m as well, which is why we have additional sets of lines on these charts. The parameters in charts need to be replaced as follows: The y-axis will now be $(c - c_\infty/K^*)/(c_i - c_\infty/K^*)$ and $m = D_{AB}/h_m K^* L$ for the general situation of $K^* \neq 1$. For the special case of $K^* = 1$, the y-axis is $(c - c_\infty)/(c_i - c_\infty)$ and $m = D_{AB}/h_m L$. Note that m is the inverse of mass transfer Biot number, Bi_m, analogous to how it was defined in heat transfer.

13.3 Transient Diffusion in a Finite Geometry— Multi-Dimensional Problems

There are mass transfer situations when we cannot approximate the geometry as one-dimensional. Two- and three-dimensional effects need to be considered, as described for heat transfer problems in Section 5.4 on page 82. In a similar way as mentioned under heat transfer, to solve for multi-dimensional diffusion mass transfer problems, a finite geometry is considered as the intersection of two or three infinite geometries. A rectangular box, for example, is considered as intersection of three infinite slabs and

13.3. TRANSIENT DIFFUSION IN A FINITE GEOMETRY—MULTI-DIMENSIONAL PROBLEMS

the concentration $c_{x,y,z,t}$ can be calculated as

$$\frac{c_{xyz,t} - c_s}{c_i - c_s} = \left(\frac{c_{x,t} - c_s}{c_i - c_s}\right)_{\substack{\text{infinite}\\\text{x slab}}} \left(\frac{c_{y,t} - c_s}{c_i - c_s}\right)_{\substack{\text{infinite}\\\text{y slab}}} \left(\frac{c_{z,t} - c_s}{c_i - c_s}\right)_{\substack{\text{infinite}\\\text{z slab}}} \quad (13.26)$$

Similarly, for a finite cylinder, the concentration $c_{r,z,t}$ can be calculated as

$$\frac{c_{r,z,t} - c_s}{c_i - c_s} = \left(\frac{c_{r,t} - c_s}{c_i - c_s}\right)_{\substack{\text{infinite}\\\text{cylinder}}} \left(\frac{c_{z,t} - c_s}{c_i - c_s}\right)_{\substack{\text{infinite}\\\text{slab}}} \quad (13.27)$$

Numerical computer-based solutions, mentioned in Section 5.3.6, can also handle very efficiently the multidimensional problems.

Example 13.3.1 Drying of Wood

Wooden slabs of 50.8 mm thick containing 30 wt% (dry basis or db) moisture are to be dried by blowing dry air over them. Forced convection (blowing) conditions can maintain the moisture at the wood surface to be in equilibrium with the moisture in the air. This equilibrium moisture content at the wood surface is 5 wt% moisture, dry basis. An effective diffusivity value for combined liquid and vapor diffusion through the wood is $1 \times 10^{-9} \text{m}^2/\text{s}$. Calculate the time needed to terminate the drying process when the center of the wooden slab reaches 10% (db) moisture.

Known: Slab of wood subjected to sudden change in the surface moisture

Find: The time for the center of slab to reach a moisture content of 10%

Figure 13.8: Section of a wood. Effective moisture diffusivities would be different along the three dimensions.

Schematic and given data:

1. Thickness of slab =50.8 mm
2. Diffusivity of moisture in the slab, $D = 1 \times 10^{-9} \text{ m}^2/\text{s}$
3. Initial moisture content = 30%
4. Final moisture content = 10%
5. Equilibrium moisture = 5%

Assumptions:

1. Thickness is much smaller than other dimensions, therefore drying from the other four faces is ignored.
2. The diffusivity value remains constant during the drying process
3. No shrinkage—dry density does not change

Analysis: For transient diffusion in a slab, the Heisler charts are to be used. The non-dimensional concentration, Y, is

$$Y = \frac{c - c_s}{c_i - c_s}$$
$$= \frac{0.10 - 0.05}{0.3 - 0.05}$$
$$= 0.2$$

The other parameters needed to read the chart are the non-dimensional distance n at the center

$$n = \frac{x}{x_1} = \frac{0}{25.4} = 0$$

and the parameter m signifying the convective condition at the surface

$$m = \frac{D}{h_m x_1} = 0$$

where $h_m \to \infty$ has been used since the surface concentration is specified (see discussion under the third kind of boundary condition in Section 11.3). With these values of Y, m, and n, value of non-dimensional time X is obtained from the chart on page 327 as

$$X = 0.75$$

To find time,

$$0.75 = \frac{Dt}{x_1^2}$$
$$= \frac{1 \times 10^{-9} t}{(0.0254)^2}$$

which gives a time $t = 483870$s, about 5.6 days.

Comments: The actual drying time could be shorter since moisture would also be lost through the other four faces that we have ignored here. However, since diffusivity drops with moisture content, drying time can also be higher.

Example 13.3.2 Diffusion of CO_2 to Retard Ripening Process

Although ripening of tomatoes can be retarded by low temperatures, storage under refrigeration can cause chilling injury and an increase of fungal contamination. Use of atmospheres rich in CO_2 and low O_2 can also slow down the ripening process of tomatoes. To determine the optimum level of CO_2 treatment, diffusivity of CO_2 through the fruits need to be known. Whole fruit without its skin was used to establish effective diffusivity of CO_2 in the internal tissue and the value found was $D_{eff} = 2.3 \times 10^{-8} \mathrm{m^2/s}$.

What exposure time is necessary for the concentration of CO_2 at the center of a spherical tomato of 8 cm diameter to change to 15% by volume of air inside the tomato? The initial concentration of CO_2 in air in the tomato is 2% by volume. In the storage room, the surface of the tomato is maintained at 18% by volume of CO_2.

Known: A spherical object subjected to sudden increase in the surface concentration of CO_2

Find: The time for the center of sphere to reach a CO_2 concentration of 15%

Schematic and given data:

1. Surface CO_2 concentration, $c_s = 18\%$
2. Initial CO_2 concentration, $c_i = 2\%$
3. Final CO_2 concentration, $c = 15\%$

Figure 13.9: Section of a tomato as an example of an inhomogeneous material. Effective moisture diffusivities are likely to vary for the different constituents.

Assumptions:

1. Assume the tomato is a perfect sphere
2. Diffusivity value provided is an average over all the material from center to surface, see Figure 13.9.

Analysis: For transient diffusion problem in a sphere, the Heisler chart on page 329 is to be used. The non-dimensional concentration, Y, is

$$Y = \frac{c - c_s}{c_i - c_s} = \frac{15 - 18}{2 - 18} = 0.1875$$

The other parameters needed to read the chart are the non-dimensional distance n at the center

$$n = \frac{x}{x_1} = \frac{0}{0.04} = 0$$

and the parameter m signifying the convective condition at the surface

$$m = \frac{D}{h_m x_1} = 0$$

where $h_m \to \infty$ has been used since the surface concentration is specified. With these values of Y, m, and n, the value of non-dimensional time X is obtained from the chart on page 329 as

$$X = 0.30$$

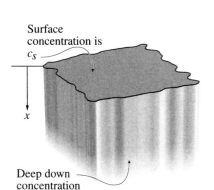

Figure 13.10: Schematic of a semi-infinite region showing only one identifiable surface.

To find time,

$$0.30 = \frac{Dt}{x_1^2}$$
$$= \frac{2.3 \times 10^{-8} t}{(0.04)^2}$$

which gives the necessary exposure time t of about 5.8 hours.

Comments The actual time to change concentration is likely to be longer since skin would pose additional resistance to diffusion.

13.4 Transient Diffusion in a Semi-Infinite Region

This section parallels the section on page 82 for heat conduction in a semi-infinite region. A semi-infinite region extends to infinity in two directions and has a single identifiable surface in the other direction, as shown in Figure 13.10. The semi-infinite region provides a useful idealization for many practical situations where we are interested in mass transfer for a relatively short period of time and/or in a relatively thick material. Examples can be short time exposure of a solid or liquid to a gas.

The governing equation for diffusion with no chemical reaction is

$$\underbrace{\frac{\partial c_A}{\partial t}}_{\text{storage}} + \underbrace{u \frac{\partial c_A}{\partial x}}_{\text{no bulk flow}}{}^{0} = \underbrace{D \frac{\partial^2 c_A}{\partial x^2}}_{\text{diffusion}} + \underbrace{r_A}_{\text{no reaction}}{}^{0} \quad (13.28)$$

which leads to the simplified governing equation as

$$\frac{\partial c}{\partial t} = D \frac{\partial^2 c}{\partial x^2} \quad (13.29)$$

where the subscript A from c_A has been dropped for simplicity. The boundary conditions at the surface and far from the surface (at infinity) are given by

$$c(x = 0) = c_s \quad (13.30)$$
$$c(x \to \infty) = c_i \quad (13.31)$$

The initial condition is given by

$$c(t = 0) = c_i \quad (13.32)$$

Following discussion on page 83, the solution for concentration c is given by

$$\frac{c - c_i}{c_s - c_i} = 1 - \text{erf}\left[\frac{x}{2\sqrt{Dt}}\right] \quad (13.33)$$

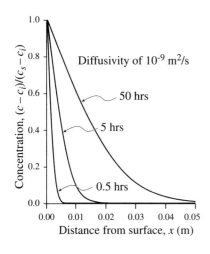

Figure 13.11: Plot of Eqn. 13.33 showing how concentration profile changes with time.

13.4. TRANSIENT DIFFUSION IN A SEMI-INFINITE REGION

The error function is tabulated on page 325. The solution given by Eqn. 13.33 is plotted in Figure 13.11 for various times. The diffusive mass flux at the surface can be calculated from concentration given by Eqn. 13.33 as (see details on page 83):

$$n_s = -D \left.\frac{dc}{dx}\right|_{x=0}$$

$$= \sqrt{\frac{D}{\pi t}}(c_s - c_i) \qquad (13.34)$$

We can see that the diffusive mass flux at the surface reduces with time. This is expected since the concentration gradient at the surface drops with time, as shown in Figure 13.11.

When can we use semi-infinite approximation? The semi-infinite approximation is true for thin materials over shorter time or thick materials over longer time. As described in Eqn. 5.29 on page 84, at distances greater than $4\sqrt{Dt}$, i.e., for

$$x \geq 4\sqrt{Dt} \qquad (13.35)$$

concentration change is less than 0.5%. Thus, any material whose thickness L is larger than $4\sqrt{Dt}$ can be approximated as semi-infinite for this purpose. This is illustrated in Figure 13.12 for a diffusivity of $10^{-8}\,\text{m}^2/\text{s}$, a value chosen in the range of gas diffusivity in liquids. The material needs to be at least as thick as the minimum thickness shown, to be able to use Eqn. 13.33.

Example 13.4.1 Oxygen Concentration and Bacterial Growth in a Silage

Many of the problems associated with the storage of silage in a bunker silo arise during the unloading phase, through the exposed face (see Figure 13.13). Upon exposure to air, yeasts, molds and aerobic bacteria begin to grow rapidly causing a loss of nutrients, a rise in pH and heating in particularly warm weather. Associated effects may include a decrease in animal production and in some cases toxicity and death. The mechanism of oxygen penetration into the silage through the exposed face have sometimes been assumed to be primarily diffusion (Pitt and Muck, 1993).

Assuming the oxygen concentration at the surface to be constant and equal to $0.21\,\text{m}^3$ of O_2/m^3 of air, calculate the O_2 concentration within the silage at a depth of 0.15 m after 1 hour. Diffusivity of O_2 in air is $1.781 \times 10^{-5}\,\text{m}^2/\text{s}$ at $0°C$. Assume the silo to be long so that a 1-D model can be used. Neglect microbial growth and CO_2 generated during respiration.

Known: A thick object subjected to sudden increase in the surface concentration of O_2

Find: The concentration of O_2 at 0.15 m after 1 hour

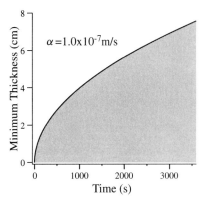

Figure 13.12: Illustration of minimum thickness of a material for which error function solution can be used for diffusive mass transfer.

Figure 13.13: Schematic of a bunker silo showing the exposed face.

Schematic and given data:

1. Surface O_2 concentration, $c_s = 0.21$
2. Initial O_2 concentration, $c_i = 0$
3. Diffusivity of O_2 in air, $D = 1.781 \times 10^{-5} \text{m}^2/\text{s}$
4. Depth of interest, $x = 0.15$ m

Assumptions: To check whether we can use a semi-infinite approximation for oxygen penetration over one hour, we can use Eqn. 13.35. The quantity $4\sqrt{Dt}$ turns out to be 1.01 m. We assume the thickness of the silage is higher than this distance.

Analysis: For transient diffusion problem in a semi-infinite region, the solution is given by Eqn. 13.33.

$$\frac{c - c_i}{c_s - c_i} = 1 - \text{erf}\left[\frac{x}{2\sqrt{Dt}}\right]$$

$$\frac{c - 0}{0.21 - 0} = 1 - \text{erf}\left(\frac{0.15}{2\sqrt{(1.781 \times 10^{-5})(3600)}}\right)$$

$$= 1 - \text{erf}(.296)$$

$$= 1 - .328 = .672$$

which leads to $c = .14 \text{ m}^3 O_2/\text{m}^3$ of air

Comments: Thus within a short period of 1 hour the top layer of 15 cm has a very high oxygen concentration of 0.14 m^3 O_2/m^3 of air. This layer is very much susceptible to bacterial and mold growth. Note, however, we have used the diffusivity value of oxygen in air whereas the diffusivity value of oxygen in the air contained in the silage can be significantly smaller depending on the tortuous paths that the oxygen would have to take in this porous media.

13.5 Chapter Summary—Transient Diffusive Mass Transfer

- **No Internal Resistance, Lumped Parameter (page 254)**

  1. For $Bi_m < 0.1$ in a finite size material, the diffusive mass transfer resistance in the solid can be ignored (the solid is considered lumped) in comparison with the convective mass transfer resistance at the solid surface.

  2. As diffusive mass transfer resistance in the solid is ignored, concentration is not a function of position, is a function of time only and is given by Eqn. 13.5 on page 256.

- **Internal Resistance Is Significant (pages 257, 262, 266)**

  1. When internal diffusive resistance is significant ($Bi_m > 0.1$), concentration is a function of both position and time.

  2. For $Bi_m > 0.1$, $Fo > 0.2$ and for infinite slab, infinite cylinder and a sphere, solutions are given by the same Heisler charts as for conduction heat transfer (pages 327-329). Concentration parameter for the charts need to be modified, as discussed on page 261.

  3. For a finite slab and a finite cylinder (page 263), the solutions are given by Eqns. 13.26 and Eqn. 13.27, respectively.

  4. Materials with thicknesses $L \geq 4\sqrt{Dt}$ are considered effectively semi-infinite (Eqn. 13.35) and the solution is given by the error function (Eqn. 13.33 on page 266).

13.6 Concept and Review Questions

1. In mass transfer, what parameter compares internal diffusive resistance to external convective resistance?

2. In lumped parameter analysis, we ignored the internal diffusional resistance. Is this physically possible? Explain.

3. Under what conditions can you approximate a slab as one-dimensional?

4. What dimension would you choose for analyzing a long cylinder?

5. When can you approximate a geometry as semi-infinite?

6. Can you use Heisler charts for any duration of the diffusion process?

13.7 Further Reading

Bartholomew, C. 1965. *Soil Nitrogen.* American Society of Agronomy, Madison, Wisconsin.

Bertola, N., A. Chaves, and N.E. Zaritzky. 1990. Diffusion of carbon dioxide in tomato fruits during cold storage in modified atmosphere. International Journal of Food Science and Technology 25:318-327.

ICT. 1933. International critical tables of numerical data, physics, chemistry and technology. Volume II. Published for the National Research Council of the United States of America by McGraw-Hill, New York.

Lissik, E.A. 1988. Mechanisms of pathogen transport through the teat canal of a dairy cow. Ph.D. Thesis, Cornell University.

Mannapperuma, J.D., R.P. Singh, and M.E. Montero. 1991. Simultaneous gas diffusion and chemical reaction in foods stored in modified atmospheres. Journal of Food Engineering 14:167-183.

Pitt, R.E. and R.E. Muck. 1993. A diffusion model of aerobic deterioration at the exposed face of bunker silos. J. Agric. Engr. Res. 55(1):11-26.

Rhee, J. and L.E. Bode. 1990. Transport model of spray droplets above and within a plant canopy. Presented at the 1990 International Meeting of ASAE, Paper No. 90-1580. ASAE, St. Joseph, Michigan.

13.8 Problems

13.8.1 Average Moisture Content in a Slab

Show that for a slab of material dried symmetrically, the average moisture of the slab as a function of time at long times is given by

$$\frac{c_{av} - c_s}{c_i - c_s} = \frac{8}{\pi^2} \exp(-D\,(\pi/2L)^2\, t)$$

where c_{av} is the volume averaged moisture content.

13.8. PROBLEMS

13.8.2 Experimental Determination of Moisture Diffusivity

As you may recall, parameters such as diffusivity are typically found from experiments. Consider a simple experiment where a slab 2 cm thick was initially weighed at 250 g. It is dried for 1 hour and weighed again to be 200 g. The weight of dry solids was known to be 70 g. The slab is thin enough in one direction compared to the other two directions so that the moisture transfer is along the thin direction only. Also, it is being dried from both faces equally, i.e., drying is symmetric. The dry air maintains a moisture content of zero at both faces of the slab. 1) Using the weight loss data, calculate the diffusivity of moisture in the slab. Hint: Use the expression for average moisture content as a function of time, c being moisture content *on dry weight basis*. 2) How would you expect the diffusivity to change as the material dries?

13.8.3 Diffusion in Agar Gel

Agar gel is a gel matrix formed from agarose that is a purified hydrocolloid isolated from agar or agar-bearing marine algae. This gel matrix is nearly ideal for diffusion and electrokinetic movement of biopolymers. Therefore, it is useful for studying processes such as electrophoresis and immunodiffusion. An agar gel cylinder of 20 mm diameter contains a uniform concentration of urea of 0.1 kmol/m^3. Consider the length of the cylinder to be much larger than 20 mm. The cylinder is suddenly immersed in water containing no urea. The water flow is rapid enough such that the surface convective mass transfer resistance can be assumed to be negligible. The diffusivity of urea in the agar is 4.5×10^{-10} m^2/s. 1) Calculate the concentration at the center of the cylinder *and* at 5 mm from the center after 20 hours. 2) If the diameter of the cylinder is doubled, what would be the concentration at the center after 20 hours?

13.8.4 Drying of a Cement Slab

A cement slab 75 mm thick at a moisture content of 75% dry basis needs to be dried to a moisture content of 25% dry basis. Air blowing over both sides of the slab keeps the surfaces at equilibrium moisture content of 15% dry basis. The average diffusivity of water in the cement is 5.6×10^{-8} m^2/s. Calculate the time for the center of the slab to reach a moisture content of 25% dry basis.

13.8.5 Drying of a Compost Pile

Consider drying of a layer of compost 30 cm thick spread on a surface that is impermeable to water vapor, i.e., no vapor can pass through the surface. Air at 20% relative humidity is blowing over the top of the compost and is leading to an average mass transfer coefficient of 0.1 m/s. The compost has an initial moisture content of 1 kg of water/kg of dry matter that is uniform throughout. The diffusivity of moisture in compost is approximately 10^{-6} m^2/s. The equilibrium moisture content of the compost is approximated, for the purpose of this problem, to be that of the paper in Figure 13.14. 1) How long does it take for the most moist location in the compost layer to reach

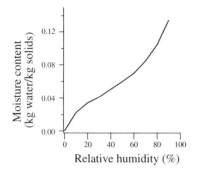

Figure 13.14: Equilibrium moisture content of paper. Moisture content is given in kg of water per kg of solids. Data from ICT, 1933.

0.3 kg of water/kg of dry matter? 2) What is the top surface moisture content (in the compost) at this time?

13.8.6 Effective Diffusivity in Drying

During a drying process for raisins, the average moisture content of raisins measured at 30 minutes and one hour are 60 wt% and 50 wt% dry basis, respectively. The air being circulated over the raisins maintains its surface at its equilibrium moisture content that is 5% for the temperature and humidity of the air. Consider an average raisin to be a slab of thickness 1 cm. Estimate the effective diffusivity of water in the raisins during this drying process. Assume that shrinkage can be neglected during this period. [Hint: Although this is a transient problem, the moisture contents are average and therefore Heisler charts on pages 327-329 cannot be used for this problem.]

13.8.7 Diffusion of Oxygen Through Alveoli

We would like to estimate how quickly the non-uniformities in gas composition in the alveoli are damped out. Consider an alveolus to be spherical of diameter 0.1 mm. Let the sphere have an uniform concentration of c_i and at a certain instant its walls are raised to a oxygen concentration of c_∞ and maintained at this value. If the oxygen diffusivity in alveolus can be approximated as that in water and is equal to 2.4×10^{-9} m^2/s, how long does it take for the *concentration change* $(c - c_i)$ at the center to be 90% of the final concentration change?

13.8.8 Ripening of Tomatoes

As shown in the text example, ripening of tomatoes can be retarded by storing in atmospheres that are rich in CO_2 that replaces the O_2 in the tissues with CO_2. After a certain time, sampling of tissue in a spherical tomato stored in a CO_2 rich environment was found to have a CO_2 concentration of 0.1736 at a non-dimensional radius of 0.80. 1) Calculate the concentration at the center at this time. 2) If a concentration of 0.16 is required at the center for the completion of the diffusion process, what *additional* time is needed? The diffusivity of CO_2 in tomato is 2.3×10^{-8} m^2/s. The radius of the tomatoes are 4 cm. The tomato surface is maintained at a CO_2 concentration of 0.18. The initial concentration of CO_2 in the tomato is 0.02.

13.8.9 Size Effect in Diffusion

In the example of diffusion of CO_2 through a tomato, how would the time of exposure to CO_2 to reach the same concentration of CO_2 at the center change for tomatoes that are half the diameter? Note: No other numerical data are necessary.

13.8.10 Transport of Medication Through Skin

For a simplified case of tranpsort through skin, assume total diffusional resistance of skin is entirely due to the stratum corneum (see Figure 5.14). This ignores penetration in sweat ducts, etc. This would be sufficiently accurate for relatively rapidly penetrating molecules of small molecular weight. Consider application of a solute to the outer surface of stratum corneum which has a thickness of 1 micron. The partition coefficient relating the membrane surface concentration to the concentration of applied material is 1. The inner surface of stratum corneum is assumed to be maintained at essentially zero concentration since the molecules are removed as soon as they reach the microcirculation by a sufficiently high peripheral blood flow through skin. Also, the viable layers below stratum corneum have high diffusivity. Assume diffusivity for stratum corneum for a low-molecular weight non-electrolyte is $10^{-6} m^2/s$. 1) How much material enters the skin up to a given time? Simplify for large times and sketch. 2) How much material leaves the stratum corneum into the viable tissue? Simplify for large times and sketch. 3) How much material remains in the stratum corneum? Simplify for large times and sketch.

13.8.11 Nitrogen Movement in Soils

Nitrogen is an extremely important element for the proper growth and maintenance of plants, since it is an essential part of the nucleic acids that plants synthesize. For that reason a large part of any fertilizer is comprised of nitrogen-based compounds. However nitrogen compounds can be extremely toxic to humans when found in drinking water, especially when the ground-water concentration of nitrogen is $>$ 10 mg/l. Plants, however, generally need a concentration of at least 20 mg/l to thrive, depending on the crop species. 1) Find the depth of soil beneath the crop layer where the nitrogen concentration is at least 10 mg/l. For the analysis, assume that the soil region is a semi-infinite plane. 2) Do you think that the fertilizer practices on this field are adversely affecting the quality of the drinking water in that area? If not now, how about if those practices continue for ten years? Consider the yearly average concentration of nitrogen in the soil directly beneath the crop layer is 20 mg/l. This is the first year that the field has been fertilized. Initial concentration of nitrogen in unfertilized soil is 6 mg/l. The diffusivity of nitrogen in soil is 1 cm^2/day.

13.8.12 Unsteady-State Diffusion in a Semi-Infinite Region

Consider a semi-infinite region to have an initial concentration of a certain solute of 0.01 kmol of A/m^3 of solid. It is suddenly exposed to a fluid that has a concentration of A of 0.10 kmol/m^3 of fluid. Assume very large mass transfer coefficient at the surface. The distribution coefficient K^* for equilibrium at surface is 2. The diffusivity of solute A through the solid is $5 \times 10^{-9} m^2/s$. 1) Calculate the concentration at the surface. 2) Calculate the concentration at 1 cm below the surface after 10 hours.

13.8.13 Evaporation from Wet Ground

Consider evaporation from wet ground (silt loam soil) at night. The ground was initially at a moisture content of 0.8 kg of water/kg of dry soil. Air is flowing over the ground at a relative humidity of 10%. The mass transfer coefficient for air flow over the surface is 0.01 m/s and the average diffusivity of moisture in the ground is $10^{-6} m^2/s$. 1) For a distance of 100 m along the wind direction, show that the diffusive resistance to moisture movement inside the ground is a lot higher than the convective resistance (in air) to the moisture movement. 2) Because of item 1, the surface of the ground can be considered to be at the moisture level in equilibrium with the flowing air. What is this equilibrium moisture at the surface in the soil? 3) For the condition given in item 2, find the expression for moisture loss per unit area considering diffusion of moisture in the ground and show that this moisture loss varies as the inverse of the square root of time. Hint: For small values of the argument ϕ, the error function erf(ϕ) can be approximated as ϕ. 4) For the condition given in item 2, find the moisture flux in g/m^2·s after 6 hours. The soil density is 900 kg/m^3.

13.8.14 Drying of Top Soil

Soil, very wet at the end of spring (20% wet basis), begins to dry from the lack of rain during summer. For purposes of calculation, consider the temperature of the top layers of the soil to be constant. 1) After one month (30 days) of dry weather when the average relative humidity of air is 40%, calculate the depth (*in centimeters*) at which the moisture levels would be down to 1/5th of the value at the end of spring. Assume surface mass transfer coefficient is infinity. Consider average diffusivity in the soil to be 10^{-7} m^2/s. Note the moisture content is wet basis and all moisture contents need to be converted to dry basis using the formula

$$\text{dry basis (fraction)} = \frac{\text{wet basis(fraction)}}{1 - \text{wet basis(fraction)}}$$

13.8.15 Diffusive Transport of Pathogens in a Teat Canal

Bacterial infection leading to mastitis is a costly problem for the dairy industry. Mastitis results from infections due to the penetration of pathogens through the teat canal into the teat cistern (see Figure 13.15). Of the many possible mechanisms, the diffusion of pathogens through the column of milk fluid (in the teat canal) into the teat cistern was considered (Lissik, 1988). Bacteria colonize the teat and subsequently grow through the teat canal. By the use of growth model, it was shown that under ideal conditions pathogens which have penetrated the teat canal can grow to sufficient numbers to fill the entire teat canal in less than 12 hours, the inter-milking period. Thus diffusion is a feasible mechanism contributing to the incidence of new mastitis infection.

1) Assuming a one-dimensional diffusion model, calculate the time required for the bacterial concentration to reach detectable amounts (8 nanogram/ml) at the entrance to the teat cistern from a high initial concentration of 250,000 nanogram/ml. Also assume that this high concentration at the teat canal entrance is a constant. Diffusivity of bacteria through milk 3.8×10^{-13} m^2/s. Length of the teat canal is 1 cm. Ratio of length of teat canal/diameter of canal is 125. 2) Is this time to infect the entire length of the teat canal more than the intermilking period (12 hours)? Which part of the anatomy is likely to be infected the most during these 12 hours?

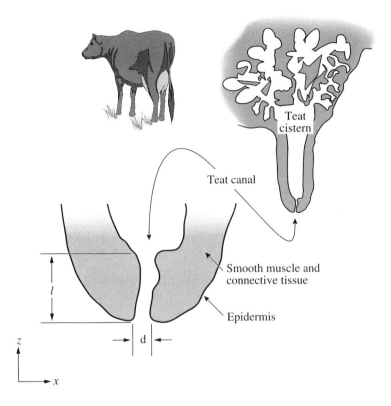

Figure 13.15: Schematic of an udder showing the length of the canal for possible infection.

Chapter 14

CONVECTION MASS TRANSFER

CHAPTER OBJECTIVES

After you have studied this chapter, you should be able to:

1. Formulate and solve for convective-dispersive mass transfer in a flowing liquid, through a porous media, and in a gas.

2. Explain the process of convective-diffusive mass transfer in a fluid over a solid surface as a function of the relatively stagnant fluid layer over the surface called the boundary layer.

3. Calculate the convective mass transfer coefficient h_m knowing the flow situation.

KEY TERMS

- convective mass transfer coefficient
- velocity boundary layer
- mass transfer boundary layer
- Sherwood number (mass transfer Nusselt number)
- Schmidt number
- mass transfer Grashof number
- natural and forced convection mass transfer

In this chapter we will study how the mass transfer in a fluid is modified by the effects of bulk flow. We will do this by developing a more general version of the governing equation that we derived in Chapter 11, by keeping the term representing the bulk flow in addition to transport due to diffusion or dispersion. The relationship of this chapter to other chapters in mass transfer is shown in Figure 14.1.

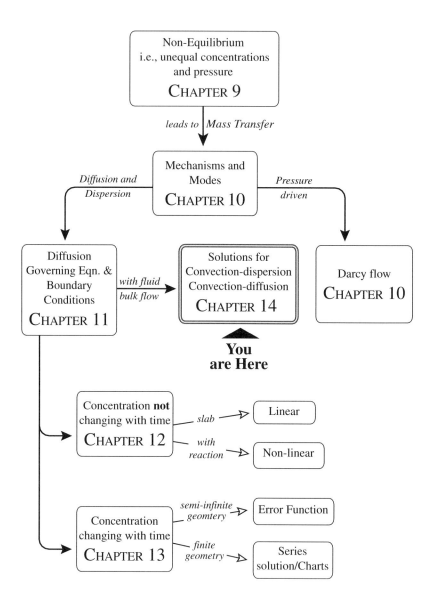

Figure 14.1: Concept map of mass transfer showing how the contents of this chapter relates to other chapters in mass transfer.

14.1. CONVECTION-DISPERSION IN AN INFINITE FLUID

When bulk flow or convection is present, the governing equation for the transport of species A is given by Eqn. 11.17 from page 229, which is repeated here.

$$\underbrace{\frac{\partial c_A}{\partial t}}_{\text{storage}} + \underbrace{u\frac{\partial c_A}{\partial x}}_{\substack{\text{bulk flow or}\\ \text{convection}}} = \underbrace{D_{AB}\frac{\partial^2 c_A}{\partial x^2}}_{\substack{\text{diffusion or}\\ \text{dispersion}}} + \underbrace{r_A}_{\text{generation}} \qquad (14.1)$$

In Eqn. 14.1, u is the velocity contributing to bulk flow. The term representing convection or bulk flow is the additional term retained in the governing equation in this chapter. We will consider four different scenarios of convective mass transfer using this equation that have important practical applications in environmental and biological transport processes. These four scenarios are 1) In an infinite fluid, 2) In a porous solid, 3) In a stagnant gas and 4) Over a surface. Remember that Eqn. 14.1 can be used for both diffusion or dispersion. To differentiate between the two, we will use the symbol D for diffusion and E for dispersion, as introduced in Chapter 10. We will consider the cases 1 and 2 in the context of dispersion, whereas cases 3 and 4 will be considered in the context of diffusion. To contrast this chapter on convective mass transfer with Chapter 6 on convective heat transfer, note that all of Chapter 6 was for flow over a surface, i.e., analogous to case 4 of the four scenarios considered here. The reason the first two cases considered here was not included in Chapter 6 is none other than that their practical applications are more abundant in mass transfer than the corresponding heat transfer situation.

14.1 Convection-Dispersion in an Infinite Fluid

One-dimensional convection-dispersion can be useful idealization for many practical transport situations, for example, pollutants being introduced in a flowing stream, as shown in Figure 14.2. Due to turbulence in the stream, pollutants will spread (disperse) in addition to being carried by the stream (bulk flow). If we can assume instantaneous mixing in the vertical and lateral directions (two directions perpendicular to the flow), the convective-dispersive transport can be approximated as one-dimensional, often referred to as a plug-flow system. In this example, the pollutants can also degrade, thus the term containing mass depletion (negative of generation) would also need to be present. Thus, for a 1D dispersion with chemical reaction in a flowing medium that is of infinite extent, the governing equation can be written as:

$$\frac{\partial c_A}{\partial t} + \underbrace{u\frac{\partial c_A}{\partial z}}_{\text{convection}} = \underbrace{E\frac{\partial^2 c_A}{\partial z^2}}_{\text{dispersion}} - k'' c_A \qquad (14.2)$$

where k'' is the rate of reaction, E is the dispersion coefficient, and u is the bulk velocity along the flow (z direction). For steady state, Equation 14.2 can be simplified to

$$u\frac{dc}{dz} = E\frac{d^2c}{dz^2} - k''c \qquad (14.3)$$

where the subscript A in c_A has been dropped for simplicity. If mass is being introduced at the rate of $\dot{m}$ only at $z = 0$ (e.g., at a discharge point), boundary conditions can be

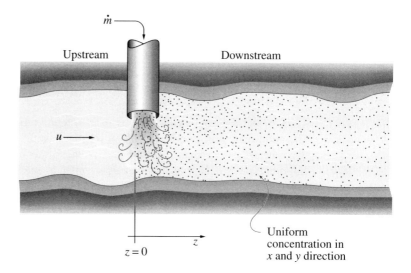

Figure 14.2: Schematic of a point source pollutant being introduced in a stream.

written for upstream and downstream points far from $z = 0$ as:

$$c(z \to \infty) = 0 \qquad (14.4)$$
$$c(z \to -\infty) = 0 \qquad (14.5)$$

Solution to Eqn. 14.3, which is a second-order, ordinary, homogeneous differential equation, has the form:

$$c = c^* e^{\lambda z} \qquad (14.6)$$

where c^* and λ are to be determined. Substituting Eqn. 14.6 in Eqn. 14.3 to find λ:

$$c^*(u\lambda - E\lambda^2 + k'')e^{\lambda z} = 0$$

which would require that

$$u\lambda - E\lambda^2 + k'' = 0$$

Solving for λ,

$$\lambda = \frac{u \pm \sqrt{u^2 + 4Ek''}}{2E}$$
$$= \frac{u}{2E}(1 \pm \psi)$$

where:

$$\psi = \sqrt{1 + \frac{4Ek''}{u^2}} \geq 1$$

Therefore, the two roots λ_1 and λ_2 satisfy the conditions:

$$\lambda_1 = \frac{u}{2E}(1 + \psi) > 0 \qquad (14.7)$$

14.1. CONVECTION-DISPERSION IN AN INFINITE FLUID

$$\lambda_2 = \frac{u}{2E}(1-\psi) < 0 \qquad (14.8)$$

The general solution is:

$$c(z) = c_1 e^{\lambda_1 z} + c_2 e^{\lambda_2 z}$$

Using the two boundary conditions (Eqns. 14.4 and 14.5), we get:

$$c(z \to \infty) = 0 \quad \text{which leads to } c_1 = 0$$
$$c(z \to -\infty) = 0 \quad \text{which leads to } c_2 = 0$$

Thus, the general solution is

$$c(z) = c_1 e^{\lambda_1 z} \quad z < 0 \qquad (14.9)$$
$$c(z) = c_2 e^{\lambda_2 z} \quad z > 0 \qquad (14.10)$$

To evaluate c_1 and c_2, note that in order for both Eqns. 14.9 and 14.10 to be satisfied at $z = 0$ (discharge point in Figure 14.2), c_1 has to be equal to c_2. Also, difference between the mass fluxes on both sides of $z = 0$ has to be due to the mass source, $\dot{m}$. For a cross-sectional area A of the stream, Au is its flow rate and the mass balance at $z = 0$ can be written as

$$\underbrace{\left[-EA\frac{dc}{dz} + Auc\right]_{z=0^+}}_{\text{Flux to the right of } z=0} - \underbrace{\left[-EA\frac{dc}{dz} + Auc\right]_{z=0^-}}_{\text{Flux from the left of } z=0} = \underbrace{\dot{m}}_{\substack{\text{source of mass} \\ \text{at } z=0}}$$

Substituting for c from Eqns. 14.9 and 14.10 and using $c_2 = c_1$,

$$-EA\, c_1\lambda_2 e^{\lambda_2 0} + Auc_1 e^{\lambda_2 0} + EA\, c_1\lambda_1 e^{\lambda_1 0} - Auc_1 e^{\lambda_1 0} = \dot{m}$$
$$-EA\, c_1\lambda_2 + Auc_1 + EA\, c_1\lambda_1 - Auc_1 = \dot{m}$$
$$EAc_1(\lambda_1 - \lambda_2) = \dot{m}$$

$$c_1 = \frac{\dot{m}}{EA(\lambda_1 - \lambda_2)}$$

Using Eqns. 14.7 and 14.8 to substitute for λ_1 and λ_2,

$$c_1 = \frac{\dot{m}}{A\,u\,\psi}$$

Solution for the general case of convection and dispersion

Substituting for c_1, we get the solution when both advection and dispersion effects are present:

$$c(z) = \begin{cases} \frac{\dot{m}}{Au\psi} e^{\frac{u}{2E}(1+\psi)z} & z < 0 \\ \frac{\dot{m}}{Au\psi} e^{\frac{u}{2E}(1-\psi)z} & z > 0 \end{cases} \qquad (14.11)$$

This is illustrated graphically in Figure 14.3, showing greater concentration along the positive z direction where dispersion and bulk flow contribute to the transport. The small concentration along the negative z direction is due to dispersion only and has to work against the bulk flow.

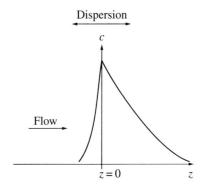

Figure 14.3: Concentration distribution due to a steady source when both convection and dispersion are present.

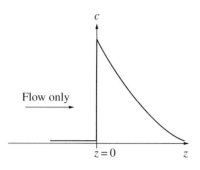

Figure 14.4: Concentration distribution due to a steady source in presence of convection only.

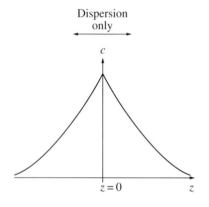

Figure 14.5: Concentration distribution due to a steady source in presence of dispersion only.

When convection is the major mechanism

For flow in a freshwater river, when it is not under tidal influence, the dispersion coefficient E is often small. For small E, $\psi \approx 1$ and

$$c(0) = \frac{\dot{m}}{Au}$$

By setting $E = 0$ in Equation 14.2 and solving:

$$c(z) = \begin{cases} 0 & z < 0 \\ \frac{\dot{m}}{Au} e^{\frac{-k'' z}{u}} & z > 0 \end{cases} \quad (14.12)$$

This is illustrated graphically in Figure 14.4 showing the effect of only bulk flow (with decay) in the positive z direction. In absence of dispersion, there is no concentration in the negative z direction.

Dispersion only

As the river approaches the sea, the dispersion coefficient E increases and the net downstream velocity u decreases. For $u \approx 0$

$$u\psi = u\sqrt{\frac{u^2 + 4k''E}{u^2}}$$
$$\approx \sqrt{4k''E}$$

Substituting in Eqn. 14.11 for $u\psi$ and $u \approx 0$, we get the solution when only dispersion is present

$$c(z) = \begin{cases} \frac{\dot{m}}{A\sqrt{4k''E}} e^{+\sqrt{k''/E}\, z} & z < 0 \\ \frac{\dot{m}}{A\sqrt{4k''E}} e^{-\sqrt{k''/E}\, z} & z > 0 \end{cases} \quad (14.13)$$

This is illustrated graphically in Figure 14.5, where the profiles are symmetric due to the absence of bulk flow.

Example 14.1.1 Movements of Pollutants in a River from Steady Dumping at Multiple Locations

In this example, we wish to develop a water quality model for a river with many BOD (Biological Oxygen Demand) sources. BOD is the amount of dissolved oxygen that is consumed by organisms for a specified length of time at $20°C$. The total oxygen consumption possible if the reactions are allowed to run to completion is called the ultimate BOD. It provides a direct measure of the most damaging effect of pollutants since the depletion of dissolved oxygen leads to the death of many aerobic organisms.

14.1. CONVECTION-DISPERSION IN AN INFINITE FLUID

Consider the steady dumping of pollutants at two points in a flowing river, as shown in Figure 14.6. The BOD contents of the two pollutant streams are shown in Table 14.1. Compute the BOD concentration profile in the river due to the two sources when the river is not under tidal influence (negligible dispersion).

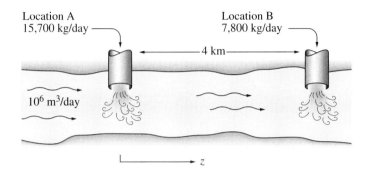

Figure 14.6: Schematic for Example 14.1.1.

Table 14.1: Data for Example 14.1.1

	Location A	Location B
Distance from A [km]	0	4
k'', reaeration rate constant [days^{-1}]	0.4	0.45
Stream flow [10^6 m^3/day]	2	2.3
Velocity [km/day]	3.5	3.8
BOD discharge [kg/day]	15,700	7,800
Initial BOD [kg/m^3]	3.0×10^{-6}	3.0×10^{-6}

Known: Rate of dumping and decay of pollutants in a stream and the flow rate of the stream

Find: Concentration of pollutants with distance along downstream

Schematic and Given Data: Schematic of the problem is shown in Figure 14.6. The given data are shown in Table 14.1

Assumptions:

1. Many of the assumptions made here are discussed in Section 14.1. Instantaneous mixing of the pollutants in the lateral and vertical directions is

one of the important ones, allowing a one-dimensional formulation of the problem.

2. The stream velocity is also assumed to be constant.

3. Dispersion is negligble

4. First order kinetic model adequately describes the degradation of the pollutants

5. Steady state

Analysis: The governing equation for this problem can be written as:

$$u \frac{\partial c}{\partial z} = -k'' c$$

where c is the BOD level, representing the concentration of pollutants. For the boundary conditions as shown earlier in this section (Eqn. 14.9 and 14.10), the solution for each of the stream of pollutants is given by Eqn. 14.12. Plugging in the input parameters, the BOD concentration profile from the first source is given by:

$$c(z) = \begin{cases} 0 & z < 0 \\ \frac{15700}{2 \times 10^6} e^{-\frac{0.4z}{3.5}} = 0.00785 e^{-0.1143z} \text{ [kg/m}^3\text{]} & z > 0 \end{cases}$$

where z is in km. Similarly, the BOD concentration profile due to the second source is given by

$$c(z) = \begin{cases} 0 & z < 4 \\ \frac{7800}{2.3 \times 10^6} e^{-\frac{0.45z}{3.8}} = 0.0039 e^{-0.1184z} \text{ [kg/m}^3\text{]} & z > 4 \end{cases}$$

The combined BOD concentration profile from the two sources can be obtained by adding the two concentration profiles for $z > 0$, together with the initial BOD in the river at 3×10^{-6} kg/m^3. This combined BOD concentration profile is plotted in Figure 14.7.

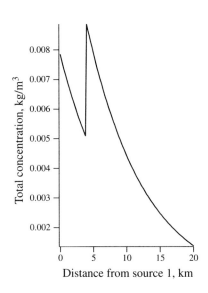

Figure 14.7: Concentration of pollutants as a function of position for Example 14.1.1.

Comments: The role of initial BOD is insignificant here.

14.2 Convection-Dispersion in a Semi-Infinite Porous Solid

In the previous section, we discussed convection-dispersion in a flowing stream where the source of dispersion was the turbulence of the stream. In this section, we consider convection-dispersion in a saturated porous solid, that is, a porous solid where the carrier fluid completely fills the interstitial region between the porous media grains. The porous media of our interest in this section is soil. Compared with dispersion in a

14.2. CONVECTION-DISPERSION IN A SEMI-INFINITE POROUS SOLID

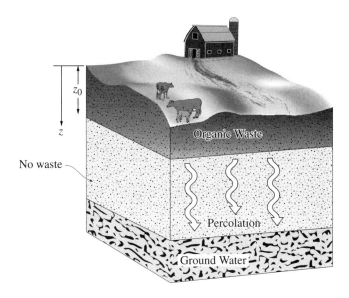

Figure 14.8: Schematic of pollutant transport through soil.

stream, dispersion in a soil is due to completely different mechanisms such as the presence of pores of varying diameter and tortuosity, as explained in Section 10.3. We will study one-dimensional convection-dispersion in a semi-infinite region, as illustrated in Figure 14.8. Note that although this figure may include initial unsaturated zone, it is assumed that with rainfall, etc., the top region will first get saturated and the wetting front will move downwards, eventually reaching groundwater. Thus, the saturated condition under which we will study transport, will be satisfied. One intended application of this section is the study of pollutant transport through soil.

The governing equation for convection-dispersion without any reaction is

$$\underbrace{\frac{\partial c_A}{\partial t}}_{\text{storage}} + \underbrace{u \frac{\partial c_A}{\partial z}}_{\text{convection}} = \underbrace{E \frac{\partial^2 c_A}{\partial z^2}}_{\text{dispersion}}$$

Here c_A is the concentration of dissolved solutes A in the liquid that is flowing through the porous media with an average velocity u. Note that u is not the average velocity in the pores (see example 10.1.1), rather it is the volumetric flux, i.e., when u is multiplied by the total area of flow, we get the volumetric flow rate through the porous media. Making a change of variable $\eta = z - ut$, as shown on page 372 in the Appendix, this equation can be transformed to:

$$\frac{\partial c}{\partial t} = E \frac{\partial^2 c}{\partial \eta^2} \tag{14.14}$$

where the subscript A from c_A has been dropped for simplicity. Note this is the familiar diffusion equation where the diffusion coefficient D has been replaced by E, the dispersion coefficient. For a semi-infinite region, this transformed equation was solved earlier (Eqn. 13.29 on page 266), which we plan to use here. However, the initial condition corresponding to Figure 14.8 is different from those in Figure 13.10. Here the initial condition is a surface layer of thickness z_0 that will have the solute concentration raised to a value of c_i as a pulse, thus representing a finite amount of dispersing species near the surface at initial time. For example, it could be a chemical that got spilled and mixed with the soil or it could be some pesticides applied to the soil. This initial condition can be written symbolically as:

$$\begin{aligned} c(z,t) &= c_i \;(z \leq z_0,\; t = 0) \\ &= 0 \;\;(z > z_0,\; t = 0) \end{aligned} \qquad (14.15)$$

Due to this important difference in boundary condition from that studied in Section 13.4, the solution is not exactly the same as in that section but, as we will see, the solution to our problem here can be developed with the help of the solution process in Section 13.4.

The problem in Figure 14.8 can be treated as a combination of two problems, each having a concentration profile of c^* and c^{**}, respectively, of the dispersing species, as shown in Figure 14.9, such that

$$c = c^* - c^{**} \qquad (14.16)$$

Note that subtraction of Figure 14.9b from Figure 14.9a does indeed lead to Figure 14.8. The first problem involving concentration c^* represents a initial concentration of material c_i extending from 0 to ∞. The governing equation is:

$$\frac{\partial c^*}{\partial t} + u \frac{\partial c^*}{\partial z} = E \frac{\partial^2 c^*}{\partial z^2} \qquad (14.17)$$

and the boundary conditions are:

$$\begin{aligned} c^*(z \to \infty) &= c_i \\ c^*(z = 0) &= 0 \\ c^*(t = 0) &= c_i \end{aligned} \qquad (14.18)$$

The second problem involving concentration c^{**} also represents an initial concentration c_i of material, but extending from z_0 to ∞. The governing equation is:

$$\frac{\partial c^{**}}{\partial t} + u \frac{\partial c^{**}}{\partial z} = E \frac{\partial^2 c^{**}}{\partial z^2} \qquad (14.19)$$

and the boundary conditions are

$$\begin{aligned} c^{**}(z \to \infty) &= c_i \\ c^{**}(z = z_0) &= 0 \\ c^{**}(t = 0) &= c_i \end{aligned} \qquad (14.20)$$

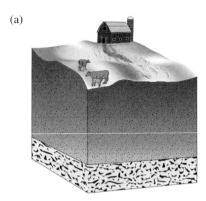

(a)

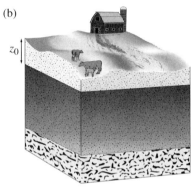

(b)

Figure 14.9: Splitting up the problem shown in Figure 14.8 into two problems a) and b).

14.2. CONVECTION-DISPERSION IN A SEMI-INFINITE POROUS SOLID

Using the transformation illustrated by Eqn. 14.14, the first problem (Eqn. 14.17) can be written as

$$\frac{\partial c^*}{\partial t} = E \frac{\partial^2 c^*}{\partial \eta^2} \qquad (14.21)$$

which, together with the following boundary conditions for η (derived from Eqn. 14.18):

$$c^*(\eta \to \infty) = c_i \qquad (14.22)$$
$$c^*(\eta = 0) = 0 \qquad (14.23)$$

is exactly equivalent to the problem in Section 13.4 with its solution given by Eqn. 13.29. Thus, the solution to the first problem is:

$$\frac{0 - c^*}{0 - c_i} = \mathrm{erf}\left(\frac{\eta}{2\sqrt{Et}}\right)$$
$$= \mathrm{erf}\left(\frac{z - ut}{2\sqrt{Et}}\right) \qquad (14.24)$$

The solution to the second problem defined by Eqn. 14.19 and Eqn. 14.20 is almost identical to the first one, except for the distance z_0. If we first transform the distance variable as:

$$z^* = z - z_0 \qquad (14.25)$$

the second problem becomes identical to the first one and its solution is:

$$\frac{0 - c^{**}}{0 - c_i} = \mathrm{erf}\left(\frac{z^* - ut}{2\sqrt{Et}}\right) \qquad (14.26)$$
$$= \mathrm{erf}\left(\frac{z - z_0 - ut}{2\sqrt{Et}}\right) \qquad (14.27)$$

Hence, the solution to the complete problem is:

$$c = c^* - c^{**}$$
$$= c_i \left[\mathrm{erf}\left(\frac{z - ut}{2\sqrt{Et}}\right) - \mathrm{erf}\left(\frac{z - z_0 - ut}{2\sqrt{Et}}\right)\right] \qquad (14.28)$$

Equation 14.28 is illustrated in Figure 14.10 at two different times, together with the initial uniform concentration distribution. Notice how the initial amount of solute flows (convects) down with time with a simultaneous dispersion (spreading) that increases with time. Since the solution assumes no reaction (such as decay), the total amount (total shaded area) remains constant with time. This solution does not include sorption whereby some of the solute "sticks" to the soil surfaces, which is described in the next section. However, this solution can represent, for example, transport of soluble salts through a soil, such as nitrate ion that stays mostly dissolved in the water.

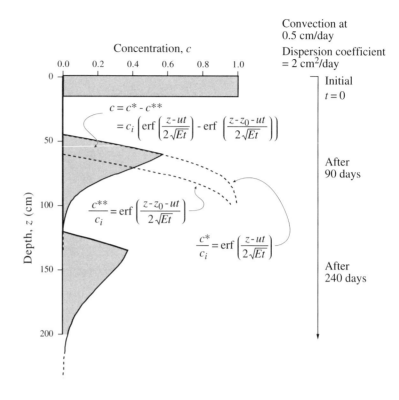

Figure 14.10: Movement and spreading of a layer of pollutant over time due to dispersion and convection (or bulk flow) as predicted by Equation 14.28. The dashed lines for the middle curve illustrate the two individual solutions, c^* and c^{**}, that make up c.

Example 14.2.1 Transport of a Pesticide Spill into the Soil

This example is a simple illustration of the solution just developed. Consider a pesticide spill that occurred where the pesticide got incorporated in the soil up to a depth of 15 cm. This leads to the initial concentration of pesticide in the soil to be 1 gm/cc. An average velocity of 0.5 cm/day due to percolation will carry this pesticide into the soil. The dispersion coefficient is available to be at 2 cm^2/day. Calculate 1) the concentration distribution in the soil after 90 and 240 days, respectively, and 2) concentration at 2 m after 240 days. Assume no sorption or decay of the pesticide.

Known: Initial concentration of pesticide in the top layer of soil

Find: Concentration in the soil as function of position and time

Schematic and Given Data: Schematic of the problem is shown in Figure 14.8. The given data are

1. Average velocity due to percolation, $u = 0.5$ cm/day
2. Dispersion coefficient in the soil, $E = 2$ cm^2/day
3. Initial depth of pesticide spill, $z_0 = 15$ cm
4. Initial concentration in the soil, $c_i = 1$ g/cc

Solution:

1. This problem is an application of Eqn. 14.28. Thus, the concentration after 90 days is given by

$$c = (1)\left[\text{erf}\left(\frac{z - (0.5)(90)}{2\sqrt{(2)(90)}}\right) - \text{erf}\left(\frac{z - 15 - (0.5)(90)}{2\sqrt{(2)(90)}}\right)\right]$$

$$= \left[\text{erf}\left(\frac{z - 45}{26.833}\right) - \text{erf}\left(\frac{z - 60}{26.833}\right)\right] \text{ g/cc}$$

and that after 240 days is given by

$$c = (1)\left[\text{erf}\left(\frac{z - (0.5)(240)}{2\sqrt{(2)(240)}}\right) - \text{erf}\left(\frac{z - 15 - (0.5)(240)}{2\sqrt{(2)(240)}}\right)\right]$$

$$= \left[\text{erf}\left(\frac{z - 120}{43.82}\right) - \text{erf}\left(\frac{z - 135}{43.82}\right)\right] \text{ g/cc} \qquad (14.29)$$

These are plotted in Figure 14.10. This calculation can be easily done on a spreadsheet where error function is available.

2. The concentration value at $z = 200$ cm after 240 days can be calculated using Eqn. 14.29 (from the same spreadsheet) as approximately 0.025 g/cc.

Comments: The data used for this problem are for illustration only. The dispersion coefficient is a strong function of the velocity due to percolation and field conditions and its value used here is low.

14.3 Convection-Dispersion in a Semi-Infinite Porous Solid: Inclusion of Sorption

In the previous section, we discussed convection-dispersion in a porous solid. We included convection and dispersion, but no reactions. Frequently, as applied to transport of contaminants through soil, contaminants can undergo sorption, where pollutants are removed from the aqueous phase (carrier fluid) and immobilized onto particles in the porous medium. Mechanisms of sorption of solutes to solid particles can be hydrophobic partitioning, electrostatic adsorption onto particle surfaces and ion exchanges (Schnoor, 1996). In this section, we will include the sorption process during the transport. In general, the effect of the sorption process, will be to slow down the transport of the solutes (since they are retained by the solid particles).

Like in the last section, we considered convection-dispersion in a semi-infinite region. As before, we will assume constant velocity of liquid flowing through the porous media, constant dispersion, and no decay. Our governing equation will now be only slightly modified to include a term due to sorption:

$$\underbrace{\frac{\partial c}{\partial t}}_{\text{Storage}} + \underbrace{u\frac{\partial c}{\partial z}}_{\text{Convection}} = \underbrace{E\frac{\partial^2 c}{\partial z^2}}_{\text{Dispersion}} - \underbrace{\frac{\partial c^{ad}}{\partial t}}_{\text{Adsorption}} \quad (14.30)$$

where c is the concentration of dissolved solute in the liquid portion of the porous media, but measured as per unit volume of the total porous media. It is related to the dissolved concentration of solute, c^*, that is measured as per unit volume of the liquid only. For transport through soil, this relationship can be written as:

$$c\left[\frac{\mu g}{cm^3 \text{ of soil}}\right] = c^*\left[\frac{\mu g}{cm^3 \text{ of water}}\right] * \theta_f \left[\frac{cm^3 \text{ of water}}{cm^3 \text{ of soil}}\right] \quad (14.31)$$

where θ_f is the maximum water content of the soil. The quantity c^{ad} is the concentration of solutes adsorbed to the solid surfaces in the porous media and is related to the dissolved solute concentration, c^*, as:

$$c^{ad}\left[\frac{\mu g}{cm^3 \text{ of soil}}\right] = c^*\left[\frac{\mu g}{cm^3 \text{ of water}}\right] K^*\left[\frac{cm^3 \text{ of water}}{g \text{ of soil}}\right] \rho_s \left[\frac{g \text{ of soil}}{cm^3 \text{ of soil}}\right] \quad (14.32)$$

where K^* is the distribution coefficient between the solute chemical and soil particles and ρ_s is the dry bulk density of soil. The total concentration of the solute chemical in the soil will be composed of the dissolved solute in the liquid and adsorbed solute on the solid surfaces, i.e., $c + c^{ad}$. Substituting c and c_{ad} into the governing equation:

$$\theta_f \frac{\partial c^*}{\partial t} + u\theta_f \frac{\partial c^*}{\partial t} = E\theta_f \frac{\partial^2 c^*}{\partial z^2} - \rho_s K^* \frac{\partial c^*}{\partial t} \quad (14.33)$$

and rearranging

$$\underbrace{\left(1 + \frac{\rho_s K^*}{\theta_f}\right)}_{R} \frac{\partial c^*}{\partial t} + u\frac{\partial c^*}{\partial z} = E\frac{\partial^2 c^*}{\partial z^2} \quad (14.34)$$

Defining R as the retardation factor, the governing equation becomes

$$\frac{\partial c^*}{\partial t} + \frac{u}{R}\frac{\partial c^*}{\partial z} = \frac{E}{R}\frac{\partial^2 c^*}{\partial z^2} \quad (14.35)$$

and the boundary conditions are:

$$\begin{aligned} c^*(z, t) &= c_i^* \ (z \leq z_0, \ t = 0) \\ &= 0 \ (z > z_0, \ t = 0) \end{aligned} \quad (14.36)$$

These governing equation and boundary conditions are identical in form to Eqns. 14.14 and 14.15, respectively. Therefore, the solution can be written from Eqn. 14.28 as:

$$c^*(z, t) = c_i^* \left[\text{erf}\left(\frac{z - (u/R)t}{2\sqrt{(E/R)t}}\right) - \text{erf}\left(\frac{z - z_0 - (u/R)t}{2\sqrt{(E/R)t}}\right)\right] \quad (14.37)$$

14.4. CONVECTION-DIFFUSION IN A STAGNANT GAS

The total concentration of solute chemical in the porous media is obtained by adding the concentration in the water to the concentration that is adsorbed on the solid surface:

$$c + c^{ad} = (\theta_f + K^* \rho_s) c^*$$
$$= \underbrace{(\theta_f + K^* \rho_s) c_i^*}_{c_i} \left[\text{erf}\left(\frac{z - (u/R)t}{2\sqrt{(E/R)t}} \right) - \text{erf}\left(\frac{z - z_0 - (u/R)t}{2\sqrt{(E/R)t}} \right) \right]$$

(14.38)

where c_i is the initial total concentration of solute chemical in the porous media. The solution given by Eqn. 14.38 will carry some features of Figure 14.10, in the sense that it also will have convection and dispersion. In contrast with the solution from the previous section, Eqn. 14.38 can represent transport of ions that do "stick" to the solid surface, for example, transport of ammonia and phosphorous ions that bind with aluminum and calcium in the soil.

14.4 Convection-Diffusion in a Stagnant Gas

In the previous two sections, we discussed convection-dispersion. We now pick up one important situation of convection-diffusion. Our goal is the same as before— to calculate concentration profiles and mass fluxes. This example will also serve to illustrate how molecular diffusion and convection in a gas are interrelated[1]. As an example, consider evaporation from a water surface and diffusion of the water vapor through the stagnant air column and eventually removed by the surrounding air from the top of the tube. At very low temperatures of water, the vapor pressure of water is low and the rate of evaporation is low. Water vapor moves slowly up the tube because of their thermal energy, that is the molecular diffusion process. If, however, the water is at the boiling temperature, the situation is completely different. The constantly generated high volume of water vapor will rush up the tube, which is clearly a pressure-driven flow process or convection, in contrast with a slow diffusion process. At intermediate temperatures, both diffusion and convection are important, which is the subject of this section. Examples of such processes are many. A bioenvironmental example of diffusion through a stagnant gas is the evaporation and loss of water vapor from the soil surface when it is covered by mulch. The mulch maintains somewhat of a stagnant layer of air through which the water vapor moves from the soil surface into the surrounding air.

Consider steady-state diffusion of vapor A through the stagnant or motionless gas B, as shown schematically in Figure 14.11. The gaseous species B is not soluble in A in any significant quantity, thus there is no net flux of B into the liquid A. At the top of the tube, a large volume of gas (air) flows, maintaining the concentration of vapor A to be zero at the top. The governing equation for this convection-diffusion problem can

---

[1] See the excellent illustration in Cussler (1997).

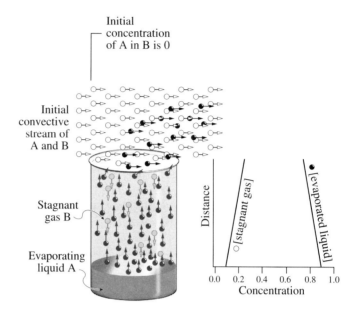

Figure 14.11: Schematic of a diffusive and convective process through a stagnant air column resulting from evaporation of a liquid.

be obtained starting from the general equation

$$\underbrace{\cancel{\frac{\partial c_A}{\partial t}}^{0}}_{\text{steady state}} + \underbrace{u\frac{\partial c_A}{\partial z}}_{\substack{\text{bulk flow or}\\\text{convection}}} = \underbrace{D_{AB}\frac{\partial^2 c_A}{\partial z^2}}_{\text{diffusion}} + \underbrace{\cancel{r_A}^{0}}_{\text{no generation}} \qquad (14.39)$$

and dropping the appropriate terms as shown. The source of the velocity u in the bulk flow term is due to continuous generation of vapor A, since the system is open at the top where vapor A is continually removed. This continuous generation of vapor "pushes" the bulk fluid upward. Also, for the gaseous species B, it's bulk flow upward (due to the "push" from vapor A) and it's diffusion downward (due to higher concentration of B near the top) balances exactly since there can be no net flux of B in liquid A (it is not soluble). This balancing of the fluxes causes B to stay stagnant. The final governing equation is

$$u\frac{dc_A}{dz} = D_{AB}\frac{d^2 c_A}{dz^2} \qquad (14.40)$$

The boundary conditions are given by

$$c_A = c_{A_1} \text{ at } z = z_1 \qquad (14.41)$$
$$c_A = c_{A_2} \text{ at } z = z_2 \qquad (14.42)$$

14.4. CONVECTION-DIFFUSION IN A STAGNANT GAS

To solve Eqn. 14.40 note that it is an ordinary, second order differential equation with constant coefficients (see discussion later on u). The equation is rewritten as

$$\frac{d}{dz}\left(-D_{AB}\frac{dc_A}{dz} + uc_A\right) = 0 \qquad (14.43)$$

Integrating,

$$-D_{AB}\frac{dc_A}{dz} + uc_A = n_A \qquad (14.44)$$

where n_A is a constant. This constant, n_A, in Eqn. 14.44 can also be physically interpreted as the total flux, since the two terms on the left side represent diffusive and convective flux, respectively. The coefficient u of the second term, the average velocity, is constant, as can be seen by approximating u under no net flux of B ($n_B = 0$), i.e., gas B is impermeable to liquid A, as

$$u = \frac{n_A + n_B}{c_A + c_B} = \frac{n_A}{c} \qquad (14.45)$$

Substituting for u in in Eqn. 14.44, we can write

$$-D_{AB}\frac{dc_A}{dz} + \frac{n_A}{c}c_A = n_A$$

or

$$-D_{AB}\frac{dc_A}{dz} = n_A\left(1 - \frac{c_A}{c}\right)$$

Rearranging and integrating between the bottom and the top of the tube

$$\int_{c_{A_1}}^{c_{A_2}} \frac{d(c_A/c)}{1 - c_A/c} = -\frac{n_A}{D_{AB}c}\int_{z_1}^{z_2} dz \qquad (14.46)$$

$$\ln(1 - c_A/c)\Big|_{c_{A_1}}^{c_{A_2}} = -\frac{n_A}{D_{AB}c}(z_2 - z_1)$$

$$\ln\frac{(1 - c_{A_2}/c)}{(1 - c_{A_1}/c)} = -\frac{n_A}{D_{AB}c}(z_2 - z_1) \qquad (14.47)$$

Since c_{A_1} and c_{A_2} are known, Eqn. 14.47 can be used to find n_A, but we will postpone this for later. Integrating Eqn. 14.46 between z_1 and any height z (where concentration is c_A), we can write an expression similar to Eqn. 14.47 as

$$\ln\frac{(1 - c_A/c)}{(1 - c_{A_1}/c)} = -\frac{n_A}{D_{AB}c}(z - z_1) \qquad (14.48)$$

Dividing the two sides of Eqn. 14.48 by the two sides of Eqn. 14.47,

$$\frac{\ln\frac{(1-c_A/c)}{(1-c_{A_1}/c)}}{\ln\frac{(1-c_{A_2}/c)}{(1-c_{A_1}/c)}} = \frac{z - z_1}{z_2 - z_1}$$

$$\ln\frac{(1 - c_A/c)}{(1 - c_{A_1}/c)} = \frac{z - z_1}{z_2 - z_1}\ln\frac{(1 - c_{A_2}/c)}{(1 - c_{A_1}/c)}$$

$$\frac{(1 - c_A/c)}{(1 - c_{A_1}/c)} = \left(\frac{(1 - c_{A_2}/c)}{(1 - c_{A_1}/c)}\right)^{\frac{z-z_1}{z_2-z_1}} \qquad (14.49)$$

294   CHAPTER 14. CONVECTION MASS TRANSFER

Equation 14.49 provides concentration c_A of the evaporating species A as a function of z, the distance from the bottom. In terms of partial pressures, $c_A = p_A/RT$ and $c = P/RT$, we can write

$$\frac{(1 - p_A/P)}{(1 - p_{A_1}/P)} = \left(\frac{(1 - p_{A_2}/P)}{(1 - p_{A_1}/P)}\right)^{\frac{z-z_1}{z_2-z_1}} \quad (14.50)$$

which provides partial pressure p_A as a function of z, the distance from the bottom. We can now use the concentration equations to calculate n_A, the flux of A, which is often of more practical interest. From Eqn. 14.47, we can write the flux as:

$$n_A = \frac{D_{AB} c}{z_2 - z_1} \ln \frac{(1 - c_{A_2}/c)}{(1 - c_{A_1}/c)} \quad (14.51)$$

In terms of partial pressures

$$n_A = \frac{D_{AB} P}{RT(z_2 - z_1)} \ln \frac{(1 - p_{A_2}/P)}{(1 - p_{A_1}/P)} \quad (14.52)$$

These two equations for the flux n_A provides the means to calculate the net loss of A through the top of the stagnant gas column.

Example 14.4.1 Evaporation in an Enclosed Area

Water evaporates from the surface in a narrow ditch where the water surface is 0.3 m below the ground level. Dry air blows over the ground. Calculate the rate of evaporative loss of water per square meter from the ditch in g/day at steady state. The temperature of water remains at 27°C.

Known: Temperature of water surface and depth of stagnant air through which water diffuses

Find: Rate of evaporative loss

Schematic and Given Data: Schematic of the process is same as that shown in Figure 14.11 with species A as water vapor and species B as air (all gases combined). The given data are:

1. Temperature of water surface = 27 °C
2. Depth of stagnant air = 0.3 m

Assumptions:

- The air column in the ditch above the water is stagnant
- The system is isothermal
- Total pressure remains at 1 atmosphere ($1.01325 \times 10^5 \text{N/m}^2$) everywhere

Analysis: Diffusive flux of water vapor as a gas through stagnant air, n_w, would be given by Eqn. 14.52:

$$n_w = \frac{D_{wa}P}{RT(z_2 - z_1)} \ln \frac{(1 - p_{A_2}/P)}{(1 - p_{A_1}/P)}$$

From Table D.3 on page 348, we get diffusivity of vapor in air, $D_{wa} = 2.538 \times 10^{-5}\,\text{m}^2/\text{s}$ at $T = 300$ K. The vapor pressure, p_{A1}, at 300 K is $0.036 \times 10^5\,\text{N/m}^2$ from Table C.10 on page 342. At the top, dry air maintains $p_{A2} = 0$. Thus, plugging in the numbers:

$$\begin{aligned} n_w &= \frac{(2.538 \times 10^{-5})(1.01325 \times 10^5)}{(8.314)(300)(0.3)} \ln \frac{(1-0)}{(1-(0.036/1.01325))} \\ &= 1.243 \times 10^{-4}\,\text{mol/m}^2\cdot\text{s} \\ &= (1.243 \times 10^{-4}\,\text{mol/m}^2\cdot\text{s})(18\,\text{g/mol})(3600 \times 24\,\text{s/day}) \\ &= 193\,\text{g/m}^2\cdot\text{day} \end{aligned}$$

Comments: Water loss would be a strong function of the temperature of water.

14.5 Convection-Diffusion Over a Surface

In contrast with the first two sections in this chapter that dealt with convection-dispersion and the third section that dealt with convection-diffusion in a stagnant gas, this section will deal with convection-diffusion over a surface, i.e., convective transport of mass from a surface into the fluid. Recall that convective mass transfer over a surface is described by the convective mass transfer coefficient, h_m, using the equation

$$N_{A_{1-2}} = h_m A(c_1 - c_2) \qquad (14.53)$$

where $N_{A_{1-2}}$ is the mass flow rate of species A from the surface (designated as 1) to the bulk fluid (designated as 2) and c is the corresponding concentrations of species A. Most of this section will deal with the details of what makes up h_m and how we can calculate it. The section will be completely analogous to the contents of Chapter 6 on convective transport of heat from a surface. As in Chapter 6 for heat transfer, mass transfer can be analyzed in terms of thin layers formed over the surface called the boundary layer. Convective mass transfer coefficients will be defined inside this mass transfer boundary layer and various formulas to predict convective mass transfer coefficient will be introduced.

14.5.1 Concentration Profiles and Boundary Layers Over a Surface

Development of a boundary layer for flow over a solid was explained in detail in Chapter 6. The fluid velocity changes from zero on the solid surface to close to free stream

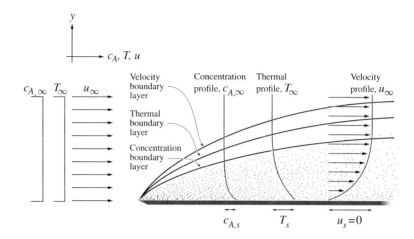

Figure 14.12: Schematic showing one example of a concentration profile and mass transfer boundary layer together with velocity and thermal boundary layers.

velocity at the edge of the boundary layer. Diffusion of a mass species near the surface is completely analogous to diffusion of energy near the surface. For example, consider dry air being blown over a wet surface. Water vapor at the wet surface first diffuses into the air and is eventually removed by the bulk flow. Like the thermal boundary layer, a concentration boundary layer will therefore develop. This concentration boundary layer is superimposed in Figure 14.12 on top of the velocity and thermal boundary layers shown in Figure 6.2 for a flat plate. Since the fluid in contact with the surface is at rest, the species in the liquid will come in equilibrium with the species in the solid. The concentration will vary from the value $c_{A,s}$ at the surface to the value $c_{A,\infty}$ in the free stream. Like the velocity and temperature variation, the change in concentration through space is also asymptotic such that the concentration is never exactly equal to $c_{A,\infty}$. The thickness of concentration boundary layer, δ_{conc}, is defined as the distance from the surface at which concentration c_A is given by:

$$\frac{c_{A,s} - c_{A,\delta_{conc}}}{c_{A,s} - c_{A,\infty}} = 0.99 \qquad (14.54)$$

which is the distance over which most of the concentration change takes place. Like the thermal boundary layer, the fluid can be considered to have two distinct concentration regions: the concentration boundary layer and outside the concentration boundary layer. In the boundary layer concentration gradients are large whereas outside the boundary layer the concentration gradients are small as the concentration at any point is within 1% of the free stream concentration. In other words, the effect of the flat plate on the mass concentration in the fluid is essentially restricted to the concentration boundary layer. Just as a thermal boundary layer exists only if there is a temperature difference between the surface and the flowing fluid, a concentration boundary layer exists only if there is a concentration difference between the surface and the flowing fluid.

14.5. CONVECTION-DIFFUSION OVER A SURFACE

Like the thermal boundary layer, the thickness of the concentration boundary layer can be related to that of the velocity boundary layer using the characteristic number Sc or Schmidt number, that is analogous to the Prandtl number, Pr (see Section 6.2 on heat transfer). Schmidt number is defined as:

$$Sc = \frac{\nu}{D_{AB}} \tag{14.55}$$

For a flat plate, the relationship between the two boundary layer thicknesses is:

$$\frac{\delta_{velocity}}{\delta_{conc}} = Sc^{1/3} = \left(\frac{\text{Momentum diffusivity}}{\text{Mass diffusivity}}\right)^{1/3} \tag{14.56}$$

or

$$\delta_{conc} = \frac{\delta_{velocity}}{Sc^{1/3}}$$

which implies that mass transfer boundary layer thickness, like other boundary layer thicknesses, decreases as velocity increases.

Example 14.5.1 Boundary Layer in Oxygen Transport from a Gas Bubble to a Cell or an Organism

As an example of a boundary layer, consider the situation where oxygen is transferred from a rising air bubble into a liquid phase containing cells (Bailey and Ollis, 1986). The oxygen must pass through a series of transport resistances, the relative magnitudes of which depend on bubble (droplet) hydrodynamics, temperature, cellular activity and density, solution composition, interfacial phenomena, and other factors. The total resistance arises from various combinations of the following resistances (Figure 14.13):

1. Diffusion from bulk gas to the gas-liquid interface.

2. Movement through the gas-liquid interface.

3. Diffusion of the solute through the relatively unmixed liquid region adjacent to the bubble into the well-mixed bulk liquid. *This is the boundary layer region just studied.*

4. Transport of the solute through the bulk liquid to a second relatively unmixed liquid region surrounding the cells.

5. Transport through the second unmixed liquid region associated with the cells. *This is the second boundary layer region.*

6. Diffusive transport into the cellular floc, mycelia, or soil particle.

7. Transport across cell envelope and to intra-cellular reaction site.

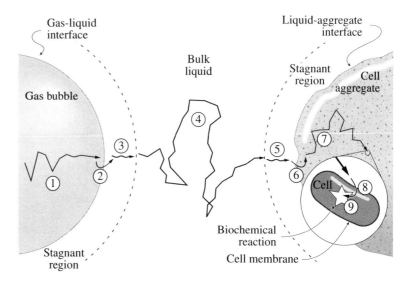

Figure 14.13: Boundary layer combined with other resistances in oxygen transfer from a bubble to a microorganism. Adapted from Biochemical Engineering Fundamentals by J.E. Bailey and D.F. Ollis. © 1986 McGraw-Hill, Inc. Reproduced with permission from McGraw-Hill, Inc.

When the organisms take the form of individual cells, the sixth resistance disappears. Microbial cells themselves have some tendency to adsorb at interfaces. Thus, cells may preferentially gather at the vicinity of the gas-bubble-liquid interface. Then, the diffusing solute oxygen passes through only one unmixed liquid region and no bulk liquid before reaching the cell. In this situation, the bulk dissolved O_2 concentration does not represent the oxygen supply for the respiring microbes.

14.5.2 Convective Mass Transfer Coefficient Defined

The convective mass transfer coefficient over a surface is defined quite similarly to the convective heat transfer coefficient. For fluid flowing over a surface, the fluid layer in contact with the surface is at rest, as illustrated in Figure 14.12. Thus, at the surface ($y = 0$), mass transfer *in the fluid* is by diffusion only and the flux is given by:

$$\text{Diffusive mass flux in the fluid} = -D_{AB} \frac{\partial c_A}{\partial y}\bigg|_{y=0,\ \text{in fluid}} \quad (14.57)$$

where c_A is the concentration of species A *in the fluid* at a distance y from the surface and D_{AB} is the diffusivity of species A *in the fluid*. However, the same mass flux can be written in terms of a surface convective mass transfer coefficient, h_m, as:

$$\begin{aligned}\text{Convective mass flux} &= h_m(c_{A,y=0,\text{in fluid}} - c_{A,\infty}) \\ &= h_m(c_{A,s} - c_{A,\infty}) \end{aligned} \quad (14.58)$$

14.5. CONVECTION-DIFFUSION OVER A SURFACE

where $c_{A,s}$ is the concentration at the surface *in the fluid* and $c_{A,\infty}$ is the concentration in the fluid far from the surface. Since Eqns. 14.57 and 14.58 represent the same mass flux:

$$-D_{AB}\frac{\partial c_A}{\partial y}\bigg|_{y=0,\text{in fluid}} = h_m(c_{A,s} - c_{A,\infty})$$

which leads to the defining equation for h_m as:

$$h_m = \frac{-D_{AB}\frac{\partial c_A}{\partial y}\big|_{y=0,\text{in fluid}}}{c_{A,s} - c_{A,\infty}} \quad (14.59)$$

The concentration profile $c_A(y)$ can be obtained by solving the governing equation for convection (Equation 14.1). Knowing $c_A(y)$, we can calculate $\partial c_A/\partial y$ in Equation 14.59 and therefore can get h_m, the mass transfer coefficient. Since $c_A(y)$ is obtained by solving diffusion and flow equations, the *convective mass transfer coefficient includes contributions from both bulk flow and diffusion*.

14.5.3 Significant Parameters in Convective Mass Transfer

To obtain the convective mass transfer coefficient h_m for many practical situations, one would like to obtain a solution to the convection-diffusion equation or simply the convection equation (Eqn. 14.1). The governing equation for convective mass transfer with its appropriate boundary conditions can be solved in much the same way as the governing equation for convective heat transfer. In order to solve this equation we need the fluid flow equations which describe the velocity u as it influences the rate of mass transfer. This solution process for convective mass transfer is completely analogous to the process described for convective heat transfer, so that we will not describe the process in the same detail. The convection equation (Equation 14.1) for steady two dimensional flow over a flat plate can be written in non-dimensional form following Section 6.5 for heat transfer as:

$$u^*\frac{\partial c_A^*}{\partial x^*} + v^*\frac{\partial c_A^*}{\partial y^*} = \frac{D_{AB}}{u_\infty L}\frac{\partial^2 c_A^*}{\partial y^{*2}} \quad (14.60)$$

By defining a non-dimensional quantity, the Schmidt number Sc as:

$$Sc = \frac{\mu}{\rho D_{AB}} = \frac{\nu}{D_{AB}} = \frac{\text{momentum diffusivity}}{\text{mass diffusivity}} \quad (14.61)$$

Eqn. 14.60 can be rewritten (like Equation 6.22) as:

$$u^*\frac{\partial c_A^*}{\partial x^*} + v^*\frac{\partial c_A^*}{\partial y^*} = \frac{1}{Re_L Sc}\frac{\partial^2 c_A^*}{\partial y^{*2}} \quad (14.62)$$

The solution to this equation, together with the momentum equation that provides for the velocities, u^* and v^*, can be written in functional form as:

$$c_A^* = f(x^*, y^*, Re_L, Sc) \quad (14.63)$$

Using the definition of the convective mass transfer coefficient (Equation 14.59), h_m can be written in terms of a non-dimensional concentration gradient as:

$$\begin{aligned} h_m &= \frac{-D_{AB} \frac{\partial c_A}{\partial y}\Big|_{y=0,\text{in fluid}}}{c_{A,s} - c_{A,\infty}} \\ &= \frac{D_{AB}}{L} \frac{\partial \left(\frac{c_A - c_{A,s}}{c_{A,\infty} - c_{A,s}}\right)}{\partial \left(\frac{y}{L}\right)}\Bigg|_{y=0} \\ &= \frac{D_{AB}}{L} \frac{\partial c_A^*}{\partial y^*}\Bigg|_{y^*=0} \end{aligned} \quad (14.64)$$

which can be rewritten using Equation 14.63 as:

$$\begin{aligned} \frac{h_m L}{D_{AB}} &= \frac{\partial c_A^*}{\partial y^*}\Bigg|_{y^*=0} \\ &= f(x^*, Re_L, Sc) \end{aligned} \quad (14.65)$$

This shows that the mass transfer coefficient h_m is a function of position, x^*, in general. The average mass transfer coefficient would not depend on the position, and would be given by:

$$\frac{h_{mL} L}{D_{AB}} = f(Re_L, Sc) \quad (14.66)$$

The subscript L in h_{mL} is generally dropped as averaging is typically assumed. This equation shows the functional relationship between the mass transfer coefficient h_m and the non-dimensional parameters Re_L and Sc. The non-dimensional quantity on the left of Equation 14.66 is defined as the Sherwood number having the symbol Sh (sometimes referred to as the mass transfer Nusselt number):

$$\begin{aligned} Sh &= \frac{h_m L}{D_{AB}} \\ &= \frac{\partial c_A^*}{\partial y^*}\Bigg|_{y^*=0} \end{aligned}$$

using Equation 14.64. Thus we see that the Sherwood number, like the Nusselt number, is a non-dimensional concentration gradient. The non-dimensional parameters defined in this section are summarized and compared with their heat transfer equivalents in Table 14.2.

14.5.4 Calculation and Physical Implications of Convective Mass Transfer Coefficient Values

Convective mass transfer coefficients over a surface can be obtained in an analogous manner to convective heat transfer coefficients by either solving the convection-

14.5. CONVECTION-DIFFUSION OVER A SURFACE

diffusion mass transfer equation or through experiments. Since the solution to the convection-diffusion mass transfer equation would result in formulas identical in form to those of heat transfer, the formulas in Section 6.6 on page 104 can be used for convective mass transfer over a surface with the substitution of appropriate dimensionless parameters for mass transfer as described in Table 14.2. Note that the Schmidt number, Sc, in that table is the equivalent in mass transfer of the Prandtl number, Pr, from heat transfer. In Schmidt number, the mass diffusivity replaces the thermal diffusivity. Schmidt number compares the rates of momentum (or viscous) diffusion with mass diffusion. The Sherwood number, Sh, in mass transfer replaces the Nusselt number, Nu, in heat transfer. As mentioned earlier, Sherwood number compares the diffusive and convective resistances *(both in the fluid)* over a surface. The diffusive resistance in Sherwood number is defined in terms of mass diffusivity, in contrast with the conductive resistance in Nusselt number defined in terms of thermal conductivity (not thermal diffusivity). The Grashof number for mass transfer is defined in terms of density difference that drives the flow in contrast with temperature difference (that drives the flow in heat transfer) used in heat transfer Grashof number. The Rayleigh number for mass transfer is defined in analogy with Rayleigh number for heat transfer.

For example, for flow over a sphere (Figure 14.14), the formulas for mass transfer coefficients can be written from Eqns. 6.51 and 6.52 on page 111 for heat transfer coefficient. Thus, for natural convection mass transfer, we have

$$Sh_D = 2 + 0.43 Ra_{mD}^{\frac{1}{4}} \quad \text{for } 1 < Ra_{mD} < 10^5, \ Sc \cong 1 \quad (14.67)$$

For forced convection mass transfer over a sphere, the correlation would be:

$$Sh_D = 2 + \left(0.4 Re_D^{1/2} + 0.06 Re_D^{2/3}\right) Sc^{0.4} \quad (14.68)$$
$$\text{for } 3.5 < Re_D < 7.6 \times 10^4 \text{ and } 0.71 < Sc < 380$$

In these equations, Nu, the Nusselt number for heat transfer, has been replaced by Sh, the Sherwood number. Likewise, $Ra = Gr \times Pr$, the Rayleigh number for heat transfer, has been replaced by $Ra_m = Gr_{AB} \times Sc$, the Rayleigh number for mass transfer and the Prandtl number, Pr, has been replaced by Schmidt number, Sc. See Table 14.2 for the complete list of non-dimensional parameters in mass transfer and their heat transfer equivalent.

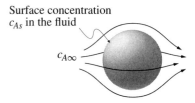

Figure 14.14: Mass transfer over a sphere.

To understand the implication of the correlations for mass transfer coefficients over a surface, we also turn to refreshing our knowledge of the correlations for heat transfer coefficients over a surface, as discussed in Section 6.6 on page 104. Identical conclusions can be made about the implications of the correlations. For example, as Re increases, the thickness of the mass transfer boundary layer decreases. Since this boundary layer is the slower moving layer that provides the resistance to mass transfer, reduced thickness will mean higher rate of mass transfer and a larger value of h_m, the mass transfer coefficient. Likewise, turbulence will increase the rate of mass transfer.

Table 14.2: Non-dimensional parameters in mass transfer and their heat transfer equivalents. To obtain the correlations for convective mass transfer coefficient, appropriate non-dimensional parameter needs to be substituted in the equations in Section 6.6 on page 104. The quantity $\Delta\rho$ in the Grashof number for mass transfer is the density difference due to concentration gradient

Mass transfer	Heat transfer equivalent
Schmidt number	*Prandtl number*
$Sc = \dfrac{\mu/\rho}{D_{AB}}$ $= \dfrac{\text{momentum diffusivity}}{\text{mass diffusivity}}$	$Pr = \dfrac{\mu/\rho}{\alpha_{fluid}}$ $= \dfrac{\text{momentum diffusivity}}{\text{thermal diffusivity}}$
Sherwood number	*Nusselt number*
$Sh = \dfrac{h_m L}{D_{AB}} = \dfrac{\frac{L}{D_{AB}}}{\frac{1}{h_m}}$ $= \dfrac{\text{diffusive resistance}}{\text{convective resistance}}$	$Nu = \dfrac{hL}{k_{fluid}} = \dfrac{\frac{L}{k_{fluid}}}{\frac{1}{h}}$ $= \dfrac{\text{conductive resistance}}{\text{convective resistance}}$
Grashof number[1]	*Grashof number*
$Gr_{AB} = \dfrac{g\rho\Delta\rho L^3}{\mu^2}$	$Gr = \dfrac{\beta g \rho^2 L^3 \Delta T}{\mu^2}$
Rayleigh number	*Rayleigh number*
$Ra_m = Gr_{AB} \times Sc$	$Ra = Gr \times Pr$

[1] The quantity $\Delta\rho$ in the mass transfer Grashof number is the magnitude of the difference between the two densities across the boundary layer.

Example 14.5.2 Maximum Oxygen Uptake of a Microorganism

Calculate the maximum possible rate of oxygen uptake at 37°C of microorganisms having a diameter of 1 μm suspended in an agitated aqueous solution. It is assumed that the surrounding liquid is saturated with O_2 from air at 1 atm abs pressure. It will be assumed that the oxygen is utilized by the microorganism much faster than it can

14.5. CONVECTION-DIFFUSION OVER A SURFACE

diffuse to it. The microorganism has a density very close to that of water. Diffusivity of O_2 in water is 3.25×10^{-9} m^2/s. *(Adapted from Geankoplis, 1993)*

Known: A spherical microorganism is suspended in water (zero velocity of water flowing over it)

Find: The rate of oxygen transport to the microorganism

Schematic and Given Data: The schematic for the problem is shown in Figure 14.15. The given data are

1. $c_{O_2,\text{surface}} = 2.26 \times 10^{-4}$ kmol/m^3 at saturation
2. $D_{O_2,\text{water}} = 3.25 \times 10^{-9}$ m^2/s
3. $D = 1 \times 10^{-6}$ m, diameter of the microorganism

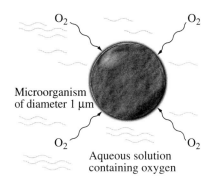

Figure 14.15: Schematic for Example 14.5.2.

Assumptions:

- Concentration of O_2 is zero at surface of the microorganism since oxygen is consumed faster than it can diffuse to the surface.
- Density of the microorganism is same as water

Analysis: We recognize this problem as one of convective mass transfer over a spherical surface. Thus, the formula for mass transfer coefficient, h_m, will be given by Eqn. 6.51, with Gr replaced by Gr_{AB} and Pr replaced by Sc (see Table 14.2):

$$\frac{h_m D}{D_{AB}} = 2 + 0.43 \, (Gr_{AB} Sc)^{\frac{1}{4}} \qquad (14.69)$$

where D is the diameter of the microorganism. Since we are assuming density of the microorganism is approximately the same as that of water, $\Delta \rho = 0$ makes $Gr_{AB} = 0$. Thus, Eqn. 14.69 reduces to

$$h_m = \frac{2 D_{AB}}{D}$$

which shows that the convective mass transfer coefficient is simply due to the contribution to transport from diffusion (limiting case of no bulk flow). Substituting numbers, we get

$$\begin{aligned} h_m &= 2 \times \frac{3.25 \times 10^{-9} [\text{m}^2/\text{s}]}{1 \times 10^{-6} [\text{m}]} \\ &= 6.5 \times 10^{-3} \text{m/s} \end{aligned}$$

The flux of O_2 to the microorganism is given by:

$$\begin{aligned} n_{O_2} &= h_m \left(c_{O_2,\infty} - c_{O_2,\text{surface}} \right) \\ &= (6.5 \times 10^{-3}[\text{m/s}])(2.26 \times 10^{-4}[\text{kmol/m}^3] - 0) \\ &= 1.47 \times 10^{-6} \text{ kmol } O_2/\text{m}^2\text{s} \end{aligned}$$

Comments: The microorganism can grow only so big before oxygen becomes limiting in a stagnant water.

14.5.5 Convection-Diffusion of Heat and Mass (Simultaneous) Over a Surface: Example of a Wet Bulb Thermometer

Heat and mass transfer are often simultaneous. In this section, we will illustrate simultaneous heat and mass transfer over a surface, using the wet bulb thermometer. A schematic of a wet bulb thermometer is shown in Figure 14.16. The mercury bulb of one of the thermometers is covered with a thin wet cloth or wick and the cloth is kept wet by a continuous supply of water. To keep a continuous flow of moving air, the wet bulb thermometer is swung about an axis. Evaporation cools the cloth and the temperature drops compared to another identical thermometer that is placed side-by-side but without the wet cloth. This second thermometer without the wet cloth measures the ambient air temperature and its temperature is referred to as the dry bulb temperature. The wet bulb temperature, together with the dry bulb temperature can provide the humidity or the relative humidity of air, as we will see shortly. Psychrometric charts, such as Figure C.9 on page 341 in the appendix can be used to obtain humidity or relative humidity from the dry and the wet bulb temperature.

We will now develop a convection heat and mass transfer analysis of the wet bulb thermometer. Heat transferred due to the temperature difference between the ambient air temperature (or the dry bulb temperature) and the wet bulb temperature causes evaporation of water in the wet bulb. At steady state, when the temperature of the wet bulb stays constant, the heat gained by the wet bulb balances the heat lost due to evaporation:

$$\begin{aligned} \text{Convected heat gain} &= \text{Evaporated heat loss} \\ &= \Delta H_{vap} \times \text{Convected mass loss} \end{aligned}$$

where ΔH_{vap} is the latent heat of evaporation of water. Substituting formulas for convective heat and mass transfer:

$$\underbrace{h(T - T_s)A}_{\substack{\text{Convected} \\ \text{Heat gain}}} = \Delta H_{vap} \underbrace{h_m(c_s - c)A}_{\substack{\text{Convected} \\ \text{mass loss}}} \tag{14.70}$$

where subscript s stands for the surface of the wet wick. Concentration of water vapor in the air, c is in kg of vapor per m³ of dry air. Humidity of the air, $\bar{H}$, is defined as kg of water vapor per kg of dry air and, therefore related to c as

$$c = \bar{H}\rho_0$$

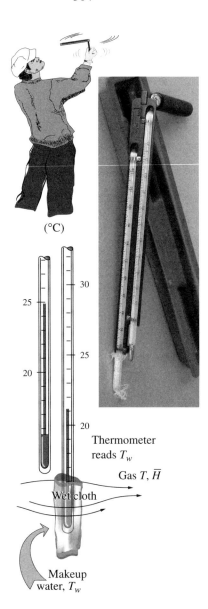

Figure 14.16: Schematic of a wet bulb thermometer (lower left figure). The picture on the right shows what the thermometer looks like and the upper left figure illustrates how the thermometer is rotated around the handle during operation.

where ρ_0 is the density of dry air in kg of dry air per m³ of dry air. Substituting in Eqn. 14.70 and simplifying

$$\frac{\bar{H}_s - \bar{H}}{T - T_s} = \frac{h}{h_m} \frac{1}{\Delta H_{vap} \rho_0} \qquad (14.71)$$

where $\bar{H}_s$ is the humidity at the wet surface, or saturation humidity. Assuming the bulb is a cylindrical geometry and crossflow occurs over it, we can write the general equations for h and h_m as (see Section 6.6.4 on page 110)

$$\frac{hD}{k_{air}} = B Re_D^n Pr^{1/3}$$

$$\frac{h_m D}{D_{wa}} = B Re_D^n Sc^{1/3}$$

where D is the diameter of the wet bulb. From these two equations, we can calculate that the ratio h/h_m as

$$\frac{h}{h_m} = \frac{k_{air}}{D_{wa}} \left(\frac{Pr}{Sc}\right)^{1/3} \qquad (14.72)$$

$$= \frac{k_{air}}{D_{wa}} \left(\frac{D_{wa}}{\alpha_{air}}\right)^{1/3} \qquad (14.73)$$

where k_{air} and α_{air} are the thermal conductivity and thermal diffusivity of air, respectively, and D_{wa} is the diffusivity of water vapor in air. Thus, substituting Eqn. 14.73 in Eqn. 14.71, we get

$$\frac{\bar{H}_s - \bar{H}}{T - T_s} = \frac{k_{air}}{D_{wa}} \left(\frac{D_{wa}}{\alpha_{air}}\right)^{1/3} \frac{1}{\Delta H_{vap} \rho_0} \qquad (14.74)$$

Note that the right hand side depends only on properties of air and diffusivity of vapor in air. It turns out that the right hand side has only a small variation over a large temperature range and can be treated approximately as a constant. Thus, Eqn. 14.74 can be used to relate humidity $\bar{H}$ to the wet bulb temperature, T_s, all other variables being known.

14.6 Chapter Summary—Convective Mass Transfer

- **Convection-Dispersion in Infinite (page 279) and Semi-Infinite Region (page 284)**

    1. For infinite region in a fluid, the solution is given by Eqn. 14.11.
    2. For semi-infinite region in a porous media, the solution is given by Eqn. 14.28.

- **Convection-Diffusion Through a Stagnant Gas (page 291)**

    1. When a gas is diffusing through another stagnant gas, it is still a diffusion-convection situation. The flux is given by Eqn. 14.52 and the concentration is given by Eqn. 14.49.

- **Convection-Diffusion Over a Surface or Convective Mass Transfer Coefficient (page 295)**

    1. The convective mass transfer resistance of fluid flowing over a solid surface is restricted to a thin layer on the surface where the fluid moves relatively slowly. This layer is called a mass transfer boundary layer.

    2. Convective mass transfer coefficient, h_m, represents the mass transfer resistance of the boundary layer.

    3. h_m includes the effect of diffusion in the fluid and flow. It therefore also depends on the flow parameter Re. Thus h_m should not be confused with a material property such as the mass diffusivity, D.

    4. In complete analogy to the heat transfer coefficient, h, the many significant variables such as the flow velocity which h_m depends on are grouped into similar dimensionless parameters. The new dimensionless parameters being Sherwood number (replacing Nusselt number), Schmidt number (replacing Prandtl number) and Grashof number for mass transfer.

    5. Since the heat and mass transfer coefficients are completely analogous, the values of h_m for a situation are given by the same formulas as for heat transfer coefficient (Section 6.6, page 104) with appropriate substitution of parameters noted in Table 14.2.

14.7 Concept and Review Questions

1. Can you have dispersion without convection or bulk flow? Explain.

2. Can you think of a heat transfer analog of the convection-dispersion in an infinite fluid? Hint: Consider thermal pollution.

3. In Section 14.1 (page 281), plot the concentration profile for the general case of convection and dispersion, when there is no decay ($k'' = 0$).

4. In convection-dispersion in a porous solid, does the dispersion coefficient depend on the magnitude of the average velocity, u? How?

5. In pollutant transport through soil, what role do macro pores play?

6. How does sorption in the porous matrix affect the transport of a species through the porous matrix? Explain qualitatively.

7. What is meant by a concentration boundary layer? How is its thickness defined?

8. Are the velocity and thermal boundary layers always thicker than the concentration boundary layer? Explain.

9. Give an example of natural convection mass transfer, identifying the two densities, the difference between which causes the natural convection.

10. What dimensionless group in natural convection mass transfer substitutes for the Reynolds number in forced convection?

11. What are the mass transfer analogs of the Grashoff and Prandtl number? Describe their physical significances.

12. The analysis of the wet bulb thermometer has several assumptions. It is important to understand these assumptions in the context of our simple analysis.

    (a) Why can you treat the surface concentration of moisture as a constant? Is this a good assumption if you keep reading the thermometer for days without adding water?

    (b) What is the reason the temperature reaches steady state?

    (c) As we rotate the psychrometer, would the dry bulb temperature change? Why or why not?

    (d) The faster we rotate, the higher is the relative velocity between the air and the thermometer. Thus, according to the formulas for the heat transfer coefficient, the rate of heat transfer should keep increasing. Would our temperature readings depend on how fast we rotate the psychrometer?

14.8 Further Reading

Bailey, J.E. and D.F. Ollis. 1986. *Biochemical Engineering Fundamentals*. McGraw-Hill, New York.

Clark, M.M. 1996. *Transport Modeling for Environmental Engineers and Scientists*. John Wiley & Sons, New York, NY.

Cussler, E. L. 1997. *Diffusion Mass Transfer in Fluid Systems*. Cambridge University Press, Cambridge, UK.

Enfield, C.G. 1982. Approximating pollutant transport to groundwater. Ground Water 20(6):711-722.

Geankoplis, C.J. 1993. *Transport Processes and Unit Operations*. P T R Prentice Hall, Inc., Englewood Cliffs, New Jersey.

Gish, T.J. and W.A. Jury. 1983. Effect of plant roots and root channels on solute transport. Trans. of the ASAE 26(2):440-444, 451.

Johnson, W.B., R.C. Sklarew, and D.B. Turner. 1976. *Urban Air Quality Simulation Modeling, in Air Pollution: Air Pollutants, Their Transformation, and Transport*. A.C. Stern, ed. Academic Press, New York.

Knox, R.C., D.A. Sabatini, and L.W. Cantor. 1993. *Subsurface Transport and Fate Processes*. Lewis Publishers, Boca Raton, Florida.

Leij, F.J., Jiri Simunek, D.L. Suarez, and M. Th. Van Genuchten. 1999. Nonequilibrium and multicomponent transport models. In *Agricultural Drainage* by R.W. Skaggs and J. van Schilfgaarde, American Society of Agronomy, Inc., Madison, Wisconsin, pp. 405-430.

Loucks, D.P., J.R. Stedinger, and D.A. Haith. 1980. *Water Resource Systems Planning and Analysis*. Prentice Hall. Englewood Cliffs, New Jersey.

Patterson, M.R. 1992. A mass transfer explanation of metabolic scaling relations in some aquatic invertebrates and algae. Science 255:1421-1423.

Pitt, R.E. and R.E. Muck. 1993. A diffusion model of aerobic deterioration at the exposed face of bunker silos. Journal of Agricultural Engineering Research 55(1):11-26.

Schnoor, J. L. 1996. *Environmental Modeling: Fate and Transport of Pollutants in Water, Air and Soil*. John Wiley & Sons, New York.

Schulin, R., P.J. Wierenga, H. Flühler, and J. Leuenberger. 1987. Solute transport through a stony soil. Journal of the Soil Science Society of America 51:36-42.

Waldschmidt, S.R. and W.P. Porter. 1987. A model and experimental test of the effect of body temperature and wind speed on ocular water loss in the lizard Uta stansburiana. Physiological Zoology 60(6):678-86.

Weber, E. 1982. Air Pollution: Assessment Methodology and Modeling. Plenum Press, New York.

14.9 Problems

14.9.1 Convection-Dispersion in a Soil

Consider a pollutant on a soil surface being dissolved in the rainfall and snowmelt and carried through the soil depth into the groundwater. Calculate the concentration profiles of the pollutant in the soil after 2 and 5 years. The bulk density of the soil is 1.2 g/cm^3, field capacity of the soil is 0.3 cm^3 of water per cm^3 of soil, partition coefficient (cm^3 of water per gram of soil) is 2, initial concentration of pollutant is 333 μg/cm^3, depth of incorporation in the soil is 15 cm, retardation factor is 9, dispersion coefficient is 2 cm^2/day and mean velocity due to percolation is 0.5 cm/day.

14.9.2 Transport of Material Injected in a Vein

Patients are sometimes administered mineral, saline water, which is injected into the veins. We would like to consider the transport of the saline solution in the vein. A simple model would be to consider the diffusion of the saline water with diffusivity

D, as the entire liquid (blood + saline water) moves with some average velocity, u. Consider a continuous supply of the saline substance of known concentration to be maintained at a location $z = 0$ on the vein. 1) Draw a schematic for this problem. 2) Setup the one-dimensional governing equation for this problem, keeping only the relevant terms. 3) Setup the boundary and initial conditions for this problem. 4) Solve the governing equation to provide concentration as a function of time and position along the vein (Hint: See Section 14.2).

14.9.3 Diffusion Through a Stagnant Gas

The water surface, in an open cylindrical tank is 7.5 meter below the top. Dry air is blown over the top of the tank and the entire system is maintained at 18°C and 1 atm total pressure. If the air in the tank is stagnant, determine the rate of water loss through the top in $g/m^2 \cdot s$. The diffusivity of water vapor through air at this temperature is $2.77 \times 10^{-5} m^2/s$. The diameter of the tank is 2 m. (Note: Some required data is in the Appendix).

14.9.4 Diffusion Without a Stagnant Gas

Consider dry air flow 18°C at 15 km/hr over a circular water surface of 2 m diameter. The diffusivity of vapor in air is given by the same value as in Problem 14.9.3. 1) Calculate the rate of water loss in $g/m^2 \cdot s$. 2) Compare the water loss just calculated with that calculated in Problem 14.9.3, and comment on which is more and why?

14.9.5 Convective Loss of Moisture from Soil Surface

Estimation of evaporation from soil is normally very complicated. For simplicity, consider dry air blowing over a very wet silt loam soil having a flat and bare surface, i.e., no grass or other cover. Calculate the moisture loss per unit area per unit time (in $g/m^2 \cdot s$) from this soil under the following conditions: Length of the field is 100 m, soil surface and air temperature is 20°C, moisture content of saturated air at 20°C is 0.0173 kg water/m^3 dry air, average wind speed is 20 km/hour, diffusivity of water vapor in air is $2.2 \times 10^{-5} m^2/s$, viscosity of air is 2×10^{-5} Pa·s and density of air is 1.14 kg/m^3.

14.9.6 Reduction of Moisture Loss Using a Cover

Consider the same very wet soil surface at 20°C as in Problem 14.9.5 but now with mulch (loose straw) of thickness 2.5 cm on it. Because the mulch maintains a layer of stagnant air over the soil surface, we cannot calculate evaporation from the soil surface in the same manner as for Problem 14.9.5. Instead, the water vapor from the soil surface is now considered to diffuse through air held stagnant by the mulch. Assume flow of dry air over the mulch maintains a water vapor concentration of zero at the top of the mulch. Vapor pressure of water at 20°C is 0.0231 atm (1 atm = $1.013 \times 10^5 N/m^2$), atmospheric pressure is 1 atm and molecular weight of water is 18 g/mol. 1) Calculate the loss of moisture per unit area of soil (in g of water per m^2 per second) for the mulched condition. 2) Compare the evaporative water loss for the mulched

Figure 14.17: Schematic for Problem 14.9.7.

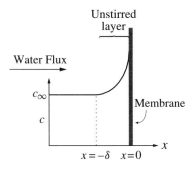

Figure 14.18: Concentration profile of solute near membrane.

condition with evaporative loss from bare soil (no mulch) calculated in Problem 14.9.5 and explain the difference.

14.9.7 Thickness of Various Boundary Layers

Consider dry air flowing at 10 cm/s over a wet sponge of size 10 cm × 5 cm where the flow is perpendicular to the longer side, as shown in Figure 14.17. Air is warmer than the sponge surface. Calculate and compare the thicknesses of the velocity, thermal and moisture boundary layers over this sponge at 5 cm (trailing edge) from where the flow hits the sponge. The average or film temperature to be used is 20°C and the properties at this temperature are given as $\mu_{air}/\rho_{air} = 15.89 \times 10^{-6} m^2/s$, $\alpha_{air} = 2.08 \times 10^{-5} m^2/s$ and $D_{vapor,air} = 2.6 \times 10^{-5} m^2/s$.

14.9.8 Solute Concentration Profile in Unstirred Layer

Water and solute transport in roots is a much-researched topic. Some researchers have indicated that the solute concentration very near the membrane (i.e. in the unstirred surface layer) is very different from the solute concentration in the bulk. This increase occurs because there is a flux of water across the membrane, but the membrane has a very low solute permeability and a very high water permeability. Therefore the solute cannot pass through the membrane as quickly as the water and there is a subsequent solute build-up very near the membrane.

1) Write the governing equation for the concentration profile of solute very near the membrane (see Figure 14.18). The profile is steady-state because, as concentration builds up in the unstirred layer near the membrane, diffusion of solute occurs in the direction opposite to convection (water flux) and eventually balances it. 2) Write the two boundary conditions. The boundary condition at the membrane surface ($x = 0$) is zero total flux (convective+diffusive). The other boundary condition is given by c_∞, concentration of solute in bulk solution, at $x = -\delta$ (the thickness of the unstirred layer). 3) Calculate the concentration profile as a function of position x for a concentration of solute in bulk solution of 2 mol/liter, bulk velocity of water at 30×10^{-9} m/s, diffusivity of solute $2 \times 10^{-10} m^2/s$ and the thickness of the unstirred layer of 8×10^{-9} m.

14.9.9 Ocular Evaporative Water Loss in Lizards

Reptiles are often able to successfully colonize arid habitats because their bodies have a very low rate of evaporative water loss (EWL). Although a reptile's outer covering is virtually impermeable to water, the wet ocular surfaces freely evaporate water to the environment. While the eye area comprises less that 0.03% of the lizard's total surface area, the water loss from these moist surfaces can be very significant.

Consider the data below for a lizard habitat and 1) Write the equation for mass transfer coefficient assuming air flow over the eye as laminar flow over a flat plate. 2) For a wind velocity that leads to an average heat transfer coefficient of 11 W/m²·K, calculate the average mass transfer coefficient for air flow over the eye. 3) Find the rate of evaporative water loss (EWL) from the eye. 4) If the total evaporative water loss (total EWL) from the eye and the body surface is 1.77 mg/hr, find the EWL from the eye as a percentage of total EWL.

Assume that the surrounding air is completely dry. Percentage of time the eyes are open is 60. The total surface area of both eyes combined is $8.5 \times 10^{-7} m^2$, diffusivity of vapor in air is $2.6 \times 10^{-5} m^2/s$, vapor concentration in the air at the eye surface corresponds to the vapor pressure of water at the lizard's surface temperature of 37°C, thermal conductivity of air is 0.027 W/m·K, density of air is 1.14 kg/m^3 and specific heat of air is 1 kJ/kg·K.

14.9.10   Evaporation from a Falling Drop of Water

Calculate the instantaneous rate of evaporation per unit surface area from a falling drop of water in dry air at an average temperature of 35°C at the instant when the diameter is 0.001 m. Diffusivity of water vapor in air at 35°C is $0.273 \times 10^{-4} m^2/s$, instantaneous velocity of drop is 3 m/s, vapor pressure of water at 35°C is 2.33×10^3 Pa. Air properties can be found in Appendix C.8.

14.9.11   Time to Completely Evaporate a Drop

Consider convective mass transfer from tiny spherical droplets of water in dry air as evaporation occurs at the droplet surface. The droplets are very small, so they are suspended in air, i.e., no relative velocity between the droplets and the air. The droplet temperature remains constant at 40°C. The diffusivity of water vapor in air at this temperature is $0.273 \times 10^{-4} m^2/s$ and the density of water is 1000 kg/m^3. 1) Equate the time rate of change of a droplet mass to the water loss through convective mass transfer at its surface. 2) Simplify the equation in step 1 and solve for the time it takes to completely evaporate a droplet of radius r. 3) Calculate the time it takes for a 0.1 mm radius drop to completely evaporate. Vapor pressure data can be found in Appendix C.10.

14.9.12   Mass Transfer to a Rising Bubble

A bubble of pure chlorine gas at 1.0 atm and 0.5 cm in diameter is rising at a velocity of 20 cm/s in pure water at 16°C. The diffusivity of chlorine gas in water at this temperature is $1.26 \times 10^{-9} m^2/s$ and the maximum amount of chlorine that can dissolve in water (saturation solubility) at this temperature is 0.823 g of chlorine per 100 g of water. Chlorine from the bubble surface is dissolved in water and diffuses away, making the bubble smaller. What is the rate (in grams per second) of absorption of the bubble by the water when the bubble is 0.5 cm in diameter? The density and viscosity of water are 1000 kg/m^3 and 1.155×10^{-3} kg/m·s, respectively.

14.9.13   Oxygen Diffusion to Sustain Microorganisms in Water

Consider a stagnant water at 25°C with a colony of microorganisms having an effective diameter of 0.1 mm. Oxygen diffuses through water to the surface of microorganisms where it is completely consumed, maintaining a zero concentration at the surface.

Oxygen in the bulk water is in equilibrium with air (21% oxygen). Henry's constant for oxygen and water is 4.4×10^4 atm/mole fraction at 25°C. Diffusivity of oxygen in water is 3.25×10^{-9} m^2/s and density of water is 1000 kg/m^3. 1) Write the governing equation for oxygen diffusion in water, considering radial variation only. 2) Write the boundary conditions needed. 3) Obtain the oxygen concentration profile as a function of radius. 4) Calculate the oxygen concentration in bulk water assuming equilibrium between air and water. 5) Calculate the rate of oxygen consumption by the microorganism colony in mol/m$^2 \cdot$s.

Part III

APPENDIX

Appendix A

SUMMARY AND ANALOGIES BETWEEN HEAT AND MASS TRANSFER

A.1 Generic Transport Equation and the Physical Meanings of the Constituent Terms

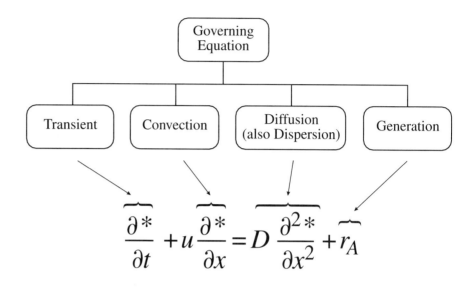

A.2 Generic Boundary Conditions and Initial Condition Needed to Solve the Transport Equation in A.1

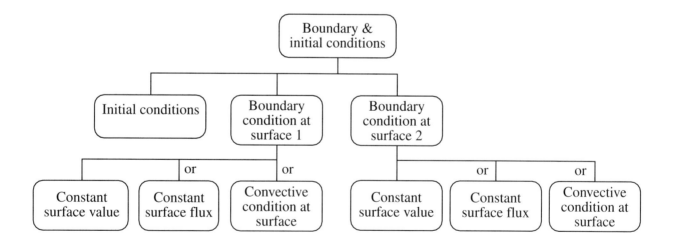

A.3 Solution Map for Heat Conduction and Mass Diffusion Problems (Without Bulk Flow)

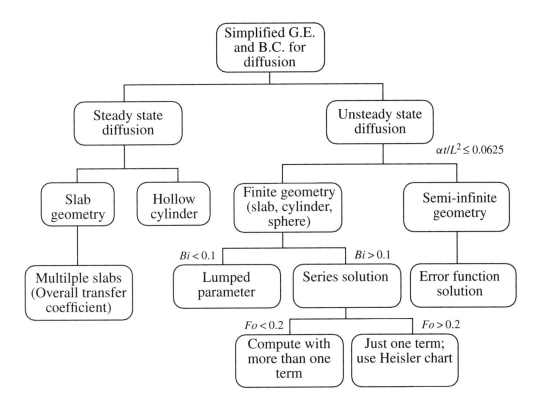

A.4 Solution Map for Mass Diffusion or Dispersion with Bulk Flow (Convection-Diffusion/Dispersion)

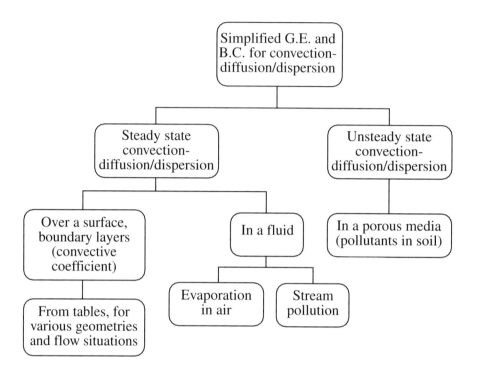

A.5 Basic Heat and Mass Transfer Parameters and Fluxes

	Heat Transfer	Mass Transfer
Quantity of Interest	$T\,[^\circ\mathrm{C}]$	$c_A\,[\mathrm{kg/m^3}]$
Equation for Diffusive Flux	$q'' = -k\dfrac{\partial T}{\partial z}\,\left[\dfrac{\mathrm{W}}{\mathrm{m^2}}\right]$ $= -\alpha\dfrac{\partial(\rho c_p T)}{\partial z}$	$j_A = -D_{AB}\dfrac{\partial c_A}{\partial z}\,\left[\dfrac{\mathrm{kg}}{\mathrm{m^2 s}}\right]$
Diffusivity	$\alpha\,[\mathrm{m^2/s}]$	$D_{AB}\,[\mathrm{m^2/s}]$
Convective Transfer Coefficient	$h\,[\mathrm{W/m^2\cdot K}] = \dfrac{q''}{\Delta T}$	$h_m\,[\mathrm{m/s}] = \dfrac{j_A}{\Delta c}$
Overall Coefficient	$\dfrac{1}{U} = \dfrac{1}{h_1} + \dfrac{\Delta L}{k} + \dfrac{1}{h_2}$	$\dfrac{1}{U_m} = \dfrac{1}{h_{m_1}} + \dfrac{\Delta L}{K^* D_{AB}} + \dfrac{1}{h_{m_2}}$
Equation for total flux	$q'' = \underbrace{-k\dfrac{\partial T}{\partial z}}_{\text{conductive}} + \underbrace{\rho c_p T u}_{\text{convective}}$	$n_{A,z} = \underbrace{-D_{AB}\dfrac{\partial c_A}{\partial z}}_{\text{diffusive}} + \underbrace{c_A u}_{\text{convective}}$

A.6 Heat and Mass Transfer Governing Equations and Boundary Conditions

	Heat Transfer	Mass Transfer
Governing Equation	$\frac{\partial T}{\partial t} + u \frac{\partial T}{\partial z}$ $= \alpha \frac{\partial^2 T}{\partial z^2} + \frac{Q}{\rho c_p}$ Q is the rate of heat generation or absorption per unit volume	$\frac{\partial c_A}{\partial t} + u \frac{\partial c_A}{\partial z}$ $= D_{AB} \frac{\partial^2 c_A}{\partial z^2} + r_A$ r_A is the rate of mass generation or disappearance per unit volume
Boundary Conditions	$T\big|_{surface} = T_s$ surface temperature is specified, e.g., *condensing steam*	$c_A\big|_{surface} = c_{A,s}$ surface concentration is specified, e.g., *later stages of drying*
	$-k \frac{\partial T}{\partial z}\big|_{surface} = q_s''$ surface heat flux is specified, e.g., *insulated (zero heat flux)*	$-D_{AB} \frac{\partial c_A}{\partial z}\big|_{surface} = n_{A,s}$ surface mass flux is specified, e.g., *impermeable (zero mass flux)*
	$-k \frac{\partial T}{\partial z}\big|_{surface}$ $= h(T\big|_{surface} - T_\infty)$ convection over surface	$-D_{AB} \frac{\partial c_A}{\partial z}\big|_{surface}$ $= h_m(c_A\big|_{surface} - c_{A,\infty})$ convection over surface

Appendix B

PHYSICAL CONSTANTS, UNIT CONVERSIONS, AND MATHEMATICAL FUNCTIONS

B.1 Physical Constants

Quantity	Symbol	Value
Universal Gas Constant	R_g	8.205×10^{-2} m$^3 \cdot$ atm/kmol$\cdot$K
		8.314×10^{-2} m$^3 \cdot$ bar/kmol$\cdot$K
		8.315 kJ/kmol$\cdot$K
Avogadro's Number	N	6.024×10^{23} molecules/mol
Planck's Constant	h	6.625×10^{-34} J$\cdot$ s/molecule
Boltzmann's Constant	κ	1.380×10^{-23} J/K$\cdot$molecule
Speed of Light in Vacuum	c_o	2.998×10^8 m/s
Stefan-Boltzmann Constant	σ	5.670×10^{-8} W/m$^2 \cdot$ K^4
Gravitational Acceleration (Sea Level)	g	9.807 m/s^2
Normal Atmospheric Pressure	p	$101,325$ N/m^2

B.2 Some Useful Conversion Factors

To Convert From	to	Multiply by
Area		
acre	m^2	$4.046\,856 \times 10^3$
acre	ft^2	$4.356\,000 \times 10^4$
ft^2	m^2	$9.290\,304 \times 10^{-2}$
m^2	ft^2	$1.076\,391 \times 10^1$
cm^2	m^2	$1.000\,000 \times 10^{-4}$
ft^2	in.^{2}	$1.440\,000 \times 10^2$
in.^{2}	m^2	$6.451\,600 \times 10^{-4}$
Energy		
Btu (international)	kJ	$1.055\,056$
Btu	kcal	$2.519\,958 \times 10^{-1}$
kcal	kJ	$4.186\,800$
kW·h	kJ	$3.600\,000 \times 10^3$
kW·h	Btu	3.413×10^3
hp·h	Btu	2.454×10^3
J	ergs	$1.000\,000 \times 10^7$
eV	J	$1.602\,19 \times 10^{-19}$
Btu	ft-lbf	$7.781\,693 \times 10^2$
ft·lbf	kJ	$1.355\,818 \times 10^{-3}$
Force		
lbf	N	$4.448\,222$
dyne	N	$1.000\,000 \times 10^{-5}$
Length		
angstrom	m	$1.000\,000 \times 10^{-10}$
caliber	m	$2.540\,000 \times 10^{-4}$
fathom	m	$1.828\,800$
ft	m	$3.048\,000 \times 10^{-1}$
inch	m	$2.540\,000 \times 10^{-2}$
mile	m	$1.609\,344 \times 10^3$
mile	ft	5.280×10^3
yard	m	$9.144\,000 \times 10^{-1}$

B.2. SOME USEFUL CONVERSION FACTORS

To Convert From	to	Multiply by
Mass		
carat	kg	$2.000\,000 \times 10^{-4}$
grain	kg	$6.479\,891 \times 10^{-5}$
lb_m (pound mass)	kg	$4.535\,924 \times 10^{-1}$
kg	lb_m	$2.204\,622$
kg	slug	$6.852\,1 \times 10^{-2}$
slug	lb_m	$3.217\,405 \times 10^{1}$
ton	lb_m	2.000×10^{3}
ton	kg	$9.071\,847 \times 10^{2}$
Mass per Unit Volume (density)		
lb_m/ft^3	kg/m^3	$1.601\,846 \times 10^{1}$
$lb_m/in.^3$	kg/m^3	$2.767\,990 \times 10^{4}$
lb_m/gal	kg/m^3	$1.198\,264 \times 10^{2}$
Power		
Btu/h	W (watt)	$2.930\,711 \times 10^{-1}$
J/s	W	$1.000\,000$
erg/s	W	$1.000\,000 \times 10^{-7}$
ft · lbf/s	W	$1.355\,818$
hp	W	$7.456\,999 \times 10^{2}$
hp	ft · lbf/s	5.50×10^{2}
hp	ft · lbf/min	$3.300\,0 \times 10^{4}$
hp (boiler)	Btu/h	$3.347\,14 \times 10^{4}$
hp (boiler)	kW	$9.809\,50$
Pressure		
std atm	kPa	$1.013\,25 \times 10^{2}$
std atm	$lbf/in.^2$	$1.469\,6 \times 10^{1}$
cm Hg (0°C)	kPa	$1.333\,22$
in. Hg (32°F)	kPa	$3.386\,389$
in. H_2O (60°F)	kPa	$2.488\,4 \times 10^{-1}$
bar	kPa	$1.000\,000 \times 10^{2}$
bar	$lbf/in.^2$	$1.450\,377 \times 10^{1}$
$lbf/in.^2$	kPa	$6.894\,757$
in. Hg (32°F)	$lbf/in.^2$	$4.911\,542 \times 10^{-1}$
ft H_2O (60°F)	$lbf/in.^2$	$4.330\,943 \times 10^{-1}$
ft H_2O (60°F)	kPa	$2.986\,08$
$dyne/cm^2$	kPa	$1.000\,000 \times 10^{-4}$

324   APPENDIX B. PHYSICAL CONSTANTS, UNIT CONVERSIONS, AND MATHEMATICAL FUNCTIONS

To Convert From	to	Multiply by
Specific Heat		
Btu/$lb_m \cdot °F$	kJ/kg·K	4.186 800
cal/g·°C	Btu/$lb_m \cdot °F$	1.000 000
Thermal Conductivity		
Btu/ft·h·°F	W/m·K	1.730 6
Btu·in./ft^2·h·°F	W/m·K	$1.442\,279 \times 10^{-2}$
cal/cm·h·°C	Btu/ft·h·°F	$6.719\,69 \times 10^{-2}$
Thermal Conductance		
Btu/ft^2·h·°F	W/m^2·K	5.674 466
Temperature		
°C	K	$T_K = T_{°C} + 273.15$
°F	°R	$T_{°R} = T_{°F} + 459.67$
K	°R	$T_{°R} = T_K \times 1.8$
°C	°F	$T_{°F} = (T_{°C} + 273.15) \times 1.8 - 459.67$
°F	°C	$T_{°C} = (T_{°F} + 459.67)/1.8 - 273.15$
Velocity		
ft/s	m/s	$3.048\,888 \times 10^{-1}$
miles/h	km/h	1.609 344
knot	m/s	$5.144\,444 \times 10^{-1}$
Viscosity		
centipoise	Pa·s	$1.000\,000 \times 10^{-3}$
Centistokes	m^2/s	$1.000\,000 \times 10^{-6}$
lb_m/ft·s	Pa·s	1.488 164
lbf·s/ft^2	Pa·s	$4.788\,026 \times 10^{1}$
lbf·s/ft^2	lb_m/ft·s	$3.217\,4 \times 10^{1}$
Volume		
acre·ft	m^3	$1.233\,483 \times 10^{3}$
barrel (42 gal)	m^3	$1.589\,873 \times 10^{-1}$
ft^3	m^3	$2.831\,685 \times 10^{-2}$
ft^3	$in.^3$	$1.728\,000 \times 10^{3}$
$in.^3$	m^3	$1.638\,706 \times 10^{-5}$
ft^3	gallons	7.480 52
liter	m^3	$1.000\,000 \times 10^{-3}$
gallons	liter	3.785 412
gallons	m^3	$3.785\,412 \times 10^{-3}$
gallons	ft^3	$1.336\,81 \times 10^{-1}$
quart (U.S. liquid)	m^3	$9.463\,529 \times 10^{-4}$
pint (U.S. liquid)	liter	$4.731\,765 \times 10^{-1}$

B.3 Error Function Tabulated

ϕ	erf(ϕ)	ϕ	erf(ϕ)	ϕ	erf(ϕ)
0.000	0.0000	0.850	0.7707	1.700	0.9838
0.025	0.0282	0.875	0.7841	1.725	0.9853
0.050	0.0564	0.900	0.7969	1.750	0.9867
0.075	0.0845	0.925	0.8092	1.775	0.9879
0.100	0.1125	0.950	0.8209	1.800	0.9891
0.125	0.1403	0.975	0.8321	1.825	0.9901
0.150	0.1680	1.000	0.8427	1.850	0.9911
0.175	0.1955	1.025	0.8528	1.875	0.9920
0.200	0.2227	1.050	0.8624	1.900	0.9928
0.225	0.2497	1.075	0.8716	1.925	0.9935
0.250	0.2763	1.100	0.8802	1.950	0.9942
0.275	0.3027	1.125	0.8884	1.975	0.9948
0.300	0.3286	1.150	0.8961	2.000	0.9953
0.325	0.3542	1.175	0.9034	2.025	0.9958
0.350	0.3794	1.200	0.9103	2.050	0.9963
0.375	0.4041	1.225	0.9168	2.075	0.9967
0.400	0.4284	1.250	0.9229	2.100	0.9970
0.425	0.4522	1.275	0.9286	2.125	0.9973
0.450	0.4755	1.300	0.9340	2.150	0.9976
0.475	0.4983	1.325	0.9390	2.175	0.9979
0.500	0.5205	1.350	0.9438	2.200	0.9981
0.525	0.5422	1.375	0.9482	2.225	0.9983
0.550	0.5633	1.400	0.9523	2.250	0.9985
0.575	0.5839	1.425	0.9561	2.275	0.9987
0.600	0.6039	1.450	0.9597	2.300	0.9989
0.625	0.6232	1.475	0.9630	2.325	0.9990
0.650	0.6420	1.500	0.9661	2.350	0.9991
0.675	0.6602	1.525	0.9690	2.375	0.9992
0.700	0.6778	1.550	0.9716	2.400	0.9993
0.725	0.6948	1.575	0.9741	2.425	0.9994
0.750	0.7112	1.600	0.9763	2.450	0.9995
0.775	0.7269	1.625	0.9784	2.475	0.9995
0.800	0.7421	1.650	0.9804		
0.825	0.7567	1.675	0.9822		

B.4 Charts for Unsteady Diffusion

B.4. CHARTS FOR UNSTEADY DIFFUSION

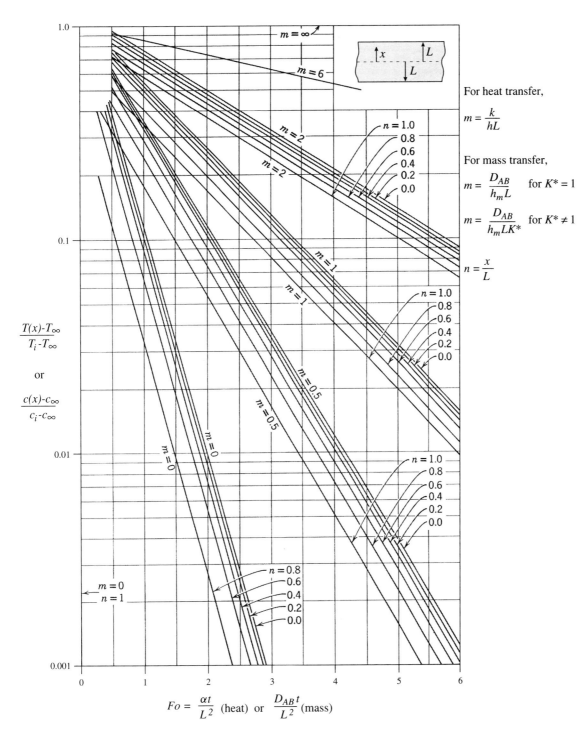

Figure B.1: Unsteady state diffusion in a large slab. Before using in mass transfer for $m \neq 0$, refer to discussion in Section 13.2.5 on page 261. From Principles of unit operations by A.S. Foust et al., © 1960 by John Wiley & Sons, Inc. Reprinted by permission of John Wiley & Sons, Inc.

328 APPENDIX B. PHYSICAL CONSTANTS, UNIT CONVERSIONS, AND MATHEMATICAL FUNCTIONS

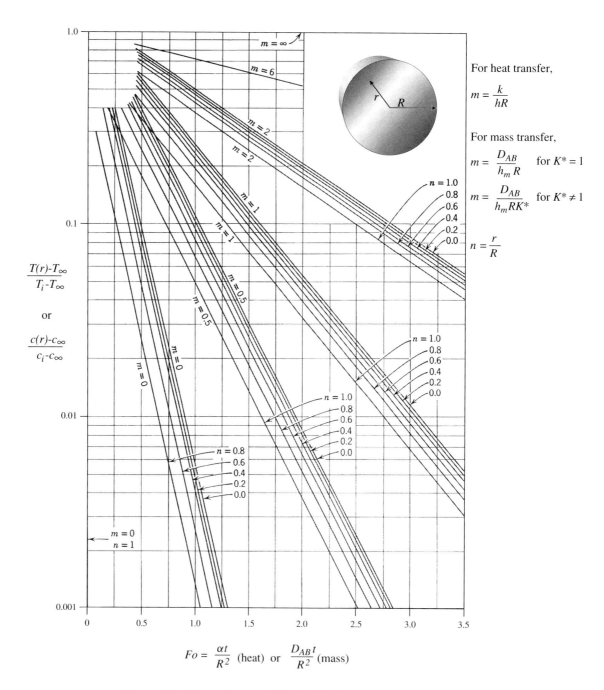

Figure B.2: Unsteady state diffusion in a long cylinder. Before using in mass transfer for $m \neq 0$, refer to discussion in Section 13.2.5 on page 261. From Principles of unit operations by A.S. Foust et al., © 1960 by John Wiley & Sons, Inc. Reprinted by permission of John Wiley & Sons, Inc.

B.4. CHARTS FOR UNSTEADY DIFFUSION

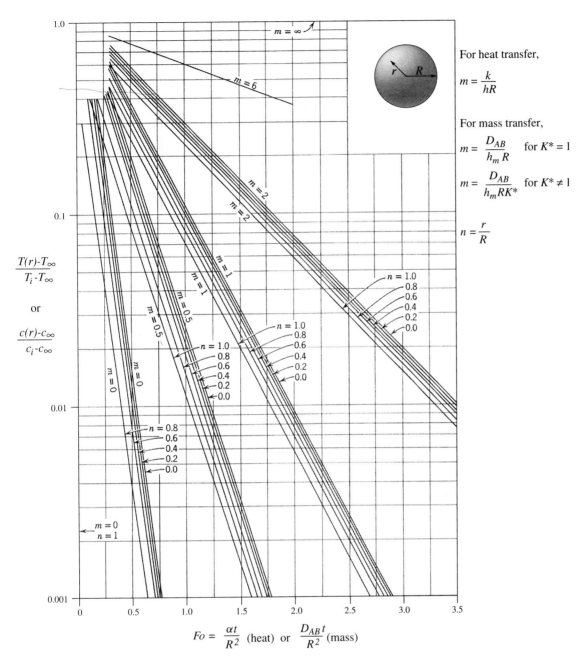

Figure B.3: Unsteady state diffusion in a sphere. Before using in mass transfer for $m \neq 0$, refer to discussion in Section 13.2.5 on page 261. From Principles of unit operations by A.S. Foust et al., © 1960 by John Wiley & Sons, Inc. Reprinted by permission of John Wiley & Sons, Inc.

Appendix C

HEAT TRANSFER AND RELATED PROPERTIES

C.1 Basal Metabolic Rate for a Few Animals

Animal	Mass [kg]	Basal Metabolic Rate [W]	[kcal/day]
Dove	0.16	0.97	20
Rat	0.26	1.45	30
Pigeon	0.30	1.55	32
Hen	2.0	4.8	100
Dog (female)	11	14.5	300
Dog (male)	16	20	420
Sheep	45	50	1050
Woman	60	68	1400
Man	70	87	1800
Cow	400	266	5500
Steer	680	411	8500

Adapted from Greenberg, L.H. 1975. *Physics for Biology and Medical Students.* W. B. Saunders Company, Philadelphia.

C.2 Typical Metabolic Rate for Various Human Activities

Activity	Metabolic rate per unit body surface area W/m^2
Sleeping	41
Reclining	33
Sitting quietly	58
Standing relaxed	70
Dressing and undressing	74
Walking on the level at 3-6 km/hr	116-221
Driving	
Car	87
Motorcycle	116
Heavy vehicle	186
House cleaning	116-198
Shopping	81-105
Carpentry, metal working, industrial painting	150
Dancing	140-256
Swimming	314
Jogging (at about 9 km/hr)	357
Walking up stairs	690

Adapted from Shitzer, A. and R.C. Eberhart. 1985. Heat generation, storage, and transport processes. 1:137-151. In: *Heat Transfer in Medicine and Biology* edited by A. Shitzer and R.C. Eberhart. Plenum Press, New York. Reprinted with permission from Plenum Publishing Corp.

C.3 Blood Flow, Oxygen Consumption, and Metabolic Rate for Various Human Organs

Organ	Mass kg	Blood Flow ml/min	Oxygen Consumption ml/min	Metabolic Rate W
Heart muscle	0.3	250-1,800	30-90	10-31
Skeletal muscle	31.0	1,200-24,000	50-1,000	17-350
Skin	3.6	400-2,800	12-85	4-30
Liver	2.6	1,500	51	18
Kidney	0.3	1,260	18	6
Brain	1.4	750	49	17

Adapted from Shitzer, A. and R.C. Eberhart. 1985. Heat generation, storage, and transport processes. 1:137-151. In: *Heat Transfer in Medicine and Biology* edited by A. Shitzer and R.C. Eberhart. Plenum Press, New York. Reprinted with permission from Plenum Publishing Corp.

C.4 Thermal Properties of Animal Materials

Description	Thermal Conductivity k W/m°C	Thermal Diffusivity $\alpha \times 10^7$ m²/s
A. Materials *in vivo*		
(values depend on blood perfusion)		
Bone, bovine and caprine	0.33-3.1	
Brain, cat	0.56-0.66	1.1-1.2
Cartilage, scapula, bovine	1.8-2.8	
Cartilage, scapula, caprine	1.4-1.9	
Kidney, sheep	0.60-1.2	2.0-4.3
Liver, canine	0.60-0.90	1.5-2.4
Muscle, canine	0.70-1.0	0.7-1.3
Skin, animal and human	0.48-2.8	0.4-1.6
B. Materials *in vitro*		
(room to body temperatures)		
Bone, fresh to several months postmortem	0.41-0.63	
Dry Bone	0.22	
Bone marrow, bovine	0.22	
Brain, bovine, cat, and human	0.16-0.57	0.44-1.4
Fat	0.094-0.37	
Heart	0.48-0.59	1.4-1.5
Kidney	0.49-0.63	1.3-1.8
Kidney, cortex, and medulla	$0.664w - 0.04$[a]	
Liver	0.42-0.57a	1.1-2.0
Liver, parenchyma	0.32	1.7-2.0
Lung		2.4-2.8
Muscle	0.34-0.68	1.8
Skin, animal and human	0.21-0.41	0.82-1.2
Spleen	0.45-0.60	1.3-1.6
Tumors, 37°C, General range	0.47-0.58	

continued

[a] w is the mass fraction of water.

Description	Thermal Conductivity k W/m°C	Thermal Diffusivity $\alpha \times 10^7$ m²/s
C. Biological fluids		
(at room to body temperatures)		
Agar gel, 1-1.75%	0.60-0.70	1.6
Blood	0.48-0.60	
Blood, human	$0.57 - 0.12H^b$	
Blood, whole		
−10°C	1.6	8.7
−20°C	1.7	10.4
−40°C	1.9	13.6
−60°C	2.1	16.9
−80°C	2.4	20.4
−100°C	2.7	23.7
Blood plasma	0.57-0.60	
Humor, aqueous and vitreous	0.58-0.59	
Milk, regular and skimmed	0.53-0.59	
Cream (double Devon)	0.31	
Cod liver oil	0.17	
Egg white	0.56	
Egg yolk	0.34-0.42	
Gastric juice	0.44	
Urine	0.56	
Water	0.59-0.63	

[b] H is the hematocrit fraction.

Adapted from Chato, J.C. 1985. Selected thermophysical properties of biological materials. 2:413-418. In: *Heat Transfer in Medicine and Biology* edited by A. Shitzer and R.C. Eberhart. Plenum Press, New York. Reprinted with permission from Plenum Publishing Corp.

C.5 Thermal Conductivities of Some Animal Hair Coats

Material	Thermal conductivity, k W/m°C
Arctic mammals	0.036-0.106
Various wild animals	0.038-0.051
Merino sheep	0.037-0.048
Newborn Merino	0.065-0.107
Cattle	0.076-0.147
Rabbit	0.038 -0.100
Kangaroo	0.043-0.064
Harp seal pups	0.047 -0.065
Penguin	0.031-0.046
Gosling	0.036-0.046
Artifical fur	0.040-0.067
Woven fabric	0.040
Dry air (for comparison) at 10°C	0.025

Adapted from Cena, K. and J.A. Clark. 1979. Transfer of heat through animal coats and clothing. International Review of Physiology, Environmental Physiology III, Vol. 20, pp. 1-42, Edited by D. Robertshaw, University Park Press, Baltimore.

C.6 Thermal Properties of Some Agricultural Materials

Material	Thermal conductivity, k W/m°C
Wood, G = 0.45 M = 12% ⊥	0.13
Wood, G = 0.45 M = 12%	0.31
Wood, G = 0.70 M = 12% ⊥	0.18
Wood, G = 0.70 M = 12%	0.44
Cell-wall substance ⊥	0.44
Cell-wall substance	0.88
Douglas-fir plywood	0.12
Concrete	0.93
Expanded polyurethane	0.02
Copper	386.74
Aluminum	201.96
Stainless steel	16.3
Sand	
Porosity - 0.4, Volumetric wetness = 0.0	0.29
Porosity - 0.4, Volumetric wetness = 0.2	1.76
Porosity - 0.4, Volumetric wetness = 0.4	2.18
Clay	
Porosity - 0.4, Volumetric wetness = 0.0	0.25
Porosity - 0.4, Volumetric wetness = 0.2	1.17
Porosity - 0.4, Volumetric wetness = 0.4	1.59
Peat	
Porosity - 0.8, Volumetric wetness = 0.0	0.06
Porosity - 0.8, Volumetric wetness = 0.4	0.29
Porosity - 0.8, Volumetric wetness = 0.8	0.50
Snow	
Porosity - 0.95, Volumetric wetness = 0.05	0.06
Porosity - 0.8, Volumetric wetness = 0.2	0.13
Porosity - 0.5, Volumetric wetness = 0.5	0.71

C.7 Thermal Properties of Food Materials (Representative Values)

Material	Thermal Conductivity W/mK	Density kg/m^3	Specific heat kJ/kgK
High moisture, frozen			
Apple[a] at −40°C	1.669	785	2.29
Beef, lean	1.42		1.68
Strawberries, tightly packed	1.1		1.14
High moisture, unfrozen			
Apple	0.418		3.60
Beef, lean	0.506		3.35
Egg white	0.558		3.88
Low moisture, non-porous			
Butter	0.197		1.38
Low moisture, porous			
Apple, dried	0.219		2.27
Apple, freeze dried (at pressure 0.2880×10^4 Pa)	0.0405		
Beef, freeze dried (at pressure of 1 atm)	0.0652		
Egg albumin gel, freeze dried (at pressure of 1 atm)	0.0393		

Data from 1998 ASHRAE Refrigeration Handbook, unless otherwise mentioned. Note that the thermal conductivity, density and specific heat are not always from the same study. Thus they should be treated as representative values. Due to the large variability in composition, thermal properties of foods are sometimes estimated using empirical correlations based on composition, such as the one in Choi, Y. 1985. Food thermal property prediction as affected by temperature and composition. PhD Thesis, Purdue University. An internet database has also become operational at http://www.nelfood.com/ that anyone can access. At least one commercial database for foods also exists (Singh, R.P. 1995. Food Properties Database 2.0 Windows. CRC press, Boca Raton, Florida).

[a]From Heldman, D.R. and R.P. Singh. 1986. Thermal properties of frozen foods. In: *Physical and Chemical Properties of Foods*, M. R. Okos, editor. ASAE, St. Joseph, Michigan.

C.8 Thermal Properties of Air at Atmospheric Pressure

Temp K	Density ρ kg/m^3	Specific Heat c_p kJ/kg·K	Viscosity $\mu \times 10^5$ kg/m·s	Thermal Conductivity k W/m·K	Thermal Diffusivity $\alpha \times 10^5$ m^2/s	Prandtl Number Pr
200	1.7690	1.0064	1.3286	0.01809	1.0161	0.739
250	1.4133	1.0054	1.5992	0.02227	1.5673	0.722
260	1.3587	1.0054	1.6504	0.02308	1.6896	0.719
270	1.3082	1.0055	1.7005	0.02388	1.8154	0.716
280	1.2614	1.0057	1.7504	0.02467	1.9447	0.714
290	1.2177	1.0060	1.7985	0.02547	2.0792	0.710
300	1.1769	1.0063	1.8465	0.02624	2.2156	0.708
310	1.1389	1.0068	1.8929	0.02701	2.3556	0.705
320	1.1032	1.0073	1.9392	0.02779	2.5008	0.703
330	1.0697	1.0079	1.9855	0.02853	2.6462	0.701
340	1.0382	1.0085	2.0302	0.02928	2.7965	0.699
350	1.0086	1.0092	2.0748	0.03003	2.9503	0.697
360	0.9805	1.0100	2.1177	0.03078	3.1081	0.695
370	0.9539	1.0109	2.1606	0.03150	3.2666	0.693
380	0.9288	1.0120	2.2018	0.03223	3.4289	0.691
390	0.9050	1.0130	2.2447	0.03295	3.5942	0.690
400	0.8822	1.0142	2.2859	0.03365	3.7609	0.689
450	0.7842	1.0212	2.4849	0.03710	4.6327	0.684
500	0.7057	1.0300	2.6703	0.04041	5.5594	0.681

Adapted from *Tables of Thermal Properties of Gases*, National Bureau of Standards Circular 564, Washington, D.C. (1955).

C.9 Psychrometric Chart

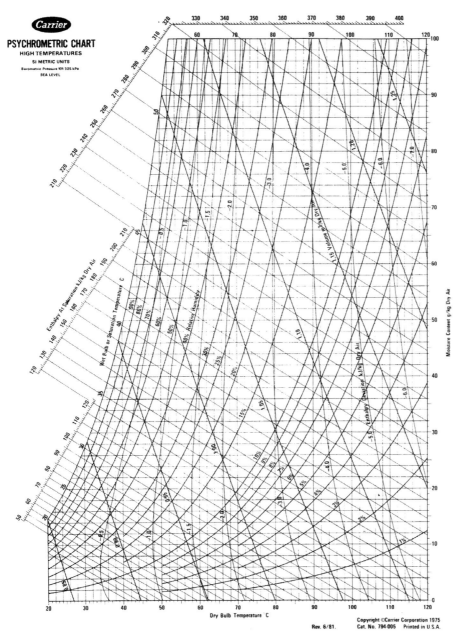

Courtsey of Carrier Corporation.

C.10 Vapor Pressure of Liquid Water from 0 to 100°C

Temp °C	Vapor Pressure mm Hg	$\times 10^{-5}$ N/m^2
0	4.58	0.00611
5	6.54	0.00872
10	9.21	0.01228
15	12.79	0.01705
20	17.54	0.02338
25	23.76	0.03168
30	31.82	0.04242
35	42.18	0.05624
40	55.32	0.07375
45	71.88	0.09583
50	92.51	0.12334
60	149.4	0.19918
70	233.7	0.31157
80	355.1	0.47343
90	525.8	0.70101
95	633.9	0.84513
100	760.0	1.01325

C.11 Steam Properties at Saturation Temperatures

Temp °C	Pressure bar	Density Liquid kg/m³	Density Vapor kg/m³	Sp. enthalpy Liquid kJ/kg	Sp. enthalpy Vapor kJ/kg	Heat of vaporization kJ/kg
0.01	0.006108	999.80	0.004847	0.00	2501	2501
1	0.006566	999.90	0.005192	4.22	2502	2498
2	0.007054	999.90	0.005559	8.42	2504	2496
3	0.007575	999.90	0.005945	12.63	2506	2493
4	0.008129	999.90	0.006357	16.84	2508	2491
5	0.008719	999.90	0.006793	21.05	2510	2489
6	0.009347	999.90	0.007257	25.25	2512	2489
7	0.010013	999.90	0.007746	29.45	2514	2485
8	0.010721	999.80	0.008264	33.55	2516	2482
9	0.011473	999.70	0.008818	37.85	2517	2479
10	0.012277	999.60	0.009398	42.04	2519	2477
15	0.017041	999.00	0.01282	62.97	2528	2465
20	0.02337	998.20	0.01729	83.90	2537	2454
25	0.03166	997.01	0.02304	104.81	2547	2442
30	0.04241	995.62	0.03037	125.71	2556	2430
35	0.05622	993.94	0.03962	146.60	2565	2418
40	0.07375	992.16	0.05115	167.50	2574	2406
45	0.09584	990.20	0.06544	188.40	2582	2394
50	0.12335	988.04	0.08306	209.3	2592	2383
55	0.15740	985.71	0.1044	230.2	2600	2370
60	0.19917	983.19	0.1302	251.1	2609	2358
65	0.2501	980.49	0.1613	272.1	2617	2345
70	0.3117	977.71	0.1982	293.0	2626	2333
75	0.3855	974.85	0.2420	314.0	2635	2321
80	0.4736	971.82	0.2934	334.9	2643	2308
85	0.5781	968.62	0.3536	355.9	2651	2295
90	0.7011	965.34	0.4235	377.0	2659	2282
95	0.8451	961.91	0.5045	398.0	2668	2270
100	1.0131	958.31	0.5977	419.1	2676	2257
110	1.4326	951.02	0.8264	461.3	2691	2230
120	1.9854	943.13	1.121	503.7	2706	2202
130	2.7011	934.84	1.496	546.3	2721	2174
140	3.614	926.10	1.966	589.0	2734	2145
150	4.760	916.93	2.547	632.2	2746	2114
200	15.551	864.68	7.862	852.4	2793	1941
250	39.776	799.23	19.28	1085.7	2801	1715
300	85.92	712.45	46.21	1344.9	2749	1404.3
374.15	221.297	306.75	306.75	2100	2100	0.0

Adapted from Irvine, T.F., Jr. and J.P. Hartnett. 1976. *Steam and Air Tables in SI Units.* Hemisphere Publishing Co., Washington.

C.12 Thermophysical Properties of Saturated Water

Temp K	Pressure $P \times 10^{-5}$ Pa	Specific Heat c_p kJ/kg·K	Viscosity N·s/m² $\mu \times 10^6$	Thermal Conduc. k W/m·K	Prandtl Number Pr	Expansion Coefficient $\beta \times 10^6$ K^{-1}
273.15	0.00611	4.217	1750	0.569	12.99	-68.05
275	0.00697	4.211	1652	0.574	12.22	-32.74
280	0.00990	4.198	1422	0.582	10.26	46.04
285	0.01387	4.189	1225	0.590	8.81	114.1
290	0.01917	4.184	1080	0.598	7.56	174.0
295	0.02617	4.181	959	0.606	6.62	227.5
300	0.03531	4.179	855	0.613	5.83	276.1
305	0.04712	4.178	769	0.620	5.20	320.6
310	0.06221	4.178	695	0.628	4.62	361.9
315	0.08132	4.179	631	0.634	4.16	400.4
320	0.1053	4.180	577	0.640	3.77	436.7
325	0.1351	4.182	528	0.645	3.42	471.2
330	0.1719	4.184	489	0.650	3.15	504.0
335	0.2167	4.186	453	0.656	2.88	535.5
340	0.2713	4.188	420	0.660	2.66	566.0
345	0.3372	4.191	389	0.668	2.45	595.4
350	0.4163	4.195	365	0.668	2.29	624.2
355	0.5100	4.199	343	0.671	2.14	652.3
360	0.6209	4.203	324	0.674	2.02	697.9
365	0.7514	4.209	306	0.677	1.91	707.1
370	0.9040	4.214	289	0.679	1.80	728.7
373.15	1.0133	4.217	279	0.680	1.76	750.1
375	1.0815	4.220	274	0.681	1.70	761
380	1.2869	4.226	260	0.683	1.61	788

Adapted from Mills, A.F. 1995. *Basic Heat and Mass Transfer*. Irwin, Chicago.

Appendix D

MASS TRANSFER PROPERTIES

D.1 Apparent Diffusivities in Solids

Solid	Diffusing material	Temp °C	Diffusivity $D \times 10^{11}$ m^2/s	Ref
Agar gel (0.79% solids)	Glucose	5	3.3	1
Cellulose acetate (5% moisture)	water	25	0.20	2
Polyethylene	Oxygen	30	6	1
Rubber	Oxygen	25	21	1
Agarose gel (2% solids)	NaCl	25	14	1
Meat muscle (fresh)	NaCl	2	22	3
Meat muscle (frozen-thawed)	NaCl	2	39	3
Starch gel (60% moisture dry basis)	water	30	15	4
Apple (air dried, 12% moisture db)	water	30	0.65	5
Apple (freeze dried, 12% moisture db)	water	30	12	5
soil, clay (water content 0.45)	chloride ion		57.87	
soil, clay (water content 0.25)	chloride ion		18.52	
soil, loam (water content 0.25)	chloride ion		40.51	
soil, silt (water content 0.248)	chloride ion		31.83	
soil, sand (water content 0.402)	chloride ion		138.89	
soil, sand (water content 0.168)	chloride ion		4.86	

[1] Geankoplis (1992).
[2] Roussis (1981).
[3] Dussap and Gros (1980).
[4] Saravacos and Raouzeos (1984); Saravacos (1967).
[5] Saravacos (1967).

D.2 Apparent Diffusivities in Liquids

Liquid	Diffusing material	Temperature °C	Diffusivity $D \times 10^9$ m^2/s	Reference
Water	acetic acid (1 kmol/m^3)	18	0.96	5
Water	chlorine (0.12 kmol/m^3)	16	1.26	5
Water	hydrogen chloride (0.5 kmol/m^3)	16	2.44	5
Water (dilute)	caffeine	25	0.63	3
Water (dilute)	carbon dioxide	25	2.00	1
Water (dilute)	ethanol	25	1.28	2
Water (dilute)	oxygen	25	2.41	1
Water (dilute)	sodium chloride	25	1.61	4
Water (dilute)	soybean protein	25	0.03	1
Water (dilute)	sucrose	25	0.56	3
Water (dilute)	urea	25	1.37	3

[1] Geankoplis (1992).
[2] Loncin and Merson (1979).
[3] Perry and Chilton (1973).
[4] Schwartzberg and Chao (1982).
[5] Welty et al. (1984).

D.3 Diffusivities in Air at 1 atm (1.013×10^5 Pa)

Temp K	Diffusivity, $D \times 10^4$ m²/s				
	H_2O	O_2	CO_2	CO	SO_2
200	0.1095	0.095	0.074	0.098	0.058
300	0.2538	0.188	0.157	0.202	0.126
400	0.4606	0.325	0.263	0.332	0.214
500	0.6983	0.475	0.385	0.485	0.326
600	0.9403	0.646	0.537	0.659	0.440
700	1.2093	0.838	0.684	0.854	0.576
800	1.5037	1.05	0.857	1.06	0.724

Adapted from Mills (1995).

D.4 Representative Values of Resistances to Water Vapor Transport Out of Leaves

Component Condition	Conductance $\times 10^3$ m/s
Boundary layer	
thin	80
thick	8
Stomata	
large area–open	19
small area–open	1.7
closed	0
mesophytes–open	4 - 20
xerophytes and trees–open	1 - 4
Cuticle	
crops	0.1 - 0.4
many trees	0.05 - 0.2
many xerophytes	0.01 - 0.1
Intercellular air spaces	
calculation	24 - 240
typical	40 - 100
waxy layer (typical)	50 - 200
Leaf (lower surface)	
crops–open	2 - 10

The values are provided in terms of conductance values, which are inverse of resistances. See Section 12.1.1 for discussion on resistances and conductances. To calculate flux, concentration differences need to be in mol/m^3 and multiplied by the conductances. Adapted from Physicochemical and Environmental Plant Physiology by P.S. Nobel, Academic Press, San Diego, 1996.

D.5 Approximate Ranges of Dispersion Coefficients in Surface Water

Condition	Dispersion Coefficient, m^2/s
Compacted Sediment	$10^{-11} - 10^{-9}$
Bioturbated sediment	$10^{-9} - 10^{-8}$
Lakes-vertical	$10^{-6} - 10^{-3}$
Large rivers-lateral	$10^{-2} - 10^{-1}$
Large rivers-longitudinal	$10^{0} - 10^{2}$
Estuaries-longitudinal	$10^{2} - 10^{3}$

Source: Adapted from Environmental Modeling by Jerald L. Schnoor, 1996.

D.6 Approximate Ranges of Dispersion Coefficients in Porous Media

Disperion coefficients used in contaminant transport in porous media such as groundwater are mostly empirical, often obtained from tracer studies. The table below can be used to obtain some order of magnitude estimates. The dispersion coefficients are a strong function of length scale involved and the velocities. The longitudinal dispersion coefficient can be calculated from

$$E = \alpha u + D \qquad (D.1)$$

where E is the longitudinal dispersion coefficient, α is the factor depending on length scale, u is the longitudinal velocity in m/s and D is the molecular diffusion coefficient. This formula also shows how the molecular diffusion is added on to the mechanical dispersion, although, the molecular diffusion component is typically much smaller and can be ignored.

Condition	Scale, m	Average longitudinal α, m
Laboratory	< 1	0.001-0.01
Field, small scale	1-10	0.1-1.0
Field Large scale	10-100	25

Source: Adapted from Environmental Modeling by Jerald L. Schnoor, 1996.

D.7 Kinetic Parameters for Various First Order Rate Processes in Biological and Environmental Systems

Application	k'' 1/min	Temp. for k'', °C	Activation energy, kJ/mole
Food pasteurization and sterilization			
E. coli O157:H8 in ground beef	4.9 - 8.86	62.8	401
L. monocytogenes in milk	3.97 - 10.47	63.3	386
S. aureus in milk	2.56	60	224
C. Botulinum B in vegetable products	0.185 - 4.7	110	267 - 390
Thiamine (vitamin B_1) in pea puree	0.00933	121	111.5
Pesticide degradation in soil			
Aldicarb	1.604×10^{-5}		
Atrazine	0.802×10^{-5}		
Carboxin	6.875×10^{-5}		

Source for microorganisms in foods: Microorganisms in foods. Microbiological specifications of food pathogens/ICMSF. Book 5. Blackie Academic and Professional Publications, London, 1996 (cited in Kinetics of Microbial Inactivation for Alternative Processing Technologies. A report of the Institute of Food Technologists for the Food and Drug Administration of the U.S. Department of Health and Human Services). Thiamine data from Karel et al. (1975).

Source for pesticides: These values are to be used only as representative data since the actual rate constant depends on pH, temperature, soil type and crop. For a comprehensive database of pesticide properties, see the document via the internet "The ARS Pesticide Properties Database" at http://wizard.arsusda.gov/acsl/ppdb.html (last accessed on May 3, 2001).

D.8 Surface Tension of Water

Temperature T °C	Surface tension γ N/m
0	0.07550
10	0.07440
20	0.07288
30	0.07120
40	0.06948
50	0.06777
60	0.06607
70	0.06436
80	0.06269
90	0.06079
100	0.05891
110	0.05697
120	0.05496
130	0.05290
140	0.05079
150	0.04868
160	0.04651
170	0.04438
180	0.04219
190	0.04000
200	0.03777

Adapted from Vargaftik (1983).

D.9 References to Data on Heat and Mass Transfer and Related Properties

Air Force Cambridge Research Laboratories (U.S.). 1961. *Handbook of Geophysics*. The Macmillan Company, New York.

ARS. 2001. Pesticide Properties Database. Document on the internet (last accessed on May 3, 2001) at http://wizard.arsusda.gov/acsl/ppdb.html.

ASHRAE. 1997. *ASHRAE Handbook: Fundamentals*. American Society of Heating, Refrigeration and Air-Conditioning Engineers, Inc., Atlanta, GA.

Calvert, S. and H.M. Englund. 1984. *Handbook of Air Pollution Technology*. John Wiley & Sons, New York, 1984.

Cena, K. and J.A. Clark. 1979. Transfer of heat through animal coats and clothing. International Review of Physiology, Environmental Physiology III, Vol. 20, pp. 1-42, Edited by D. Robertshaw, University Park Press, Baltimore.

Construction of a database of physical properties of foods. 2000. EU Project ERB FAIR CT96-1063. Document on the internet, http://www.nel.uk/fooddb/ (last accessed June 19).

Chato, J.C. 1985. Selected thermophysical properties of biological materials. 2:413-418. In: *Heat Transfer in Medicine and Biology* edited by A. Shitzer and R.C. Eberhart. Plenum Press, New York.

Choi, Y. 1985. Food thermal property prediction as affected by temperature and composition. PhD Thesis, Purdue University.

Dussap, G. and J.B. Gros. 1980. Diffusion-sorption model for the penetration of salt in pork and beef muscle. In: *Food Process Engineering* edited by P. Linko, Y. Malkki, J. Olkku, and J. Larinkari. Applied Science Publishers, London.

Foust, A.S., L.A. Wenzel, C.W. Clump, L. Maus, and L.B. Anderson. 1960. *Principles of Unit Operations*. John Wiley & Sons, Inc.

Geankoplis, C.J. 1992. *Transport Processes and Unit Operations*. Prentice Hall, New Jersey.

Greenberg, L.H. 1975. *Physics for Biology and Medical Students*. W. B. Saunders Company, Philadelphia.

Heldman, D.R. and R.P. Singh. 1986. Thermal properties of frozen foods. In: *Physical and Chemical Properties of Foods*, M. R. Okos, editor. ASAE, St. Joseph, Michigan.

Incropera, F.P. and D.P. Dewitt. 1990. *Introduction to Heat Transfer*. John Wiley & Sons, New York.

Irvine, T.F., Jr. and J.P. Hartnett. 1976. *Steam and Air Tables in SI Units*. Hemisphere Publishing Co., Washington.

Karel, M., O.R. Fennema, and D.B. Lund. 1975. *Physical Principles of Food Preservation*. Marcel Dekker, Inc., New York.

Loncin, M. and R.L. Merson. 1979. *Food Engineering*. Academic Press, New York.

Mills, A.F. 1995. *Basic Heat and Mass Transfer*. Irwin, Chicago.

O'Brien, W.J. 2000. Biomaterials Properties Database. Document on the internet, http://www.lib.umich.edu/dentlib/Dental_tables/toc.html. (last accessed June), Univ. of Michigan, Ann Arbor.

Perry, R.H. and C.H. Chilton. 1973. *Chemical Engineers Handbook*. McGraw-Hill, New York.

Rao M.A. and S.S.H. Rizvi. 1986. *Engineering Properties of Foods*. Marcel Dekker, Inc., New York.

Roussis, P.P. 1981. Diffusion of water vapor in cellulose acetate: 2. Permeation and integral sorption kinetics. Polymer, 22:1058-1063.

Schnoor, J.L. 1996. *Environmental Modeling: Fate and Transport of Pollutants in Water, Air and Soil*. John Wiley & Sons, New York.

Schwartzberg, H.G. and R.Y. Chao. 1982. Solute diffusivities in the leaching processes. Food Technology, 36(2):73-86.

Saravacos, G.D. 1967. Effect of drying method on the water sorption of dehydrated apple and potato. Journal of Food Science, 32:81-84.

Saravacos, G.D. and G.S. Raouzeos. 1984. Diffusivity of moisture in air drying of starch gels. In: *Engineering and Food*, 1:499-507. Edited by B.M. McKenna, Elsevier Applied Science Publishers, London.

Saravacos, G.D. and Z.B. Maroulis. 2001. *Transport Properties of Foods*. Marcel Dekker, New York.

Shitzer, A. and R.C. Eberhart. 1985. Heat generation, storage, and transport processes. 1:137-151. In: *Heat Transfer in Medicine and Biology* edited by A. Shitzer and R.C. Eberhart. Plenum Press, New York.

Singh, R.P. 1995. Food Properties Database 2.0 Windows. CRC press, Boca Raton, Florida.

Vargaftik, N.B. 1983. *Handbook of Physical Properties of Liquids and Gases*. Hemisphere Publishing Corporation, Washington, DC.

Welty, J.R., C.E. Wicks, and R.E. Wilson. 1984. *Fundamentals of Momentum, Heat, and Mass Transfer*. John Wiley & Sons, New York.

Appendix E

MISCELLANEOUS ENVIRONMENTAL DATA

E.1 Atmospheric Temperature, Pressure, and Other Parameters as Function of Altitude

Altitude m	Temp K	Pressure mb	Gravity m s^{-2}	Density kg m^{-3}	Mean Free Path m
-4000	314.18	1.5960×10^3	9.8190	1.7698	4.5904×10^{-8}
-3000	307.67	1.4297×10^3	9.8159	1.6189	5.0181×10^{-8}
-2000	301.16	1.2778×10^3	9.8128	1.4782	5.4959×10^{-8}
-1000	294.66	1.1393×10^3	9.8097	1.3470	6.0310×10^{-8}
0	288.16	1.01325×10^3	9.8067	1.2250	6.6317×10^{-8}
1000	281.66	8.9876×10^2	9.8036	1.1117	7.3079×10^{-8}
2000	275.16	7.9501×10^2	9.8005	1.0066	8.0710×10^{-8}
3000	268.67	7.0121×10^2	9.7974	9.0926×10^{-1}	8.9347×10^{-8}
4000	262.18	6.1660×10^2	9.7943	8.1935×10^{-1}	9.9151×10^{-8}
5000	255.69	5.4048×10^2	9.7912	7.3643×10^{-1}	1.1032×10^{-7}
6000	249.20	4.7217×10^2	9.7882	6.6011×10^{-1}	1.2307×10^{-7}
7000	242.71	4.1105×10^2	9.7851	5.9002×10^{-1}	1.3769×10^{-7}
8000	236.23	3.5651×10^2	9.7820	5.2578×10^{-1}	1.5451×10^{-7}
9000	229.74	3.0800×10^2	9.7789	4.6706×10^{-1}	1.7394×10^{-7}
10,000	223.26	2.6500×10^2	9.7759	4.1351×10^{-1}	1.9646×10^{-7}
20,000	216.66	5.5293×10^1	9.7452	8.8909×10^{-2}	9.1374×10^{-7}
40,000	260.91	2.9977	9.6844	4.0028×10^{-3}	2.0296×10^{-5}
60,000	253.68	2.5657×10^{-1}	9.6241	3.5235×10^{-4}	2.3056×10^{-4}
80,000	165.7	1.008×10^{-2}	9.564	2.120×10^{-5}	3.831×10^{-3}
100,000	199.0	2.138×10^{-4}	9.505	3.743×10^{-7}	2.171×10^{-1}

Adapted from *Handbook of Geophysics*, The Macmillan Company, New York, 1961.

E.2 National (U.S.) Primary Ambient Air Quality Standards

Pollutant	Averaging Time	Primary Standards[a]
Sulfur oxides	Annual arithmetic mean	80 μg/m^3 (0.03 ppm)
	24 hr	365 μg/m^3 (0.14 ppm)
Particulate matter	Annual geometric mean	75 μg/m^3
	24 hr	260 μg/m^3
Carbon monoxide	8 hr	10 mg/m^3
	1 hr	40 mg/m^3 (35 ppm)
Ozone (corrected for NO$_2$ and SO$_2$)	1 hr	240 μg/m^3 (0.12 ppm)
Hydrocarbons (corrected for methane)	3 hr	160 μg/m^3 (0.24 ppm)
Nitrogen oxides	Annual arithmetic mean	100 μg/m^3 (0.05 ppm)
Lead	3 months	1.5 μg/m^3
Ozone	1 hr	235 μg/m^3 (0.12 ppm)

[a] Except for annual means, standards are not to be exceeded more than once a year.

Adapted from *Handbook of Air Pollution Technology*, Calvert, S. and H.M. Englund, John Wiley & Sons, New York, 1984.

Appendix F

EQUATIONS OF MOTION IN VARIOUS COORDINATE SYSTEMS

F.1 The Equation of Motion in Rectangular Coordinates (x, y, z)

x-component

$$\rho \left(\frac{\partial v_x}{\partial t} + v_x \frac{\partial v_x}{\partial x} + v_y \frac{\partial v_x}{\partial y} + v_z \frac{\partial v_x}{\partial z} \right) = -\frac{\partial p}{\partial x}$$
$$+ \mu \left(\frac{\partial^2 v_x}{\partial x^2} + \frac{\partial^2 v_x}{\partial y^2} + \frac{\partial^2 v_x}{\partial z^2} \right) + \rho g_x \quad \text{(F.1)}$$

y-component

$$\rho \left(\frac{\partial v_y}{\partial t} + v_x \frac{\partial v_y}{\partial x} + v_y \frac{\partial v_y}{\partial y} + v_z \frac{\partial v_y}{\partial z} \right) = -\frac{\partial p}{\partial y}$$
$$+ \mu \left(\frac{\partial^2 v_y}{\partial x^2} + \frac{\partial^2 v_y}{\partial y^2} + \frac{\partial^2 v_y}{\partial z^2} \right) + \rho g_y \quad \text{(F.2)}$$

z-component

$$\rho \left(\frac{\partial v_z}{\partial t} + v_x \frac{\partial v_z}{\partial x} + v_y \frac{\partial v_z}{\partial y} + v_z \frac{\partial v_z}{\partial z} \right) = -\frac{\partial p}{\partial z}$$
$$+ \mu \left(\frac{\partial^2 v_z}{\partial x^2} + \frac{\partial^2 v_z}{\partial y^2} + \frac{\partial^2 v_z}{\partial z^2} \right) + \rho g_z \quad \text{(F.3)}$$

F.2 The Equation of Motion in Cylindrical Coordinates (r, θ, z)

r-component

$$\rho\left(\frac{\partial v_r}{\partial t} + v_r\frac{\partial v_r}{\partial r} + \frac{v_\theta}{r}\frac{\partial v_r}{\partial \theta} - \frac{v_\theta^2}{r} + v_z\frac{\partial v_r}{\partial z}\right) = -\frac{\partial p}{\partial r}$$

$$+\mu\left(\frac{\partial}{\partial r}\left(\frac{1}{r}\frac{\partial}{\partial r}(rv_r)\right) + \frac{1}{r^2}\frac{\partial^2 v_r}{\partial \theta^2} - \frac{2}{r^2}\frac{\partial v_\theta}{\partial \theta} + \frac{\partial^2 v_r}{\partial z^2}\right) + \rho g_r \quad (F.4)$$

θ-component

$$\rho\left(\frac{\partial v_\theta}{\partial t} + v_r\frac{\partial v_\theta}{\partial r} + \frac{v_\theta}{r}\frac{\partial v_\theta}{\partial \theta} + \frac{v_r v_\theta}{r} + v_z\frac{\partial v_\theta}{\partial z}\right) = -\frac{1}{r}\frac{\partial p}{\partial \theta}$$

$$+\mu\left(\frac{\partial}{\partial r}\left(\frac{1}{r}\frac{\partial}{\partial r}(rv_\theta)\right) + \frac{1}{r^2}\frac{\partial^2 v_\theta}{\partial \theta^2} + \frac{2}{r^2}\frac{\partial v_r}{\partial \theta} + \frac{\partial^2 v_\theta}{\partial z^2}\right) + \rho g_\theta \quad (F.5)$$

z-component

$$\rho\left(\frac{\partial v_z}{\partial t} + v_r\frac{\partial v_z}{\partial r} + \frac{v_\theta}{r}\frac{\partial v_z}{\partial \theta} + v_z\frac{\partial v_z}{\partial z}\right) = -\frac{\partial p}{\partial z}$$

$$+\mu\left(\frac{1}{r}\frac{\partial}{\partial r}(r\frac{\partial v_z}{\partial r}) + \frac{1}{r^2}\frac{\partial^2 v_z}{\partial \theta^2} + \frac{\partial^2 v_z}{\partial z^2}\right) + \rho g_z \quad (F.6)$$

F.3 The Equation of Motion in Spherical Coordinates (r, θ, ϕ)

In the following equations,

$$\nabla^2 = \frac{1}{r^2}\frac{\partial}{\partial r}\left(r^2\frac{\partial}{\partial r}\right) + \frac{1}{r^2\sin\theta}\frac{\partial}{\partial \theta}\left(\sin\theta\frac{\partial}{\partial \theta}\right) + \frac{1}{r^2\sin^2\theta}\left(\frac{\partial^2}{\partial \phi^2}\right) \quad (F.7)$$

The three equations of motion are:

r-component

$$\rho\left(\frac{\partial v_r}{\partial t} + v_r\frac{\partial v_r}{\partial r} + \frac{v_\theta}{r}\frac{\partial v_r}{\partial \theta} + \frac{v_\phi}{r\sin\theta}\frac{\partial v_r}{\partial \phi} - \frac{v_\theta^2 + v_\phi^2}{r}\right)$$

$$= -\frac{\partial p}{\partial r} + \mu\left(\nabla^2 v_r - \frac{2}{r^2}v_r - \frac{2}{r^2}\frac{\partial v_\theta}{\partial \theta} - \frac{2}{r^2}v_\theta\cot\theta\right.$$

$$\left. - \frac{2}{r^2\sin\theta}\frac{\partial v_\phi}{\partial \phi}\right) + \rho g_r \quad (F.8)$$

F.3. THE EQUATION OF MOTION IN SPHERICAL COORDINATES (R, θ, ϕ)

θ-component

$$\rho\left(\frac{\partial v_\theta}{\partial t} + v_r\frac{\partial v_\theta}{\partial r} + \frac{v_\theta}{r}\frac{\partial v_\theta}{\partial \theta} + \frac{v_\phi}{r\sin\theta}\frac{\partial v_\theta}{\partial \phi} + \frac{v_r v_\theta}{r} - \frac{v_\phi^2\cot\theta}{r}\right)$$
$$= -\frac{1}{r}\frac{\partial p}{\partial \theta} + \mu\left(\nabla^2 v_\theta + \frac{2}{r^2}\frac{\partial v_r}{\partial \theta} - \frac{v_\theta}{r^2\sin^2\theta} - \frac{2\cos\theta}{r^2\sin^2\theta}\frac{\partial v_\phi}{\partial \phi}\right) + \rho g_\theta \quad \text{(F.9)}$$

ϕ-component

$$\rho\left(\frac{\partial v_\phi}{\partial t} + v_r\frac{\partial v_\phi}{\partial r} + \frac{v_\theta}{r}\frac{\partial v_\phi}{\partial \theta} + \frac{v_\phi}{r\sin\theta}\frac{\partial v_\phi}{\partial \phi} + \frac{v_\phi v_r}{r} + \frac{v_\theta v_\phi}{r}\cot\theta\right)$$
$$= -\frac{1}{r\sin\theta}\frac{\partial p}{\partial \phi} + \mu\left(\nabla^2 v_\phi - \frac{v_\phi}{r^2\sin^2\theta} + \frac{2}{r^2\sin\theta}\frac{\partial v_r}{\partial \phi}\right.$$
$$\left.+ \frac{2\cos\theta}{r^2\sin^2\theta}\frac{\partial v_\theta}{\partial \phi}\right) + \rho g_\phi \quad \text{(F.10)}$$

Appendix G

SOME USEFUL MATHEMATICAL BACKGROUND

G.1 Series Solution to the One-Dimensional Heat Equation

In this section, the detailed steps in obtaining a series solution to the diffusion equation (Eqn. 5.9 on page 76 for heat diffusion and Eqn. 13.9 on page 258 for mass diffusion) is described. To simplify solution, temperature T in Eqn. 5.9 is transformed using a non-dimensional temperature θ defined as:

$$\theta = \frac{T - T_1}{T_0 - T_1} \tag{G.1}$$

Using this non-dimensional temperature θ, the derivatives in Eqn. 4.2 are transformed as:

$$\frac{\partial \theta}{\partial t} = \frac{1}{T_0 - T_1} \frac{\partial T}{\partial t}$$
$$\frac{\partial^2 \theta}{\partial x^2} = \frac{1}{T_0 - T_1} \frac{\partial^2 T}{\partial x^2}$$

Substituting these derivatives back in Eqn. 5.9, the transformed governing equation and boundary conditions are:

$$\text{G.E.} \quad \frac{\partial \theta}{\partial t} = \alpha \frac{\partial^2 \theta}{\partial x^2} \tag{G.2}$$

B.C. $\left.\dfrac{\partial \theta}{\partial x}\right|_{x=0, t>0} = 0$ (G.3)

$\theta(L, t > 0) = 0$ (G.4)

I.C. $\theta(x, t = 0) = 1$ (G.5)

To solve by separation of variables, first assume that the solution $\theta(x, t)$ can be written as product of a function in x and a function in t as:

$$\theta(x, t) = F(x)G(t)$$

Substituting into Eqn. G.2,

$$F(x)G'(t) = \alpha F''(x)G(t)$$

Separating functions in x and t as below, we note functions in two unrelated variables x and t can be equal only if each of them is equal to a constant. Let $-p^2$ be this negative constant. We will later find out that this constant has to be negative, therefore, we start with a negative value to avoid difficulty later.

$$\dfrac{F''(x)}{F(x)} = \dfrac{G'(t)}{\alpha G(t)} = -p^2 \qquad (G.6)$$

The two equations from Eqn. G.6 are:

$$F''(x) + p^2 F(x) = 0 \qquad (G.7)$$
$$G'(t) + \alpha p^2 G(t) = 0 \qquad (G.8)$$

A general solution to Eqn. G.7 for F is

$$F(x) = A \sin(px) + B \cos(px)$$

To calculate the coefficients A and B, we need boundary conditions for F. The first boundary condition is derived from Eqn. G.3 as

$$\theta'(0, t) = 0$$
$$F'(0)G(t) = 0$$

This leads to $F'(0) = 0$ at the centerline ($x = 0$). At the surface,

$$\theta(L, t) = 0$$
$$F(L)G(t) = 0$$

which leads to $F(L) = 0$ on the surface. Using the first condition $F'(0) = 0$,

G.1. SERIES SOLUTION TO THE ONE-DIMENSIONAL HEAT EQUATION

$$Ap\cos(px)|_{x=0} - Bp\sin(px)|_{x=0} = 0$$
$$-Ap = 0$$
$$A = 0$$

Using the second condition $F(L) = 0$,

$$B\cos(px)|_{x=L} = 0$$
$$B\cos pL = 0$$

Now B cannot be equal to zero simply because if it is so, we get zero as the trivial solution. Thus,

$$\cos(pL) = 0$$
$$pL = (2n+1)\frac{\pi}{2} \quad \text{where } n = 0, 1, 2, \ldots$$
$$p = (2n+1)\frac{\pi}{2L}$$

Therefore,

$$F_n(x) = B_n \cos\frac{(2n+1)\pi}{2L}x, \quad n = 0, 1, 2, \ldots \tag{G.9}$$

Now let us solve Eqn. G.8. Noting that p can take multiple values, we put a subscript to p to make it p_n and rewrite Eqn. G.8 as:

$$G'(t) + \alpha p_n^2 G(t) = 0$$

The general solution to this equation is given by:

$$G(t) = Ce^{-\alpha p_n^2 t}, \quad \text{where } p_n = (2n+1)\frac{\pi}{2L}$$

Denoting each solution of G (for each p_n) by G_n,

$$G_n(t) = C_n e^{-\alpha t\left(\frac{(2n+1)\pi}{2L}\right)^2}, \quad n = 0, 1, 2, \ldots \tag{G.10}$$

Combining Eqns. G.9 and G.10, we get the differential equation for θ as

$$\theta_n(x,t) = F_n(x)G_n(t)$$
$$= D_n \cos\frac{(2n+1)\pi x}{2L} e^{-\alpha t\left(\frac{(2n+1)\pi}{2L}\right)^2} \tag{G.11}$$

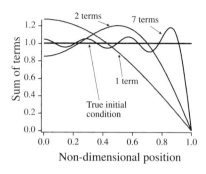

Figure G.1: As more terms are added to the series in Eqn. G.12, the summation can better approximate the initial condition of constant temperature.

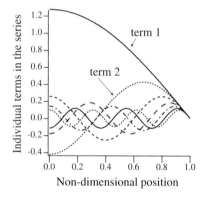

Figure G.2: Relative magnitudes of individual terms in the series solution.

This satisfies the governing equation and the boundary conditions for θ. However, it does not satisfy the initial condition given by Eqn. G.5. Also, $D_n = B_n C_n$ is an unknown in the above equation. Thus, we have an unknown and we have not used one of the conditions. We now use the initial condition to find D_n. Let us consider a solution of the form:

$$\begin{aligned}\theta(x,t) &= \sum_{n=0}^{\infty} \theta_n(x,t) \\ &= \sum_{n=0}^{\infty} D_n \cos\frac{(2n+1)\pi x}{2L} e^{-\alpha t \left(\frac{(2n+1)\pi}{2L}\right)^2}\end{aligned} \quad \text{(G.12)}$$

Why this step? The initial condition looks like a square wave. Our initial solution (Eqn. G.12) gave us a cosine wave! Thus, we will add infinite number of cosine waves to get the square wave as shown in Figure G.1. Note that although we add an infinite number of terms, the magnitude of each term drops quickly, as shown in Fig. G.2

Using the initial condition $\theta(x,0) = 1$ in Eqn. G.12, we get

$$\sum_{n=0}^{\infty} D_n \cos\frac{(2n+1)\pi}{2L}x = 1$$

To solve for D_n, both sides of the above equation are multiplied by $\cos(2n+1)\pi\frac{x}{2L}$ and integrated to obtain:

$$\begin{aligned}D_n &= \frac{2}{L}\int_0^L 1\cdot\cos\frac{(2n+1)\pi}{2L}xdx \quad \text{by orthogonality} \\ &= \frac{2}{L}\frac{2L}{(2n+1)\pi}\sin\frac{(2n+1)\pi x}{2L}\Big|_0^L \\ &= \frac{4}{(2n+1)\pi}\sin\frac{(2n+1)\pi}{2} \\ &= \frac{4}{(2n+1)\pi}(-1)^n, \quad n = 0, 1, 2, \ldots.\end{aligned}$$

Substituting D_n into Eqn. G.12 and transforming θ back to temperature T, we get the final solution of temperature T as a function of position x and time t:

$$\frac{T-T_1}{T_0-T_1} = \sum_{n=0}^{\infty} \frac{4(-1)^n}{(2n+1)\pi}\cos\frac{(2n+1)\pi x}{2L}e^{-\alpha\left(\frac{(2n+1)\pi}{2L}\right)^2 t} \quad \text{(G.13)}$$

G.2 Similarity Transformation of the Heat Equation

For certain situations, the heat equation, which is normally a partial differential equation in x and t, can be transformed to an ordinary differential equation. The final form of the equation developed here is used in Section 5.5.

$$\eta = \frac{x}{2\sqrt{\alpha t}}$$

$$t: \frac{\partial \eta}{\partial t} = \frac{x}{2\sqrt{\alpha}} \cdot t^{-3/2}(-1/2) = -\frac{1}{2t}\frac{x}{2\sqrt{\alpha t}}$$

$$\frac{\partial T}{\partial t} = \frac{\partial T}{\partial \eta} \cdot \frac{\partial \eta}{\partial t} = \frac{\partial T}{\partial \eta}\left(-\frac{1}{2t}\frac{x}{2\sqrt{\alpha t}}\right)$$

$$x: \frac{\partial \eta}{\partial x} = \frac{1}{2\sqrt{\alpha t}}$$

Therefore,

$$\frac{\partial T}{\partial x} = \frac{\partial T}{\partial \eta} \cdot \frac{\partial \eta}{\partial x} = \frac{\partial T}{\partial \eta} \cdot \frac{1}{2\sqrt{\alpha t}}$$

$$\frac{\partial^2 T}{\partial x^2} = \frac{\partial}{\partial x}\left(\frac{\partial T}{\partial \eta} \cdot \frac{1}{2\sqrt{\alpha t}}\right)$$

$$= \frac{\partial}{\partial \eta}\left(\frac{\partial T}{\partial \eta} \cdot \frac{1}{2\sqrt{\alpha t}}\right)\frac{\partial \eta}{\partial x}$$

$$= \frac{\partial^2 T}{\partial \eta^2} \cdot \frac{1}{2\sqrt{\alpha t}}\frac{1}{2\sqrt{\alpha t}} = \frac{1}{4\alpha t}\frac{\partial^2 T}{\partial \eta^2}$$

Substituting into the governing equation:

$$\frac{\partial T}{\partial t} = \alpha \frac{\partial^2 T}{\partial x^2}$$

$$-\frac{1}{2t} \cdot \frac{x}{2\sqrt{\alpha t}} \cdot \frac{dT}{d\eta} = \alpha \frac{d^2 T}{d\eta^2} \cdot \frac{1}{4\alpha t}$$

$$\frac{d^2 T}{d\eta^2} + \frac{x}{\sqrt{\alpha t}}\frac{dT}{d\eta} = 0$$

$$\frac{d^2 T}{d\eta^2} + 2\eta \frac{dT}{d\eta} = 0$$

Let

$$\theta = \frac{T - T_i}{T_s - T_i}$$

In terms of the new variable θ, the governing equation can be written as:

$$\frac{d^2 \theta}{d\eta^2} + 2\eta \frac{d\theta}{d\eta} = 0$$

The new boundary conditions (now only in η) can be derived from the old boundary conditions. Using $T(x \to \infty) = T_i$ and noting $\eta \to \infty$ when $x \to \infty$,

$$\theta(\eta \to \infty) = 0$$

Note that this condition also incorporates the initial condition $T(t = 0) = T_i$ since $\eta \to \infty$ when $t = 0$. Likewise, using $T(x = 0) = T_s$ and noting $\eta = 0$ when $x = 0$,

$$\theta(\eta = 0) = 1$$

Solution:

Let

$$\frac{d\theta}{d\eta} = Z$$

Substituting into the equation for θ, we get:

$$\frac{dZ}{d\eta} + 2\eta Z = 0$$

Integrating,

$$\int \frac{dZ}{Z} = \int -2\eta \, d\eta$$
$$\ln Z = -\eta^2 + c_1$$
$$Z = e^{-\eta^2 + c_1} = e^{c_1} \cdot e^{-\eta^2} = c_2 e^{-\eta^2}$$

Substituting for Z and integrating again,

$$\frac{d\theta}{d\eta} = c_2 e^{-\eta^2}$$
$$\int_1^\theta d\theta = \int_0^\eta c_2 e^{-\eta^2} d\eta$$
$$\theta - 1 = c_2 \int_0^\eta e^{-\eta^2} d\eta$$

where $\theta(\eta = 0) = 1$ has been used. Using initial condition, $\theta(\eta \to \infty) = 0$,

$$-1 = c_2 \cdot \frac{\sqrt{\pi}}{2} \quad \text{since} \quad \int_0^\infty e^{-\eta^2} d\eta = \frac{\sqrt{\pi}}{2}$$
$$c_2 = -\frac{2}{\sqrt{\pi}}$$

Substituting back

$$\begin{aligned} \theta &= 1 - \frac{2}{\sqrt{\pi}} \int_o^\eta e^{-\eta^2} d\eta \\ &= 1 - \mathrm{erf}(\eta) \\ &= 1 - \mathrm{erf}\left[\frac{x}{2\sqrt{\alpha t}}\right] \end{aligned}$$

where

$$\mathrm{erf}(\eta) = \frac{2}{\sqrt{\pi}} \int_o^\infty e^{-\eta^2} d\eta$$

Therefore, the solution is

$$\frac{T - T_i}{T_s - T_i} = 1 - \mathrm{erf}\left[\frac{x}{2\sqrt{\alpha t}}\right]$$

G.3 Lumped Parameter Analysis as Related to the Energy Equation

Here we show how the lumped parameter equation for energy (Eqn. 5.2 on page 71) is is in fact a variation of the general energy equation

$$\frac{\partial T}{\partial t} = \frac{k}{\rho c_p} \nabla^2 T$$

that we had derived (Eqn. 3.24 on page 41, in the absence of heat generation and convection). Integrating over the volume of the material:

$$\iiint \frac{\partial T}{\partial t} dV = \frac{k}{\rho c_p} \iiint \nabla^2 T \, dV$$

Using Green's Theorem, with n as the direction normal to surface

$$= \frac{k}{\rho c_p} \iint \frac{\partial T}{\partial n} dA$$

using the boundary condition $-k\frac{\partial T}{\partial n} = h(T - T_\infty)$

$$\begin{aligned} &= \frac{-h}{k} \frac{k}{\rho c_p} \iint (T - T_\infty) dA \\ &= -\frac{h}{\rho c_p} \iint (T - T_\infty) dA \end{aligned}$$

When the spatial variation of temperature is insignificant:

$$\frac{\partial T}{\partial t} \iiint dV = -\frac{h(T - T_\infty)}{\rho c_p} \iint dA$$

$$\frac{\partial T}{\partial t} V = -\frac{h(T - T_\infty)}{\rho c_p} A$$

$$\frac{\partial T}{\partial t} = \frac{hA}{\rho V c_p}(T - T_\infty)$$

$$= \frac{hA}{m c_p}(T - T_\infty)$$

which is the lumped parameter equation.

G.4 Transformation of the Convection-Diffusion Equation to the Diffusion Equation

The convection-diffusion equation

$$\frac{\partial c}{\partial t} + u\frac{\partial c}{\partial z} = D\frac{\partial^2 c}{\partial z^2} \tag{G.14}$$

was transformed to a diffusion equation and used in Section 14.2. To see how this transformation can be done, we define two new variables as:

$$x = z - ut$$
$$\eta = t$$

From these definitions, we can find the partial derivatives as

$$\frac{\partial x}{\partial t} = -u \qquad \frac{\partial \eta}{\partial t} = 1$$
$$\frac{\partial x}{\partial z} = 1 \qquad \frac{\partial \eta}{\partial z} = 0$$

Using the chain rule,

$$\frac{\partial c}{\partial t} = \frac{\partial c}{\partial x}\frac{\partial x}{\partial t} + \frac{\partial c}{\partial \eta}\frac{\partial \eta}{\partial t} = -u\frac{\partial c}{\partial x} + \frac{\partial c}{\partial \eta}$$

$$\frac{\partial c}{\partial z} = \frac{\partial c}{\partial x}\frac{\partial x}{\partial z} + \frac{\partial c}{\partial \eta}\frac{\partial \eta}{\partial z} = \frac{\partial c}{\partial x},$$

$$\frac{\partial^2 c}{\partial z^2} = \frac{\partial}{\partial z}\left(\frac{\partial c}{\partial z}\right) = \frac{\partial}{\partial x}\left(\frac{\partial c}{\partial z}\right)\frac{\partial x}{\partial z} + \frac{\partial}{\partial \eta}\left(\frac{\partial c}{\partial z}\right) \cdot \frac{\partial \eta}{\partial z}$$

$$= \frac{\partial}{\partial x}\left(\frac{\partial c}{\partial x}\right) \cdot 1 + \frac{\partial}{\partial \eta}\left(\frac{\partial c}{\partial x}\right) \cdot 0 = \frac{\partial^2 c}{\partial x^2}$$

Substituting in Eqn. G.14:

$$\frac{\partial c}{\partial t} + u\frac{\partial c}{\partial z} = -u\frac{\partial c}{\partial x} + \frac{\partial c}{\partial \eta} + u\frac{\partial c}{\partial x} = \frac{\partial c}{\partial \eta}$$

and using $\eta = t$, we get the diffusion equation:

$$\frac{\partial c}{\partial t} = D\frac{\partial^2 c}{\partial x^2}$$

G.5 Solution to Steady One-Dimensional Diffusion in a Slab with Chemical Reaction

The governing equation for steady one-dimensional diffusion with a first order reaction (Eqn. 12.10 on page 244) is

$$\frac{d^2 c_A}{dx^2} - m^2 c_A = 0$$

The transformed boundary conditions are:

$$c_A(x=0) = c_{A0} \qquad (G.15)$$
$$c_A(x=L) = 0 \qquad (G.16)$$

We assume a solution of the form

$$c_A = c_1 e^{-mx} + c_2 e^{mx} \qquad (G.17)$$

where c_1 and c_2 are constants to be determined using the boundary conditions. Substituting the boundary conditions, we get

$$c_{A0} = c_1 + c_2$$
$$0 = c_1 e^{-mL} + c_2 e^{mL}$$

Solving for c_1 and c_2:

$$c_1 = -c_2 \, e^{-2mL} \qquad c_2 \left(1 - e^{-2mL}\right) = c_{A,0}$$

$$c_2 = \frac{c_{A,0}}{1 - e^{-2mL}} \qquad c_1 = \frac{-c_{A,0} \, e^{-2mL}}{1 - e^{-2mL}}$$

Substituting into Eqn. G.17:

$$c_A = c_{A,0} \left[\frac{-e^{-2mL}}{1 - e^{-2mL}} \cdot e^{mx} + \frac{1}{1 - e^{-2mL}} \cdot e^{-mx} \right]$$

$$= \frac{c_{A,0}}{e^{mL} - e^{-mL}} \left[-e^{-mL} \cdot e^{mx} + e^{mL} \cdot e^{-mx} \right] \qquad (G.18)$$

Rearranging:

$$\frac{c_A}{c_{A,0}} = \frac{-e^{-mL} \left(e^{mx} - e^{-mx}\right) + e^{-mx} \left(e^{mL} - e^{-mL}\right)}{e^{mL} - e^{-mL}}$$

$$\frac{c_A}{c_{A,0}} = \frac{-e^{-mL}}{e^{mL} - e^{-mL}} \cdot \left(e^{mx} - e^{-mx}\right) + e^{-mx} \qquad (G.19)$$

INDEX

agar gel, 271
agitation in liquids to enhance heat transfer, 89, 121
agriculture, *see* plant, soil, water, industrial processing
air
    pollution, 213
    properties, 340
air-vapor mixture, *see* psychrometric chart
albedo, *see* soil
animals
    basal metabolic rate, 332
    fur thickness and insulation, 53
    oxygen diffusion in meat tissue, 235
    thermal conductivities of hair coats, 337
    thermal properties of animal materials, 335
Arrhenius's law, *see* kinetics
arteries and veins
    heat transfer in, 36
atmosphere
    greenhouse effect, 149
    ozone layer absorption, 158
    radiation from, 159
    solar radiation
        extraterrestrial, 158
        on earth surface, 158
    transmissivity, 149
atmospheric air
    properties, 358
    quality standards, 359
atmospheric pressure (normal), 321
Average (spatial) temperature
    defined, 78
Avogadro's Number, 321

bacterial clump
    sterilization of, 88
balance
    energy, 5
    species mass, 179
Beer-Lambert law, *see* radiation
biofilm, 239
bioheat equation, 36
biomaterials
    anisotropy in, *see* wood
    as capillary porous materials, 208
    emissivity, 156
Biot number, *see* resistance, 105
blood
    flow rates in human organs, 334
    heat transfer due to, 36
    thermal properties, 335
boiling, 135, 138
boiling point elevation, 138
Boltzmann's constant, 152, 321
boundary conditions (heat)
    surface convection, 34
    surface heat flux, 34
    surface insulated (zero heat flux), 35
    surface temperature, 34
    symmetry, 35
boundary conditions (mass)
    surface concentration, 230
    surface convection, 232
    surface impermeable (zero mass flux), 231
    surface mass flux, 231
    symmetry (zero mass flux), 232
boundary layer thickness
    concentration, 297
    thermal, 97

velocity, 98
Brownian motion, *see* diffusion (mass)

cell
    dehydration in freezing, 129
    semi-permeable membrane in, 129
    water transport in freezing, 129
cell membrane
    as a porous material, 209
characteristic length
    heat transfer, 73
    mass transfer, 257
chlorophyl
    energy absorption in, 148
circulatory system, 36
Clausius-Clapeyron equation, 184
clothing
    heat transfer through, 65
comfort
    humidity and, 9
    radiation and, 165
    temperature and, 9
composting
    heat transfer in, 67
concentration
    mass, 177
    perfect gases, 177
    volume, 177
conduction, 15
    derivation of G.E., 29
conduction (steady-state)
    bio-fin, 60
    cylinder(1D), 50
        comparison with slab, 52
    fins, 58
    slab(1D composite), 48
    slab(1D), 47
        with heat generation, 55
    thermoregulation and, *see* thermoregulation
conduction (transient), 69
    2D geometries, 82
    chart for long cylinder, 328
    chart for slab
        equation for, 78
    chart for slab (1D), 327
    chart for sphere, 329
    charts, limitations of, 79
    convective surface resistance included, 80
    derivation of solution for slab (1D), 365
    internal resistance neglected, 71
    lumped parameter, *see* internal resistance neglected
        alternate derivation from the general G.E., 371
        derivation of G.E., 71
        temperature change, 73
        when to use, 73
    semi-infinite region, 82
        derivation of solution, 369
        when to use, 83
    slab (1D), 75
        changes with position, 77
        changes with size, 78
        changes with time, 77
        general solution, 76
        spatial average, 77
conductivity, *see* porous media
conservation
    energy, 5
    species mass, 179
convection, 22
    boundary layer
        flat plate, 99
            thickness of thermal, 98
            thickness of velocity, 97
    boundary layers, 97
    derivation of G.E., 29, 97
    film coefficient, 22
    forced, 22
    free, 22
    heat transfer coefficeint, 22
    Newton's law, 22
convection (mass), 214, 277
    boundary layer
        flat plate, 295
            thickness of concentration, 297
    contrasted with advection, 217
    contrasted with diffusion, 217
    diffusion through stagnant gas, 291

dispersion and
    in an infinite fluid, 279
    in soil, 284
  film coefficient, 215
  governing equation, 279
  mass transfer coefficient, 215
conversion factors, 322
cooling, *see* evaporative cooling
cryosurgery, *see* mammalian

Darcy's law, *see* porous media
density
  air, 340
  bulk, 21
  steam, 343
dialysis, 243
dielectric property
  water during freezing, 127
diffusion, 17
  molecular, 203
diffusion (mass)
  Brownian motion, 204
  comparison with capillary flow, 216
  comparison with dispersion, 216
diffusion coefficient, 204
diffusion of heat (steady-state), 45
diffusion of mass (steady-state), 237
  linear profile in a slab (1D), 239
  slab (1D composite), 240
  slab (1D), 238
    solution for reaction, 373
    with reaction, 244
diffusion of mass (transient), 253
  2D geometries, 262
  charts, limitations of, 260
  convective surface resistance included, 261
  internal resistance neglected, 254
  lumped parameter, 254
    derivation of G.E., 254
    solution, 256
    when to use, 256
  semi-infinite region, 266
    when to use, 267
  slab (1D)
    average concentration, 270
    changes with position, 259
    changes with size, 260
    general solution, 257
    governing equation, 257
    spatial average, 259
  transformation of advection-diffusion G.E., 372
diffusivity (mass)
  defined, 204
  determination from experimental data, 271
  gases, 206
  in porous media, *see* porous media
  interpretation, 204
  liquids, 207, 347
  moisture in wood, 210
  myoglobin in water, 205
  proteins, 208
  solids, 346
    Knudsen, 208
    porous and non-porous, 207
    variation with concentration, 210
    variation with temperature, 210
  Stokes-Einstein relation, 207
  typical values, 206
dispersion, 213
  coefficient values
    porous media, 351
    surface water, 350
  analogy with diffusion, 213
  coefficient
    much higher than diffusivities, 214
  comparison with molecular diffusion, 217
  hydrodynamic, 213
  in porous media, 213
dispersion coefficient, 214
distribution coefficient, 185
  use in a composite slab, 241

earth
  energy balance, 160
  radiation from, *see* radiation
  surface average temperature, 159

eggs, in incubator, 74
elephant
    heat loss from, 61
    heat loss from pinna, 118
    pinna as a bio-fin, 60
emissivity, *see* radiation
equilibrium
    bends in diving, 181
    Henry's law, 180
    interphase, 179
        gas and liquid, 180
        gas and solid, 182
        solid and liquid, 185
    moisture content of solid, 182
    moisture isotherms, 182
    steady state and, 5, 45
    thermal, 5
error function
    tabulated, 325
evaporation
    from plant and soil surfaces, 137
    from wet surfaces, 136
    Penman's equation, 137
    boiling, 135
    from enclosed area, 294
    inside porous bodies, 138
    of solutions, 138
    process explained, 135
    vapor pressure, 136
evaporative cooling, 137
evapotranspiration, 137
    Penman's equation, 137
exercise
    heat transfer in, 68

Fick's law, 204
    modified by including convection, 224
film coefficient, *see* convection (mass)
fish
    oxygen availability in water, 193
    temperature equilibration, 66
flow
    laminar, 99
    turbulent, 99
flux

energy
    diffusion, 18
    through a wall, 19
    total, 226
mass
    diffusion only, 225
    total, 226
food, *see* industrial processing
    materials, specific heat, 339
    materials, thermal conductivity, 339
Fourier number, 77
Fourier's law, 18
    analogy to electricity, 49
free convection, *see* natural convection
freezing
    cryopreservation, 125
    eutectic, 129
    fast and cracking, 130
    fast vs. slow, 130
    foods, 125
    freezing point depression and, 128
    latent heat, 128
        biomaterial, 135
    mushy zone in a biomaterial, 133
    Plank's solution, 131
    time needed for a biomaterial, 133
    time needed for a cylinder, 142
    time needed for a slab, 131
    time needed for a sphere, 142
    tissues, *see* tissues
freezing point depression, 128
frost heaves, *see* water
Frozen biomaterial properties, *see* water
    blood, 336
    food, 339
fur
    thickness and insulation, 53

governing equation
    bioheat transfer, 36
    heat
        cartesian, 29, 41
        conduction term, 31
        convection term, 31
        cylindrical, 38, 41

generation term, 29, 31
source term, 31
spherical, 41, 233
storage term, 29
mass, 229
cartesian, 233
convection term, 229
cylindrical, 233
derived, 227
diffusion term, 229
generation term, 229
ignoring terms, 229
sink terms, 185
source terms, 185
storage term, 229
governing equations
of motion, 361
Grashof number, 105
gravitational acceleration (Sea Level), 321
greenhouse
radiation through glass, 171
greenhouse effect, 10, 149
CO_2 concentration and, 10
chlorofluorocarbons and, 149
gases in, 10
temperature rise in, 10
ground
cooling of, 93

half life, *see* kinetics
heat and mass transfer analogies, 300
basic parameters, 319
governing equations and boundary conditions, 320
heat equation, *see* governing equation
heat transfer (phase change), *see* freezing
heat transfer coefficient
air flow over humans, 112
as a temp. gradient, 100
cylinder
flow through tube, 104
forced convection, 104
horizontal, 104
natural convection, 104
vertical, 104
functional relationship, 104
laminar vs. turbulent, 99, 111
overall, 48
plate
forced convection, 104
horizontal, 104
natural convection, 104
vertical, 104
radiative, 163
significant parameters, 101
sphere
forced convection, 104
natural convection, 104
variation along flow, 105
Heisler chart, 327–329
alternatives, 81
hemodialysis, 243
Henderson's equation, 183
Henry's law, *see* equilibrium
human body
air flow over, 112
bends in diving, *see* equilibrium
blood oxygen, 192
diagnostics using emitted radiation, 156
dialysis, 243
metabolic rates for various organs, 334
oxygen diffusion in alveoli, 272
oxygen transport in cornea, 251
oxygen transport in tissue, 251
radiation emitted from, 156, 170
skin
diffusion through, 211
drug delivery through, 211, 272
thermal conductivity of parts, 335
thermal diffusivity of parts, 335
wind chill, 112
human organs
thermal properties, 335
hydraulic conductivity, *see* porous media
hygroscopic, 184

ice, *see* water

ice, thermal properties, *see* water
iceberg
    heat transfer in towing, 119
industrial processing
    biofilm mass transfer, 239
    compost drying, 271
    food freezing, 125
    food sterilization with ultrasound, 88
    infrared heating, 150
    microwave heating, 93, 150
    moisture transport through food packages, 251
    silage and bacterial growth, 267
    slowing down of ripening using CO_2, 264
isotherms, *see* equilibrium

kinetics, 185
    nth order, 189
    activation energy values, 352
    rate constant values, 352
    Arrhenius's law, 189
    first order, 187
    half life, 187
        pesticides, 188
    reaction rates, 186
    temperature effect, 189
    thermodynamics and, 185
    zeroth order, 186
Knudsen diffusion, *see* diffusivity (mass)

laminar flow, 99
latent heat
    of fusion, 128, 131
living system
    cryopreservation and, 125
lizards
    water loss from eye, 310
lumped parameter, *see* conduction (transient), *see* diffusion of mass (transient)

mammalian
    cooling of heart, 91
    cryosurgery, 125

metabolic rates for activities, 333
    skin burn injury, 85
    skin emissivity, 156
    skin section, 85
    thermal comfort and radiation, 165
mammalian system
    pathogens through teat canal, 274
mass and heat transfer analogies, 300
    basic parameters, 319
    governing equations and boundary conditions, 320
mass balance, 179
mass transfer
    significant parameters, 300
mass transfer coefficient
    as a concentration gradient, 298
    overall, 240
    table, *see* heat transfer coefficient
mean square displacement, *see* diffusivity
meat
    enthalpy-temperature in freezing, 134
membrane
    transport, 209
metabolic rates
    animals, 332
    human activities, 333
    human organs, 334
microorganism
    oxygen transport and boundary layer, 297
    oxygen uptake, 302
modes
    mass transfer
        capillary, 201
        convection, 214
        diffusion, 203
        dispersion, 213
        osmotic, 202
modulus of elasticity
    biomaterial during freezing, 127
multi-dimensional heat transfer, 82

natural convection, *see* heat transfer coefficient

non-dimensional parameters
  equivalence, 302
Numerical methods, 81
Nusselt number, 104, 105
  mass transfer, 300

osmotic flow, 202
oxygen consumption
  human organs, 334

packaging material
  moisture transport through, 251
partition coefficient, 185
permeability, *see* porous media
pesticides, *see* kinetics
  health advisory level, 188
photosynthesis, 247
Planck's constant, 152, 321
Plank's solution, 131
plant
  aeration, 252
  diffusive resistances in stomata and cuticle, 349
  photosynthesis and radiation, 148
  radiative freezing, 163, 164
  root zone solute transport, 310
  stomatal transport of CO_2, 247
  transpiration, 247
  water rise due to capillarity, 221
plume, 213
poikilotherms, *see* thermoregulation
pollution
  groundwater, 289
  in a river, 282
Porosity
  defined, 21
porous media
  as bundle of tubes, 198
  capillary diffusivity, 211
    how different from molecular diffusivity, 213
  cell membrane, 209
  conductivity, 197
  Darcy velocity, 197
  Darcy's law, 197
  diffusivity, effective, 211
  dispersion in, 284
  hydraulic conductivity, 197
  matric potential, 197
  permeability, 198
  Pouseuille's flow, 198
  pressure potential, 197
  saturated, 201
  unsaturated, 201, 211
  wood, 208
Pouseuille's flow, *see* porous media
Prandtl, 98
Prandtl number, 105
psychrometric chart, 341

R-value, *see* resistance
radiation, 23, 145
  absorption and absorptivity, 147
  absorption in biomaterials, 150
  atmospheric gases, 159
  Beer-Lambert law, 150
  earth's emission of, 159
  electromagnetic spectrum, 147
  emissive power, 150
    monochromatic, 151
  emissivity of a surface, 154
  emissivity of materials, 156
  energy between two wavelengths, 153
  exchange between bodies, 160
  freezing of a leaf, 163, 164
  from human body, 156, 170
  global energy balance, 160
  heat transfer coefficient due to, 163
  ideal body (blackbody), 151
  infrared, 147
  microwave, 147
  penetration depths in biomaterials, 151
  Planck's law, 152
  radio frequency, 147
  reflection and reflectivity, 147
  solar, 158
  spectroscopy and, 156
  Stefan-Boltzmann constant, 153, 321
  Stefan-Boltzmann law, 23

table for different temperatures for a blackbody, 155
thermal, 147
thermal comfort and, 165
transmission and transmissivity, 147
view factor, 160
wavelength of peak emission, 152
Wien's law, 152
rate laws, 6
of dispersion, 213
of heat conduction, 18
Rayleigh number, 109
resistance
Biot number, 72
Biot number(mass), 256
electrical analogy, 49
external vs. internal, 72
R-value, 50
thermal, 49
Reynolds number, 105

saturation pressure of steam, 343
Schmidt number, 300
semi-infinite region
when to use, heat, 83
when to use, mass, 267
sensible heat, 131
Sherwood number, 300
size
heat transfer and, 78, 90
sky
radiation from, 159
soil
albedo (reflection of radiation), 149
emissivity, 156
equilibrium moisture, 183
flow through packed sand, 198
matric potential, 197, 202
moisture diffusivity, 202
nitrogen movement in, 273
warming from sun, 89
Solution procedure for transport problems, 42
specific heat
air, 340
apparent, 133

food materials, 339
water, 344
water during freezing, 127
speed of light in vacuum, 321
steam
properties, 343
Stefan-Boltzmann constant, *see* radiation
sterilization
heat transfer in, 79
Stokes-Einstein relation, *see* diffusivity (mass)
stratum corneum, *see* skin
summary
boundary conditions, 316
heat and mass transfer analogies, 319, 320
solutions for convection, 318
solutions for diffusion, 317
transport equation, 316

temperature
animal comfort and, 9
greenhouse effect and, 10
human comfort and, 9
limits in living systems, 6
limits of human body, 7
wet bulb, 304
thawing
contrasted with freezing, 143
thermal conductivity
air, 340
animal body parts, 335
animal hair coats, 337
biomaterials, 20
clay, 338
concrete, 338
food materials, 339
gases, 17
liquids, 17
metallic solids, 17
metals, 338
non-metallic solids, 17
sand, 338
snow, 338
water, 344

water during freezing, 127
wood, 338
thermal diffusivity
    animal body parts, 335
    biomaterials, 21
thermodynamics
    equilibrium, 5
    first law, 5
    kinetics and, 185
    non-equilibrium, 6
    second law, 5
thermoregulation
    absence of (poikilotherms), 56
    at various temperatures, 57
    defined, 56
    heat fatigue and, 57
    hypothermia and, 57
    panting or sweating and, 56
    regulation of blood flow and, 56
    shivering and, 56
    warm blooded and, 56
time
    non-dimensional, see Fourier number 77
tissues
    freezing, 129
    heat transfer, 36
transpiration, 137
transport
    membrane, 209
turbulent flow, 99

ultrasound for enhancing heat transfer, 88
units, xxix
units conversion, 322

vapor pressure
    water, 136, 342
velocity
    due to bulk flow, 225
    individual species, 225
    mass average, 225
    relative to bulk flow, 225
view factor, 160
viscosity

air, 340
water, 344
water
    bound to solid, 184
    clean, from iceberg, 119
    dielectric property changes during freezing, 127
    evaporative loss from a reservoir, 215
    expansion during freezing, 127
    frost heaves from freezing, 127
    mechanical property changes during freezing, 127
    properties, 344
    specific heat changes during freezing, 127
    steps in freezing, 126
    surface tension data, 353
    thermal conductivity changes during freezing, 127
    time-temperature changes in freezing, 133
    vapor pressure, 342
water activity, 184
wind chill, see human body
    Antarctic expedition and, 119
    boundary layer and, 112
    equation, 113
wood
    anisotropy in, 209
    drying of, 263
    equilibrium moisture, 183
    moisture diffusivity in, 210
    pores, electron micrograph of, 209